BAJA BUGS & BUGGIES

By Jeff Hibbard

Contents

Photos: Jeff Hubbard; others noted
Front-cover photos by Jere Alhadeff, courtesy of Dune Buggies and *Hot VWs Magazine*; Back-cover photos by Linda Hibbard and Tom Monroe.

HPBooks
Published by The Berkley Publishing Group, a member of Penguin Putnam Inc.,
200 Madison Avenue, New York, NY 10016

Printed in the U.S.A.

27 26 25 24 23 22 21 20 19

The Putnam Berkley World Wide Web site address is http://www.berkley.com

The Library of Congress has catalogued the first edition of this title as follows:

Hibbard, Jeff.
Baja bugs & buggies / by Jeff Hibbard ; editor, Ron Sessions;
photos, Jeff Hibbard—Tucson, AZ : H.P. Books, c1982.
160p. : ill. ; 28 cm.
Includes index.
ISBN 0-89586-186-0 (pbk).
1. Dune buggies. 2. Volkswagen automobile—Modification.
I. Sessions, Ron. II. Title. III. Title: Baja bugs and buggies.
TL236.7.H53 1982 629.2′222—dc19 82-83039
AACR 2 MARC

INTRODUCTION

On through the night on the fast roads of Baja. This Class-10 single seater carried your's truly, Jeff Hibbard, to my second Baja 1000 win and a third overall in the 1984 running of this annual event. Photo by Trackside.

During World War II, Uncle Sam put thousands of GIs around the world in jeeps, tanks and half-tracks with little or no roads on which to drive. A funny thing happened. Many of these GIs discovered that driving on white-lined paved roads didn't come close to the fun of driving "off-road."

After the war, these ex-GIs, bolstered by increases in disposable income and leisure time, formed the original core of off-road enthusiasts. They used surplus jeeps, cut-up cars and stripped-down motorcycles.

Soon, these off-roader's discovered that by adding little more than a skid plate, they could get a stock Volkswagen to go almost anywhere. They used VWs not only for driving in the boonies, but for all sorts of things: hunting, fishing and camping, and in the sand, mud and snow. The Volkswagen could be made into a versatile and inexpensive off-road car.

I don't know how many times I have seen an off-road car with some of the best equipment here and little or nothing there. I'm always curious whether this is the result of an over-fix or the work of some hotshot salesman.

Avoid building a "cosmetic" off-road car. Don't start with highly visible components such as trick wheels and tires, and forget the suspension. Many people like visible, easy, bolt-on fixes. This approach is fine as long as you keep in mind that super-heavy-duty trick parts won't make an assembly any stronger than the weakest part in that system. Understand that some expensive parts are never worth the money unless used along with a full complement of heavy-duty or modified parts.

The information in this book was put together with the help of friends, competitors, manufacturers involved in off-road racing and my own experience. This book is intended to be used with and supplement VWs factory-service manual. It is not intended to replace it.

There have always been different points of view on ways to go—that's why each car is a little different. I'll try to present all the ideas and points of view so you can make your own decisions. Just keep in mind that the information presented is based on experience that did not come easily or cheaply. Most of it was accumulated through trial and error over thousands of miles of off-road driving and hundreds of off-road races.

Don't expect these modifications, driving tips and solutions to off-road problems to make you a veteran off-roader overnight. But you will be much better prepared than most off-roaders after digesting the contents of this book.

Good luck. Try to keep the shiny side up!

chapter 1 ENGINE

Dust on Parker Dam course makes driving difficult. While it may be temporary irritation to you, it will devastate an unprotected engine. Air cleaner on roof is in relatively clean air, but would be safer under the roll cage for rollover protection. Photo by Tom Monroe.

I'll start this book with the engine because it's the most expensive assembly in your car. It can be damaged very quickly the first time you go off the pavement. That is, unless you understand the problems and provide needed protection.

The Volkswagen engine is basically very simple—so simple that many people consider themselves engine builders the first time they see one apart. What sets the flat-opposed, four-cylinder VW engine apart from nearly all other automobile engines is air cooling. It's more easily compared to a motorcycle engine than to the typical water-cooled car engine.

The first thing you must recognize is that *dirt* and *heat* are the engine's greatest enemies. Dirt is an enemy because of the VW's small crankcase capacity and lack of an oil filter. A steady stream of dirt or dust entering the lubrication system or carburetor can grind an engine to death in a few hundred miles.

Heat is a problem because air is not a very good cooling medium. Air cooling requires relatively high engine or road speeds to provide enough airflow to handle temperatures generated by high engine loads. When you get stuck, engine load is high while engine speed is low and road speed is zero. Do this often enough and the result will be burned valves and scored pistons.

On the other hand, you'll be amazed how reliable these little four-bangers can be—when set up properly and used with a little care. The Volkswagen engine won't last as long *without service* as a water-cooled engine. You *must* change the oil, keep the valves adjusted, and perform regular checks for oil leaks, loose sheet metal, frayed fan belt and so on. But considering the ease and cost of working on a VW, the work and attention are worth the effort.

Unless your plans include racing, durability is far more important than all-out performance. For this reason, most recreational off-roaders are better off leaving the engine stock or semi-stock, and concentrating on suspension modifications. Why tear apart an engine that's working well?

In my opinion, the best buys are a good off-road air filter and breathers, accessory oil filter and oil cooler, upswept exhaust, skid plate, cooling-fan inlet screen and rock guard, and generally sealing the engine. But if you are building an engine especially for off-road use, here are some tips.

WHAT ENGINE TO BUILD?

Don't waste your money building anything smaller or older than a 1966 50-HP 1258cc (1300) engine. Earlier 36-HP and 40-HP engines are OK for recreational use in stock form. But precious little special equipment is available for the 36-HP and 40-HP engines. Even stock replacement parts are getting hard to find for the 36-HP engine.

The '67—'69 53-HP 1500cc Beetle engine is a worthwhile place to invest performance dollars. Best of all is the '71-and-later 1600cc engine.

Of all VW engines, the Type-4 pro-

Definitely not full race—but a good start. This sedan engine addresses the dust and dirt problem with an accessory oil filter, an in-line fuel filter, and a cotton-gauze K&N air filter.

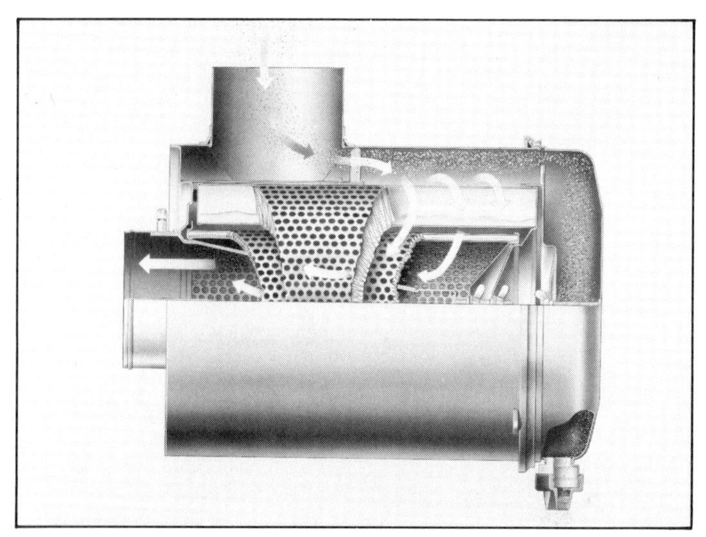

Cutaway of Cyclopac centrifugal filter. Heavy particles are flung to the outside. Rubber valve dumps these particles. Photo courtesy of Donaldson Co.

duces the most power. However, parts are expensive and hard to come by. And unless you have unlimited funds and are devoted to the brand, don't even think of running Porsche or Corvair power off-road.

AIR FILTERS

How much air does a VW engine suck through its carburetors? At wide-open throttle (WOT), a typical engine will need about 90% of its displacement for every two rotations of the crankshaft. That means a 1.6-liter engine will ingest about 1-1/2 quarts of air for two crankshaft revolutions. At 5000 rpm, this engine eats about 127 cubic feet of air per minute (cfm)—or, hold onto your hat—950 gallons of air each minute! Now that you know how much air the VW engine consumes, you can understand why a good filter is important.

Stock Oil-Bath—The stock oil-bath air filter is adequate only if most of your off-road driving is in areas with light dust, such as damp terrain, snow, mud, and so on. With this type of use, the oil at the bottom of the filter should be changed at every crankcase oil change. Off-road use requires more frequent changes. On steep inclines, the oil may leave the bottom of the filter housing. This moves the oil away from the swirling airflow and impairs filter efficiency.

If you decide to use the stock filter, fashion another mount from the air-filter body to the fan shroud to keep the filter from shaking loose from the carburetor. If you leave the air filter off for any length of time, the engine will not last very long.

Centrifugal—An effective air filter still found on a few race cars is the *centrifugal* type offered by Tri-Phase and Cyclopac. These sturdy units are designed for use on earth-moving machines that spend their lives clattering away at dusty construction sites. A replaceable dry-paper element is used inside a metal canister specially shaped to create a centrifugal airflow.

As dirty air swirls around the inside of the canister, heavier particles are *centrifuged,* or flung to the outside and out of the air going to the carburetor. This eliminates much of the dirt before it gets to the paper element. This lets the paper element last for years on a play car and several races on a competition car. Dirt removed by centrifugal action, along with any dirt shaken off the element, drops out of a one-way rubber valve at the bottom.

Because of its bulk and weight, the centrifugal filter is usually installed inside the vehicle. Its weight makes it hard to mount securely near the carburetor, so it is connected to the carburetor with a length of hose and mounted remotely. The shorter this hose, the better! A long connecting hose restricts airflow.

Single-Stage—Single-stage means one filter element. Of these, the only one you should consider for dusty off-road conditions is the cotton-gauze type offered by K&N. It has a tremendous capacity to trap dirt and still flow more air than most other foam or paper types—if you keep it oiled. Forget to oil the K&N filter and its effectiveness drops below that of a common dry-paper filter.

Dual Stage—Dual-stage means two concentric elements in a single housing—a filter within a filter. Dual-stage filters are the most popular types used off-road. Benefits are high filtering efficiency, low cost and compact size. Most designs feature a wet, open-cell foam inner filter that can be washed and reoiled.

The outer filter design varies with the manufacturer. Most have a replaceable dry-paper outer element similar to that used in street vehicles. Pro-Comp II dual-stage filter uses a *dry,* open-cell foam outer element. This allows heavy dirt to shake loose and fall away, keeping most smaller pieces away from the wet inner filter. Unlike the paper element, it can be easily removed, cleaned and replaced, thus saving money.

If you do much off-road driving, consider replacing the outside paper element with a K&N reusable gauze element. The K&N element is more expensive than paper or foam, but with frequent use it will more than pay for itself.

Wet-Sock—Still another type of off-road filter utilizes an open-cell polyfoam "sock" stretched over a wire-mesh form. The open end of the sock is clamped to a short hose or to the top of the carburetor. As with other foam filters, the sock type can be cleaned in

Centifugal air filters installed on a dual-Weber setup. Radiator hoses connect air filters to air boxes. Photo by Ron Sessions.

Pro-Comp II is a reusable, dual-stage filter with two open-cell foam elements. The inner element is lightly oiled and outer element left dry. Heavy dirt particles stopped by outer element shake free in normal operation.

Dirty on the outside and clean on the inside. It's been said that K&N cotton-gauze elements work better dirty than when clean—if you keep them oiled. Forget to oil them and they filter worse than dry-paper elements. Photo courtesy of K&N Engineering, Inc.

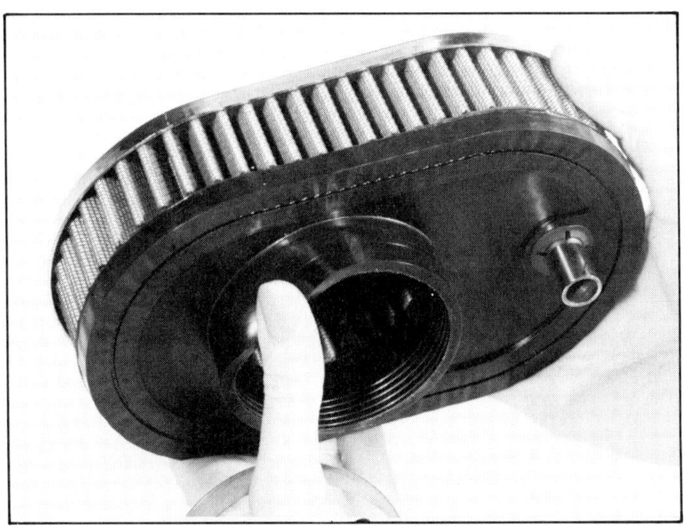

Single-stage K&N cotton-gauze element. No single-stage filter is more efficient. This one is a low-profile model for use in sedans with air-filter-clearance problems. Just oil it and watch it catch dirt. Photo courtesy of Race-Trim.

Potent combination: K&N reusable outer filter replaces a non-reusable dry-paper element in this dual-stage air cleaner. Wet-foam inner filter completes the job.

Wet-sock filters are available in various configurations to fit where other filters won't. While they're OK for street use and an occasional trek off-road, I don't use them in a race car. Photo courtesy of Bugpack.

Base for dual-element filter is simple and lightweight, making it easy to mount in a number of locations.

Filter mounted over the carburetor with a short piece of rubber hose connecting the two. Bolt at left side secures it to the shroud.

dishwashing detergent and reoiled with 20W non-detergent oil or a special wetting agent.

Sock filters are made to fit straight up, at a right angle, and in *dual-T* configurations. They are the least-expensive type of filter and will fit where the more bulky setups won't. These reusable, wet, single-stage filters have less surface area than most other filter elements. I recommend them only for street and light off-road use. Because they're unshrouded, they'll clog more easily in most off-road situations.

Mounting—Regardless of the filter type you choose, mounting location is very important. The simplest installation is a single filter mounted directly over the carburetor. When space is limited, when dual carbs are used, or when multiple filters are desired, a remote-filter setup may be used.

When mounting a single filter over the carburetor, connect the filter base to the carburetor with a short piece of rubber hose—about 4-in. long—and clamps. Install a second filter-housing mount to keep the filter from vibrating loose.

Remote-filter locations always involve some compromise. It is best to place the air filter(s) high, inboard and ahead of the rear wheels, and out of the dust stream. But it's also important to keep the hose connecting the filter(s) and carburetor(s) as short and secure as possible.

The most popular type of hose is Gates *Green-Stripe* radiator hose, available at most truck and trailer supply houses. Exhaust tubing with radiator-hose connections is another alternative.

A long hose or one with sharp bends will restrict airflow to the carburetor(s). A thin-wall hose may collapse. Never use corrugated, flex-type radiator hose—it is very restrictive.

Another consideration is filter vulnerability in a rollover. Don't mount the filter at the highest point of the roof or cage.

On the VW sedan, you have the added luxury of installing the filter inside the car, where the air is a little cleaner. Inside, the filter is protected from being broken off and allowing dirt to enter the engine in a rollover. The problem with this installation is the need for a long filter-to-carburetor air hose. Keep it as short as possible.

In this location, a low-cost solution is a large-diameter air-cleaner assembly from a junked late-model Chevy, Ford or whatever. Use it with a high-efficiency single-stage K&N gauze filter for good filtration.

Some racers prefer the extra capacity of multiple filters to avoid filter changing in a long race. Multiple-filter installations use a remote log-type air manifold, or *airbox,* custom fabricated for the specific application. On buggies, the manifold can be mounted between the hoops of the roll cage.

It's important to prevent air leaks. Even a *very* small leak downstream of the filter allows enough dirt to pass to destroy an engine. Worm-drive hose clamps keep the connections tight. Use heavy grease or silicone sealer around the connections as insurance.

Plumbing the crankcase and valve-cover vents to the air cleaner(s) is a popular method on low-rpm, recreational-buggy engines. But if you've added big 92mm barrels and other serious engine mods, plumb the vent hoses into a separate breather box. A high-performance VW engine, when hot, can pump oil-laden vapors out of the vents. If plumbed to the air cleaner, the vapors reduce air/fuel-mixture octane and can cause detonation and engine damage.

FUEL FILTERS

Another way dirt can enter the carburetor is through the fuel line. A fuel filter between the fuel pump and carburetor will do the job. Small plastic see-through filters are durable and you can see the buildup of contaminants inside. At about $1, they're effective, cheap protection.

Don't use glass-bodied filters. A rock can shatter the filter casing, bathing your engine compartment in gasoline—instant fire.

A tight-mesh wire screen in the fuel outlet of every VW fuel tank helps to protect the fuel pump from sediment that gets into the tank. But since this screen was designed with pavement and clean gas stations in mind, don't count on it—add a fuel filter in the line between the tank and fuel pump. Locate the filter near the fuel pump to discourage any vapor-lock problems.

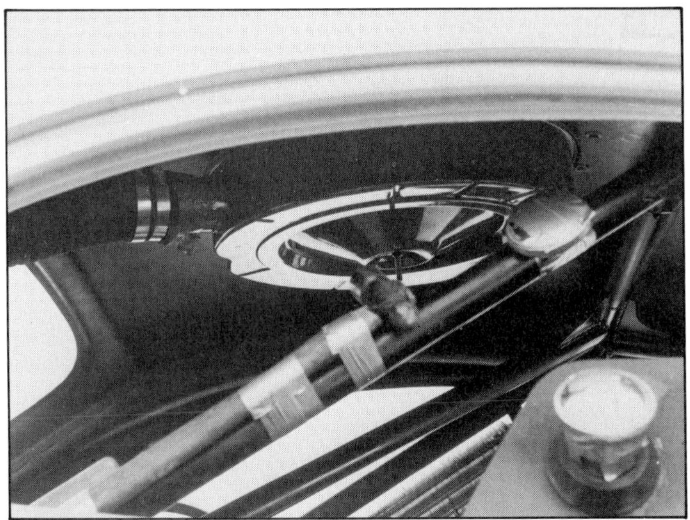

Filter mounted inside the car for protection. Huge K&N cotton-gauze element lives inside this '70 Chrysler-383 filter housing. Photo by Ron Sessions.

Plenum box over carburetor has built-in velocity stack to even out intake-airflow distribution.

Filter units mounted to an airbox or "log." This setup is placed ahead of the rear wheels and out of harm's way. Spring between the wing nuts will keep the filters from vibrating loose. Small canister at right contains filter element for crankcase breather.

Another dual-filter setup for extra filtering capacity. This one uses a Unique Metal Products "T"-type airbox and K&N elements. Photo by Ron Sessions.

A simple solution to an off-road problem. Hose slipped down over the float-bowl-vent tube keeps fuel from sloshing out of the bowl and flooding the carburetor throat in rough terrain. Hose also equalizes air pressure in float bowl with intake air.

Simple, inexpensive in-line fuel filter between pump and carb is effective and won't get broken by rocks. Fuel-pressure regulator (upper right) controls fuel pressure so it won't overcome the float needle valve and flood the engine in rough terrain.

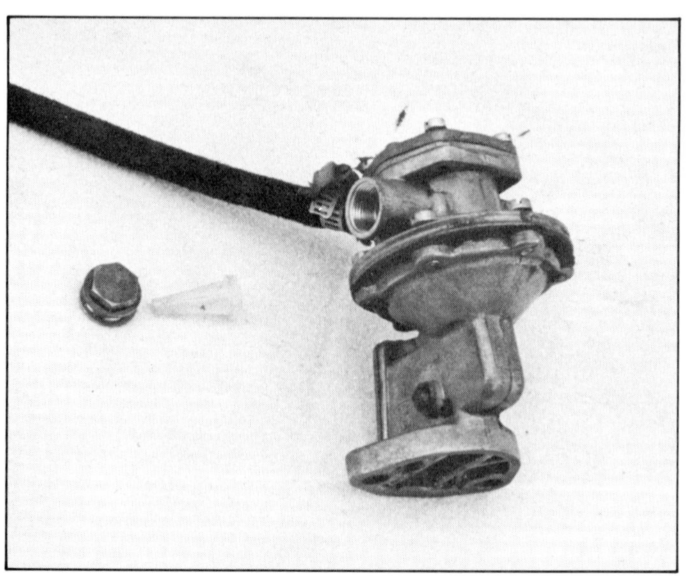

Don't forget the screen inside the stock fuel pump. This pump is a '66—'70 model. On '61—'65 models and '71—'74 models, the screen is under the pump cover.

Protect a glass-body fuel filter by wrapping it in foam. Many of the glass-body filters can be disassembled, cleaned and reused. Photo by Ron Sessions.

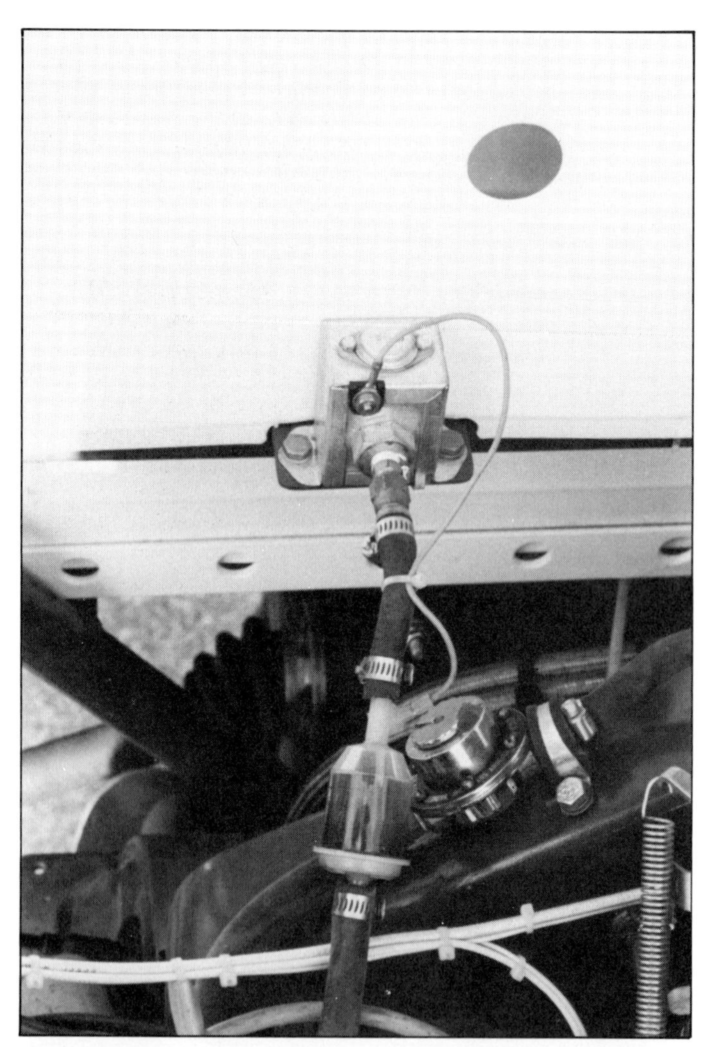

Facet vibrator-type fuel pump is used as a backup to the stock fuel pump. If the stock pump fails, the driver flicks a switch and the electric pump takes over. Neat, huh? Photo by Ron Sessions.

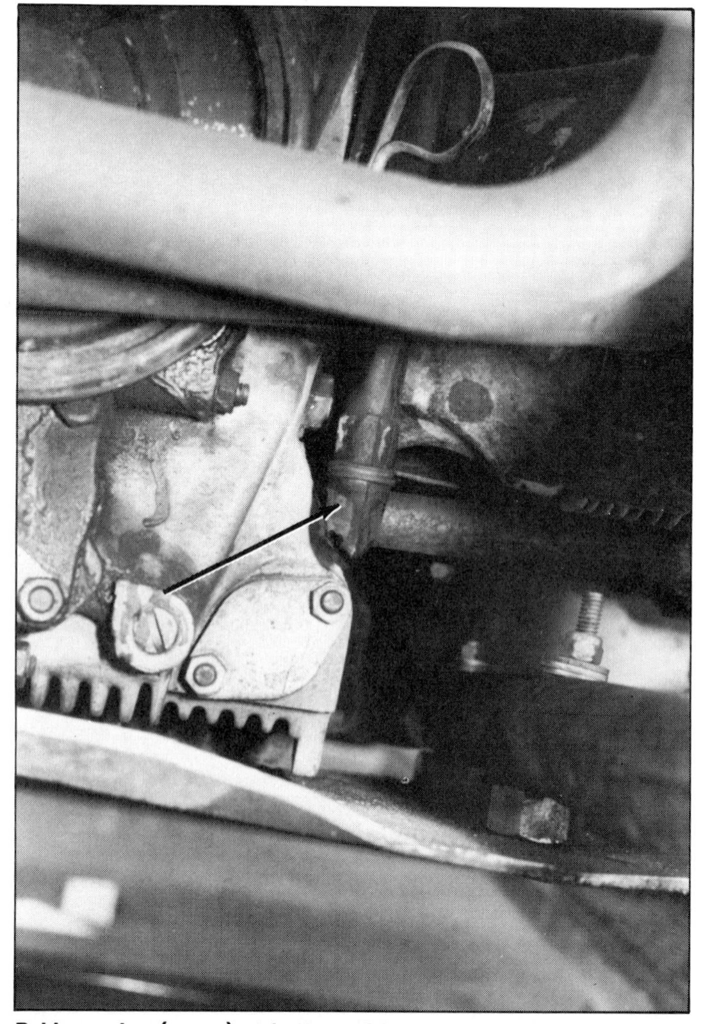

Rubber valve (arrow) at bottom of lower vent on Type-1 engine gets brittle with age and breaks off. Check it frequently. Photo by Ron Sessions.

Make sure you plumb-in an auxiliary tank upstream from the filter so fuel will pass through the filter no matter which tank you're using.

Don't forget the small filter inside the stock mechanical fuel pump. It sometimes gets clogged. On some models, it's a thimble-shaped screen under a hex plug in the side of the pump. On others, it's a flat screen, serviced by removing the pump cover. Clean the screen in solvent with an old toothbrush and reuse it. Install the cover or plug securely or its gasket may leak. Use Teflon tape or pipe dope on the plug threads. Be sure that no tape gets into the fuel system.

K&N makes this breather filter. It works well on small-bore engines which don't require additional crankcase venting. Photo by Ron Sessions.

ELECTRIC FUEL PUMP

Use the stock fuel pump because it's simple, reliable and works great with the typical carburetors used off-road.

An electric fuel pump is one more electric component to worry about. It's only as good as the electrical system's ability to feed it 12 volts. Suffer a charging-system problem or an intermittent short and you get no fuel. No fuel, no go! It's as simple as that. And electric pumps are more sensitive to problems caused by dirt or water in the fuel.

When used at all, the electric pump you'll see on most off-road race cars is the Facet pump. It is a solid-state solenoid, or *vibrator,* pump. It has no moving parts and is only about the size of a pack of cigarettes. Sometimes, electric pumps are used in twos, one as a back-up. Often only one is used, but as a back-up for the stock mechanical fuel pump.

Regardless of which pump you use, all electric pumps have some things in common. They are best at pushing fuel, so mount an electric pump so it doesn't have to do much lifting from the tank. The closer to the tank, the better.

Unlike mechanical fuel pumps, electric pumps are not self-regulating. You have to install a pressure regulator between it and the carb(s). Otherwise, high pump pressure may overcome the float needle—resulting in flooding.

BREATHERS

On most upright-fan engines, the oil-filler tube has openings for two crankcase-breather vents. The upper one, used on '63-and-later models, is next to the oil cap. The upper vent routes vapors to the air-filter assembly via a rubber hose.

The lower vent, used even before 1963, runs next to the case, much like a road-draft tube. These breathers vent crankcase vapors, which include blowby, steam from condensed water, gasoline vapors; and so on.

Usually there is positive pressure in the crankcase and little chance of dirt getting in. But when the engine is cold and the oil thick, air in the vents may pulsate back and forth with the movement of the pistons, especially when the engine is cranked by the starter. During this short startup period, the engine may draw in whatever dirt has collected in the vents.

As previously mentioned, the upper breather hose connects to the air filter. On the stock oil-bath filter, a fitting is provided for this hose. This is OK for a low-compression play buggy.

But if you decide to use an engine with compression higher than 7:1, or have racing aspirations, routing these vapors to the air cleaner may cause detonation problems. The best way to vent blowby is to route all vent hoses to a breather with its own filter.

On stock upright-fan engines, the lower breather tube has a sock-like rubber valve at its bottom. This valve may crack and rot off.

For off-road use, the valve should be wrapped in foam and clamped or *tie-wrapped*—cable-tied—in place. Use a light polyurethane foam like that used in upholstery work. Some people plug this vent, but in case of a broken ring or some other compression leak in the engine, excess blowby gets pumped into the air cleaner.

Valve-cover vents—Any engine with excessive blowby or one with 92mm-or-larger barrels and pistons should use valve-cover vents. The idea is to reduce crankcase pressure so the engine will not blow oil past the seals or crankcase vent in a foam or heavy mist.

Add one vent to each valve cover. Route a vent hose from each cover to an accessory breather box. Cut a piece of open-cell foam rubber to size and place it into the box as a baffle to catch oil mist. Oil mist added to the air/fuel mixture reduces octane, and could lead to destructive detonation on pump gas.

Put the vent on top of each valve cover. This puts the opening above the oil level in the covers. There are two possibilities for stock sheet-metal valve covers: Weld or braze 1/2-in.-ID tubes onto which a hose can be clamped, or use a bolt-on vent tube with an O-ring and nut. Bolt-on vents are offered through most off-road parts outlets.

For cast-aluminum covers, bolt-on vents can also be used if the material is not too thick. If the material is thick, try drilling and tapping each cover for a pipe-thread fitting, or TIG-welding an aluminum tube into the top of each cover.

OIL

As mentioned, dirt is a primary enemy of the Volkswagen engine. Even with the most elaborate pre-

Stock-VW-breather tube has the lower vent wrapped in foam (arrow). This prevents dirt from being drawn in during startup.

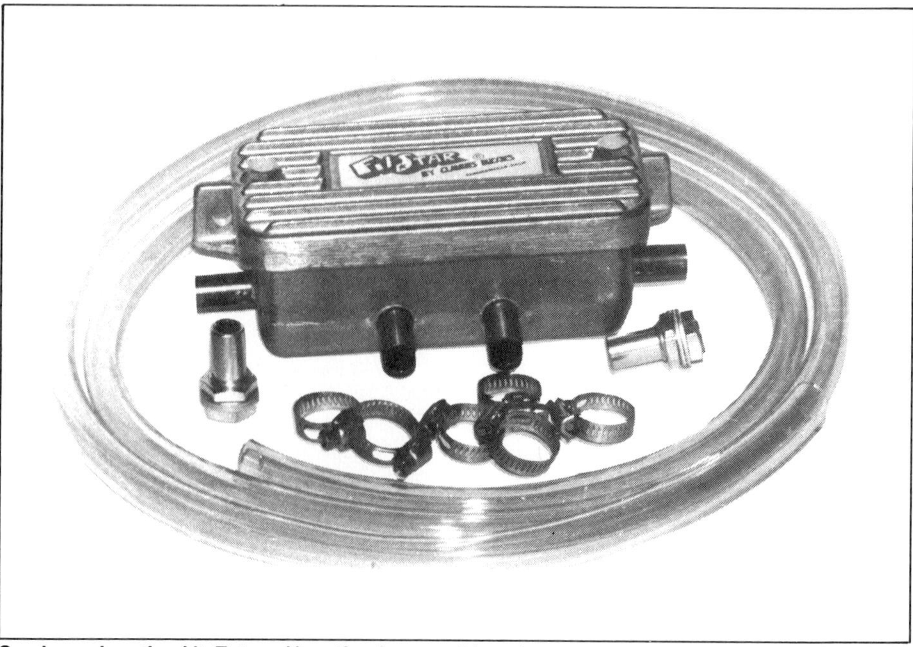

Crankcase breather kit: External breather box provides a large vent area to ensure against pressure buildup in engine. Photo courtesy of C.B. Enterprises.

ventive measures some will get in. One of the jobs of a good detergent oil is to hold dirt in suspension. Your job is to change the oil regularly.

Because the VW crankcase only holds about three quarts, an oil change isn't a big expense. It should be done often; as often as every time out on big-money engines. There's no reason to clean the screen with *every* oil change, but check it frequently. Some racers insist on looking at the screen at every oil change because they've found metal particles indicating piston-ring, piston or lifter failures.

OIL PUMP

Numerous special oil pumps are made for the VW. But don't be snowed by impressive charts and figures. The proof is what works and wins.

The stock VW oil pump is adequate for most sedans and pleasure buggies, even if you install an oil filter and oil cooler. For the rigors of racing, you should increase oil flow (volume) by adding a heavy-duty (HD) pump.

Increased oil flow can help reduce engine temperatures. This assumes that there is sufficient flow capacity in the stock cooler or in an accessory cooler to cool the oil.

Increased pressure may be needed to make up for the pressure drop created by an inefficient external filter and cooler fed from the original oil-cooler

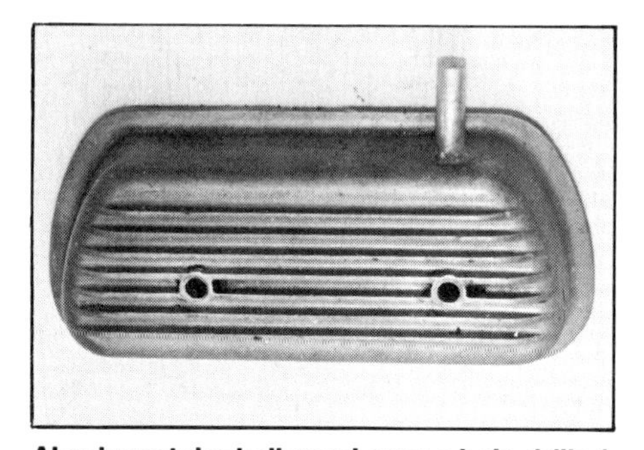

Aluminum tube heliarced over a hole drilled through the top side of this aftermarket valve cover. A hose can now be routed to an accessory breather box.

mounting. Another cause of low oil pressure is worn bearings or a case that is not holding the bearings in correct position, close to the crankshaft main-bearing journals.

In '68, VW enlarged the diameter of the oil galleries in some 1600cc engines. They also made a heavy-duty oil pump with matching 10mm ports. This pump, and all earlier pumps, are designed to be driven by the "slotted" camshaft. The heavy-duty pump can be used in earlier cases with 8mm oil galleries, although not quite as effectively. All of the restricted engine classes in off-road racing allow the early-case oil galleries to be enlarged to 10mm, which is the oil-gallery size in all replacement VW universal cases.

In '71, VW changed to a dished camshaft and changed the oil-pump drive to match it. The later pump and camshaft are not interchangeable with the earlier units. All manufactured performance cams are of the earlier slotted type. Cam grinders use the stock cam gear and replace the rivets with three bolts. So, when you install a racing cam in a '71-or-later engine, you'll need to replace the oil pump also. For more volume than the '68—70 heavy-duty VW pump provides, most racers use a Melling oil pump.

Removing/Installing—On Type-1, 2/1600 and Type-3 engines, use an extractor/puller to remove the old pump body. The pump body for Type-2/1700,-2/1800,-2/2000 and Type-4 engines has a pair of ears that permits you to pry it out with a pair of screwdrivers—if you are careful.

You may have to loosen the case bolt below the pump and pry the pump out. However, you may *lose* the case seal if you don't retorque the bolt to original specs. The new pump body can be tapped into place with a old pump cover and a mallet.

Oil-Pressure Boost Kits—Until mid-

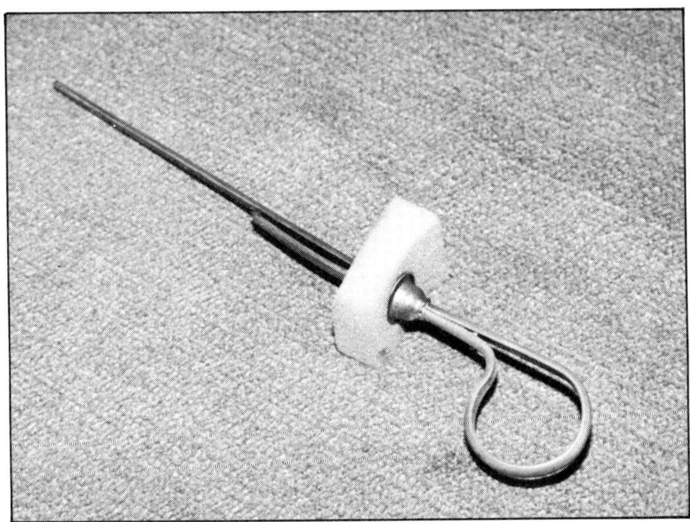

Easy, cheap and effective way of preventing dust from entering your engine through the dipstick hole.

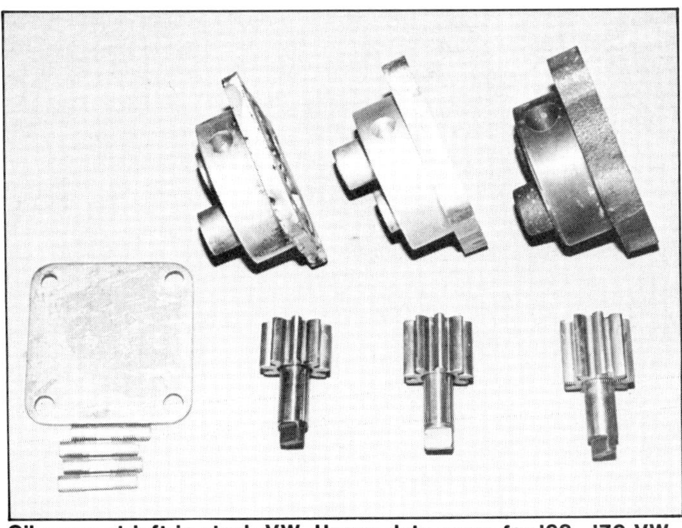

Oil pump at left is stock VW. Heavy-duty pump for '68—'70 VW engines (322-225-207A HD) is next. Pump at right is a Melling. Longer studs have to be used with the heavy-duty-VW and Melling pumps.

At left is early-type "slotted" camshaft used in '70-and-earlier engines. Three bolts hold gear to camshaft. At right is "dished" camshaft used after 1970. Four rivets retain gear to cam. Got it? Both cam types use a slot to drive oil pump, but slot on later type is recessed. Cams and pumps are not interchangeable.

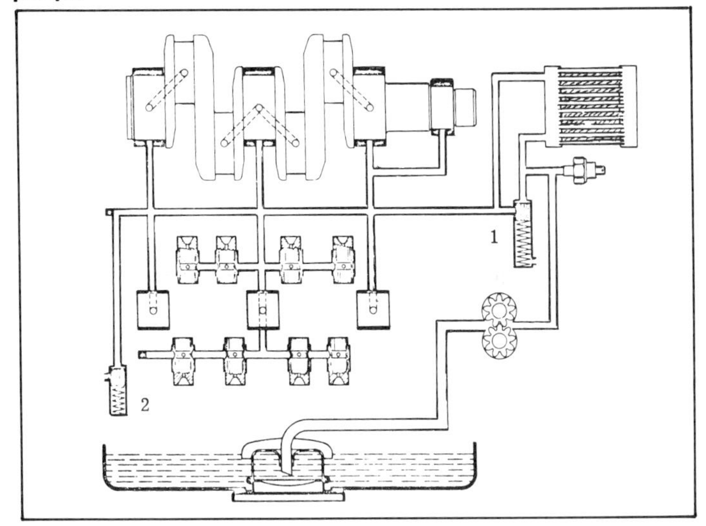

Lubrication circuit of the '70-and-later models and all universal cases. Oil-pressure-relief valve (1) routes oil to the main bearings or to the oil cooler. Oil-pressure-control valve (2) keeps pressure at the mains at a constant 28 psi. Valve (2) opens only when pressure exceeds 28 psi to allow oil to flow back to sump. Drawing courtesy Volkswagen of America, Inc.

'70, all VW cases used a single, combined pressure-regulator/relief valve located near the oil pump. In mid-'70, a separate regulator valve was added at the flywheel end of the case. This valve is subsequently used in all universal cases. Basically, these valves balance oil pressure against a spring force.

With the single-valve system, oil leaving the pump arrives at the spring-loaded regulator/relief valve. When the oil is thick and cold, it overcomes this valve and goes directly to the main bearings. When the oil warms up and thins, the valve spring closes the main-bearing passage and routes the oil through the oil cooler first.

With the dual-valve system, the relief valve works just as described above for the single-valve system. The second valve is located at the other end of the main-bearing gallery and maintains about 28 psi of constant oil pressure. If pressure rises above this value, the valve opens to bleed off excess pressure and return some oil to the sump.

The advantage of the dual-valve setup is its ability to compensate for oil-pressure losses caused by worn bearings. Why? Because it reads oil pressure downstream of the bearings.

There is a case to be made that the single relief-valve-and-spring assembly provides better oil control throughout the total rpm range by pushing more oil to the rear main and cam bearings. This is a point for engine builders to ponder. I think that both relief systems work fine. The size of the spring behind the relief valve(s) determines where the oil goes and how soon it goes there. It is therefore possible to boost pressure and direct oil to the cooler and filter sooner by using a stiffer spring. But using this technique is not without its problems.

With cold weather and/or high-viscosity oil, the added pressure will either blow out the filter seal, blow out the stock cooler seals, blow off an oil line or blow out one of the three aluminum oil-gallery plugs. The one that usually pops out is the one behind the flywheel. To fix this, you must pull the engine out and remove the flywheel. A lot of engine builders will install screw-in, Allen-type plugs to prevent this, especially on Type-4 engines.

On newly rebuilt engines, stick with the stock relief spring(s). To make up for any pressure losses caused by an external oil cooler and filter, add an aftermarket HD pump. On an engine with some oil-pressure loss due to bearing wear, a pressure-boost kit with a stiff relief spring may help. But boosting pressure will only relieve the symptom, not the cause, of the problem, with oil leaks as an undesirable side effect.

Although not the prettiest installation, this filter is up out of the way from rocks kicked up by the rear wheels and is easy to change. Cable tieing the lines makes things a bit neater and protects against the cooler fittings being cracked due to hose flexing. Cooler and filter could be damaged in a rollover.

OIL FILTER

The Type-1, -2/1600, and -3 engines do not have oil filters. Running any engine without a filter is just asking for trouble, especially in the extremely dusty conditions found off-road. Installing an aftermarket oil filter will keep the oil cleaner longer and prolong engine life. Another side benefit is oil capacity is increased from three to about four quarts.

There are more than a few ways to plumb a filter to the VW engine. Each has its pros and cons.

Filter On The Pump—A number of parts outlets offer a pump-and-filter combination. The filter mounts right off the oil-pump cover. This gives a *full-flow* filter setup—explained under "Remote Full-Flow Filter"—but should be avoided even on a street-driven car.

The filter body, made of very thin sheet metal, is under the rear of the engine. This exposes the filter to rocks kicked up by the tires. Even rocks or debris not hitting the filter directly can ricochet off other parts of the car or bounce off the skid plate and puncture the filter. A pinhole leak gone unnoticed for only a few minutes can quickly drain the VW engine of oil.

Remote Full-Flow Filter—In a full-flow oil-filter system, *all* oil leaving the pump goes through the oil filter before it gets to the bearings. Some impressive figures have been published on the wear reductions with this system as compared to a bypass-filter system.

With a bypass filter, all dirt, bearing particles, bits of carbon, etc. in the oil circulate through the engine until it warms up to open the passage to the cooler—and accessory filter.

To install a remote full-flow filter: remove the oil pump, tap and plug one passage, and drill, tap and plumb an external oil line to the case's main oil gallery. The advantage of this system is 100% oil filtration.

With the full-flow system, oil is fed from the pump to the cooler, then through the filter and back into the case. All of the oil is filtered and cooled. If the stock cooler is removed, the block-off plate over the cooler mount should include a notch for oil flow.

Because an external oil cooler is plumbed *in series* with the filter, all oil is pumped through the cooler—even when it is cold and thick. Warmup time increases. If the car is driven only for short trips, harmful sludge and varnish deposits will develop.

There is a problem with this arrangement: With a too-big pump, and a pressure-boost kit, heavy oil or cold weather, pressure can overcome the internal pressure relief and blow the filter seal. It can blow out the aluminum oil-gallery plugs and destroy the stock cooler if it is still in place. Routing oil through the cooler first doesn't sound right until you remember that the cooler creates a pressure drop. This pressure drop may "save" the filter. To avoid blowing the filter Gene Berg Enterprises offers a pump with a pressure-relief valve built into the cover.

If the weather has suddenly turned cold or you have changed to a heavier oil for some reason, be careful to let the engine warm up before burying your foot in the throttle. If you drive in cooler climates, you might look into a thermostat to bypass the oil cooler. Thermostats are available from a number of companies in the aftermarket.

Most off-roaders plumb the oil filter and external cooler with common neoprene hose and worm-drive hose clamps. Cold, thick oil could possibly blow a hose off: Beware! If you can afford it, use teflon-lined, steel-braided hose secured with aircraft-quality screw-on fittings.

If you decide that a full-flow filter setup is what you want, the best way is to do it the next time the engine is torn down. If you don't want to wait till then, Gene Berg claims that it is possible to drill and tap the oil gallery without getting metal shavings into an assembled and installed engine. Here is his plan:

Remove the oil pump and cover the hole in the case with a rag. Take out the oil-pressure sender. Install a pipe fitting where the sender came out. Attach a regulated air supply to this fitting.

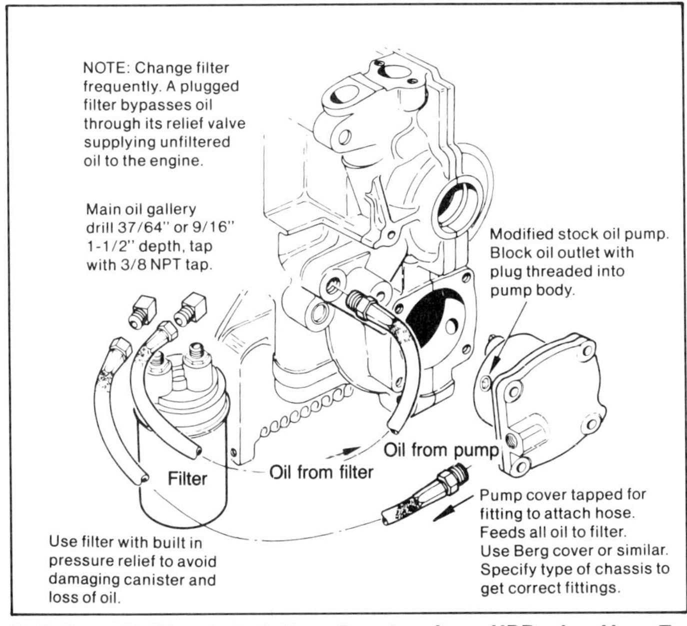

Full-flow oil filter installation. Drawing from HPBooks *How To Hotrod VW Engines*, by Bill Fisher.

There are many places to mount a filter on a buggy. This bracket is welded to the frame. Although a minor point, the filter should have been installed upright. Oil will spill from this filter when its seal is broken. And a new filter can't be filled with clean oil prior to installation unless the filter bracket is unbolted.

With air blowing out of the area you are working on, proceed to drill. When the drill breaks through, the air blows the chips out. Coat the tap with grease to catch as many chips as possible. When finished with the tapping, remove the pressure-relief cup and spring. Then spray solvent into the oil-pressure-relief opening until everything is very clean. Wipe everything very carefully. Re-install the oil pump, pressure-relief valve and oil-light sender and plumbing.

Remote Bypass Filter—Although there is no denying the wear-protection qualities of the full-flow oil-filter system, I recommend an easier way to plumb an oil filter—a remote-mounted bypass filter.

This setup allows the engine to warm up more rapidly—as intended. But it does not provide the wear protection of the full-flow system.

Atop the engine case, beneath the cooling tin, are a pair of oil passages for the stock cooler. To install an external oil cooler, remove the stock cooler and install an adapter fitting over these oil passages. Once the plumbing is in place, it's easy to route the oil through an accessory filter before it reaches the external cooler.

This oil-filter/cooler setup is referred to as a *bypass system*. When the oil is thick and cold, or when resistance to flow is great due to the system trying to pump oil through several feet of hose and through a filter and cooler, it will *bypass* most of the oil, sending it directly to the main bearings or the crankcase.

After the oil reaches operating temperature and is flowing easily, most of it goes through the filter and cooler, then to the bearings. The oil that doesn't go this route is bled off to control pressure. Considering the ease of this modification and its use on the vast majority of off-road competition cars, I see little reason to do it any other way.

ENGINE COOLING

Competing with dirt as the VW engine's worst enemy is heat. Heat is dissipated in two ways. It is transferred directly to the air blown over the engine by the engine fan. Heat is also transferred from the oil to the outside air by the oil cooler.

Cooling Fan—The cooling fan forces air over the cylinders and heads and out the bottom of the engine. On upright-fan engines—the kind used in all Beetles, Karmann Ghias, Type 181 Things and pre-'72 buses—the fan is turned by the fan belt through the generator/alternator shaft. If a belt breaks, the fan stops. On "suitcase" or "pancake" engines—used in all Type-3 Squarebacks and Fastbacks, Type 4s, and late-model buses—the fan mounts to the crankshaft.

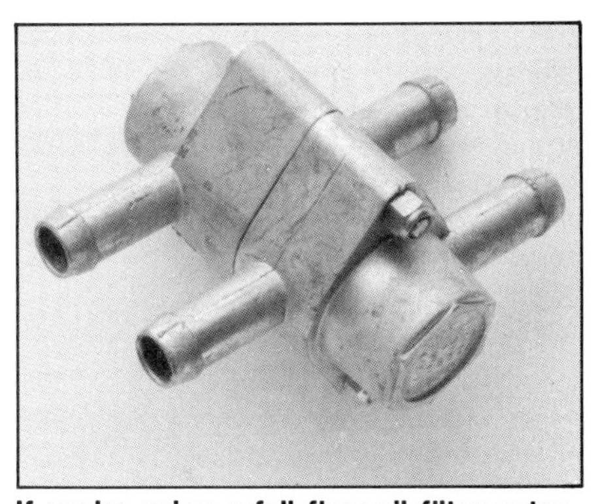

If you're using a full-flow oil-filter setup, install an oil-temperature thermostat. This Oil-Stat by Earl's allows oil to pass through the filter, but bypasses the cooler until the oil warms up. Photo courtesy of Earl's Supply.

The fan delivers a mind-boggling amount of air to the engine—over 1250 cfm at 4000 rpm. Keeping this in mind, it's no wonder that dirt, leaves, brush or grass can be sucked through the fan intake and lodged in the cooling fins of the heads and cylinders under the sheet metal. This greatly reduces cooling efficiency.

Although not a big problem on full-bodied sedans, a screen over the air

Remove the stock cooler and install an adapter fitting in its place for a remote bypass oil filter. With the stock cooler removed, you'll have to plumb for an external oil cooler as well.

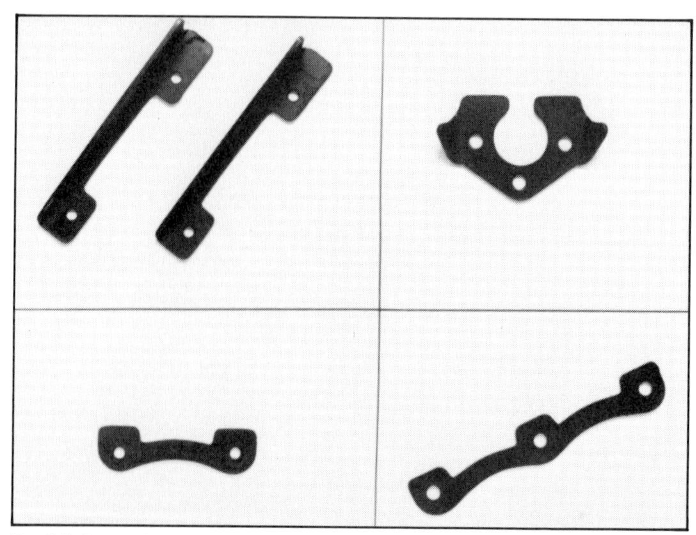

Prefabricated mounting tabs for the oil filter, oil cooler, ignition coil and other components are available from Unique Metal Products. Tabs can be welded to the frame or body in a convenient location. Photo courtesy of Unique Metal Products, Inc.

intake on a buggy or Baja Bug is a must. You can either buy one or make your own out of an old kitchen strainer or wire mesh. The screen holes should be about 1/4-in. square and the screen should be basket-shape, about 4-in. deep. A flat screen doesn't have as much surface area, so will clog faster and restrict airflow.

If you have a severe problem with this screen clogging due to grass, brush and mud, braze or bolt a short length of light chain to the fan housing just above and centered on the opening. The chain will swing back and forth across the screen, cleaning off debris.

Thermostatic Control—On all '64-and-later engines, a bellows-type thermostat operates a pair of air flaps at the sides of the engine. The flaps are closed when the engine is cold to speed warmup.

At about 150—160F(56—71C), the bellows expands to open the flaps. This directs cooling air to the cylinder heads and barrels. On earlier 1200cc and all 1100cc engines, airflow is controlled by a thermostatic ring on the fan inlet that restricts air to the fan.

Most desert off-roaders disable this system and remove the flaps. The reasoning—one more thing to break and possibly restrict airflow.

This is not a wise move for street-driven machines or off-road vehicles operating in cold climates. During the lengthened warmup time, engine wear increases rapidly due to oil dilution by gasoline from an excessively rich air/fuel mixture. Sludge and varnish form in the crankcase. If you decide to retain the thermostat and flaps or thermostatic ring, make sure they operate.

Cooling Tin—One of the items most people mistakenly remove when building a Baja Bug is the engine sheet metal or cooling tin. This shrouding directs cooling air over critical parts of the engine. It is essential for engine cooling.

Most off-roaders remove all of the tin except the fan shroud and the two large pieces that fit over top of the cylinders. This creates a problem. As the cooling air flows over the cylinders, it is diluted with hot air near the undersides of the heads and cylinders. Hot spots develop on the bottom and sides of the cylinders and cylinder-head temperature soars.

VW developed special cooling-air deflectors for the Type-3 engine to cure overheating problems, VW part 113 119 451. They can be adapted for off-road. The sheet metal clips onto the head studs from below, above the pushrod tubes. This keeps cooling air flowing around the cylinders and heads. These deflectors fit all 40-HP, 1300cc, 1500cc, and 1600cc engines.

These replace the smaller original air deflectors. If your engine has 8mm cylinder studs, you will have to crimp the tabs to hold the deflectors in place. Some off-roaders drill the tin and wire it to the studs. This eliminates any possibility of the deflectors falling down against the pushrod tubes and wearing a hole in them.

You must use the front and rear nut plates (the one in the front attaches to the heater box). Trim these pieces so the part that goes around the exhaust flange on the bottom is retained. This preserves the stock mount that prevents the fan housing and upper cylinder shrouds from vibrating loose.

Oil Cooler—Another way heat is dissipated is through the oil. Hot engine oil is pumped out of the crankcase through a cooler, where it is cooled, and then back into the engine. The desired oil-temperature range is 175F—225F(80—107C). The absolute limit is 240F(116C).

On the upright-fan and Type-3 suitcase engines, the stock oil cooler is on the left side inside the fan shroud. This cooler is adequate for normal use but when you start to flog these engines in the summer heat, the stock cooler's shortcomings quickly appear.

The cooler in the upright-fan engine looks like a one-quart milk carton sitting upright in the shroud. It's secured at the bottom only, so rough road makes it vibrate. This eventually weakens the mounting, allowing the cooler to leak at the seals. To minimize the chance of a high-pressure oil leak, *double-nut* and Loctite the mounting nuts.

The 1970-and-earlier Type-1 engines do not have enough cooling-air supply, especially to number-3 and -4 cylinders. Consequently, these two

Air-intake screen can prevent many problems: debris from collecting under the sheet metal on the cooling fins, or a rock or stick knocking the fan out of balance. Ready-made screens are available or you can make your own. I've seen a wire basket for washing vegetables used. The handle was cut off, basket was clamped on and it worked fine.

Phil's Inc. in Evanston, IL swears by these large cooling-fan screens. Mud frequently encountered in the East, can quickly clog a small screen. Air is drawn in from the sides even when the area directly in front of the fan is clogged. Heat from the engine dries the mud out, and bumps eventually knock the mud off. Neat! Photo courtesy of Phil's Inc.

cylinders run about 30F(17C) hotter than the rest; perhaps 50—75F(28—42C) hotter under full load. Number-3 and -4 exhaust valves don't live as long in this environment.

The 1971-and-later Type-1s use a cooler offset to the flywheel end of the engine, mostly out of the airflow to cylinders 3 and 4. The fan housing has a different shape, referred to as a *doghouse housing*. The 0.2-in.-wider fan pumps 100 cfm more air than the earlier fan.

The late-type cooler, fan and sheet metal will cure number 3 and 4 overheating. As an added bonus, the doghouse-type cooler is much less susceptible to vibration failure than the earlier coolers.

For racing and very serious off-road use, you should always remove the stock cooler and mount an oil cooler outside the engine compartment in the airstream. Otherwise, the dog-house cooler setup is the best all-around cooler for the VW.

It is possible to overcool the oil. If you live in a cold climate, consider using an oil-cooler thermostat. As mentioned earlier, the in-line thermostat routes oil around the cooler to maintain oil temperature at 165—170F (75C).

Mounting an External Oil Cooler and Remote Filter—This is not as much trouble as it seems. On upright-fan engines, loosen the fan belt, generator strap, throttle cable and shroud screws. Lift off the shroud and generator to expose the stock oil cooler. On Type-3 suitcase engines, the generator and distributor will have to be removed to get the left-side sheet metal off for access to the stock cooler.

Remove the stock cooler, bolt on the adapter, run 3/8-in. or 1/2-in.-ID oil lines—whichever matches cooler tube OD. Cut the shroud to clear the lines. Make sure the shroud has no jagged edges that can cut into the lines. If you have the doghouse-type shroud, and insist on an external cooler, don't cut it up because it's worth money. It shouldn't be hard to find any VW enthusiast who will trade you his old-style fan shroud, cooler and sheet metal plus about $100 for yours. Install the shroud and generator. Use grommets or silicone sealer to protect the lines from cuts and prevent loss of cooling air around the lines.

On Type-3 engines, seal the hole in the sheet metal where the edge of the cooler was, or precious cooling air will leak out. Sheet metal and a few Pop rivets will work here.

Route the hot oil through the oil filter and then to the cooler. Besides carrying more dirt in suspension, hot oil will pass through the filter with less resistance than cold oil.

When mounting the *filter*, protect it from rocks and brush. Don't mount it in a vulnerable location. When mounting the *oil cooler*, a protected location is desirable, but more important is airflow. An external oil cooler with little or no air flowing through it will be considerably less effective than the stock cooler, even though it may have twice the capacity of the stock cooler.

On a sedan, two good cooler locations are over the air vents, right below the rear window, or on the roof just above the rear window.

On a race-only Baja Bug, the oil filter and cooler are usually mounted inside the car. Window glass is usually not used, so airflow to the cooler is sufficient. Both it and the filter are protected in the event of a rollover.

If you use a roof-mounted scoop, remember that the cooler is vulnerable in a rollover. A scoop really makes a cooler work well, provided that the cooler is sealed into the shroud. All air entering the scoop is forced to pass through the cooler.

On a buggy, the cooler can be mounted anywhere you can weld on a pair of mounts. Usually, the cooler is mounted to the frame, high up, flat side to the airflow. As with sedans, most buggy coolers use solid mounts, without rubber grommets to absorb road shock and vibration. This does not mean that rubber-mounting a cooler is a bad idea. But exposed to the elements, rubber will compress, crack, harden and decompose with time. Check a rubber-mounted cooler often; better yet, mount it solid.

A lot of coolers are mounted in the engine compartment, on the back—flywheel—side of the fan shroud. The thinking here is that by mounting it over the cooling-air intake, air is

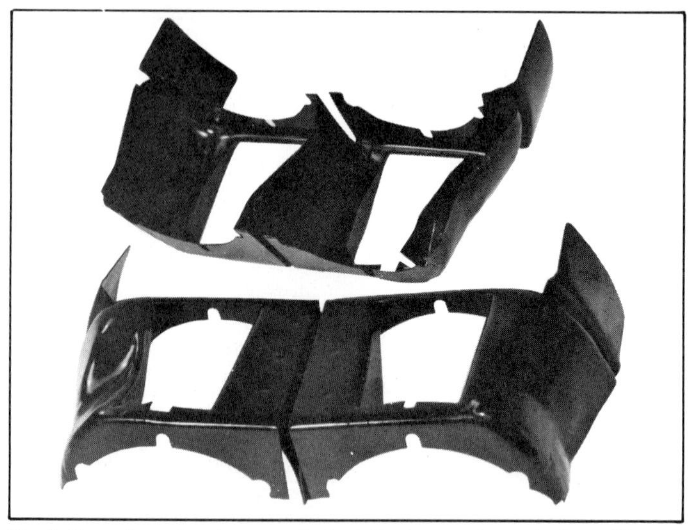

Stock Type-1 engine has small air deflectors under cylinders. These special Type-3 air deflectors bolt in place of the Type-1 units and aid cooling-air flow to Type-1 engine. Photo courtesy of Bugpack.

Stock-VW oil cooler mounts under the fan shroud. This is the one used in '70-and-earlier models.

On '71-and-later models, Volkswagen designed a new shroud with a "dog house" to move the cooler farther forward—toward the flywheel end—and out of the airflow to number-3 and-4 cylinders.

If you must use an external oil cooler on a sedan, the roof is a good location. Heat dissipation relies mostly on radiation with this setup.

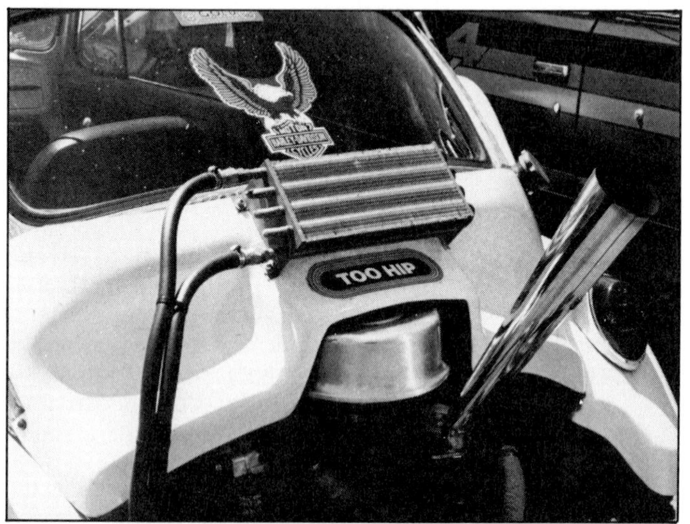

Another good location for a sedan is below the rear window. Airflow isn't quite as good, but its appearance is less obtrusive than on the roof.

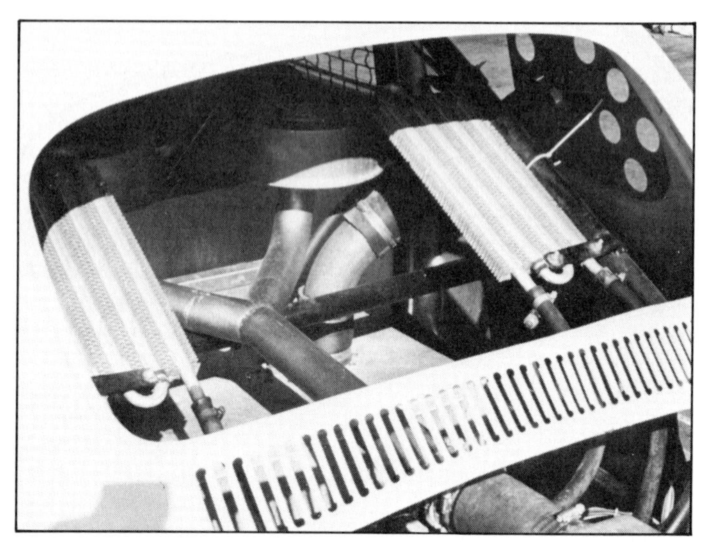

Because this race car has no window glass, coolers can be mounted inside the car. Airflow is good and the coolers are protected.

Baja-Bug roof scoop houses an auxiliary oil cooler. Cooler works great when the car is moving—as long as the car remains upright. Photo by Tom Monroe.

Poor location for a cooler. Sure the inrush of cooling air through part of the cooler removes heat from the oil. But air blown over the cylinders is then preheated. It's like robbing Peter to pay Paul.

pulled through it by the fan. This is an inefficient location for the cooler because air is not forced to flow all around the cooler. Only the part over the fan opening gets any real cooling.

Cable-tie or tape the oil lines coming off the cooler fittings. If the oil-laden lines are not secured, the cooler fittings can fatigue and crack at the hose connections, causing a high-pressure leak.

Another bad oil-cooler location. It makes no sense to mount an auxiliary oil cooler inside the engine compartment where airflow is nil and heat is high. The stock cooler would've been better.

POWER PULLEY

Upright-fan engines can be equipped with a *power pulley* to replace the crank pulley. The power pulley is about 5-3/4 in. in diameter compared to the 6-3/4 in. stock pulley. The smaller OD of the power pulley slows the cooling fan from about 1.6 times engine speed to about 1.35 times engine speed.

Test results with the smaller pulley show a savings of about 4 HP at 5000 rpm. This sounds great until you realize the fan turns slower—hence less air and less cooling. It may be a good trade-off if you have added a large external cooler. If not, stick with a stock-size pulley, particularly if your car will be driven on the street and will see a lot of idling.

SAND SEAL

A *sand seal* is a special double-lip seal (National 321460) designed to keep sand and grit from entering the crankcase behind the crank pulley at the number-4 main bearing. The sand seal got its start years ago when people started using Type-3-engined off-road cars in the dunes.

Stock crankshaft pulley is at left, power pulley at right. Power pulley's smaller diameter saves horsepower by turning fan and generator more slowly. But it delivers less cooling air as well—not good on an off-road car. Many off-road cars are equipped with a stock-size aluminum pulley because stock sheet-metal pulley sometimes fails.

Originally, installing a sand seal required machining the case. Sand seals used with a special pulley need no machining. These seals and pulleys are readily available from off-road supply houses.

Although its importance is not universally accepted, I think that the sand seal is cheap insurance, considering the high cost of building a high performance VW engine. Get your case cut to accept a seal when you have the rest of the machine work done.

DEEP SUMPS AND ACCUSUMP

Deep sumps reduce ground clearance and add another fragile component. Though they can help prevent loss of oil pressure on long steep hillclimbs and side slopes, they require an elaborate skid plate for protection. Consequently, off-roaders don't use them.

Extra oil capacity and engine protection is more effectively gained with a Mecca ACCUSUMP accessory. An ACCUSUMP is a pressurized, reserve-oiling system. It uses an accumulator to inject supplementary motor oil into the lubrication circuit if oil pressure momentarily drops below a predetermined value. The Mecca ACCUSUMP is available from Auto World in Scranton, PA.

WINDAGE TRAY

A windage tray is a baffle designed to keep oil in the sump where it belongs and away from moving engine parts. Usually, the moving part you want to keep out of the oil is the crankshaft. But in the VW engine, the crankshaft sits high in the case, away from the oil.

A windage tray is a popular item with pavement and circle-track racers. The centrifugal force encountered in that type of racing tends to push the oil up against one side of the sump, occasionally exposing the oil pickup tube. Zero oil pressure is a definite danger to main bearings, and a bright red oil-pressure warning light will distract your attention at the worst possible times.

DRY-SUMP LUBRICATION

A dry-sump lubrication system uses a remote oil reservoir holding several gallons. Oil passes through the engine sump on its way out of the engine but is not stored there. Two oil pumps are used; a non-restricted scavenge pump to suck the sump dry and send oil to the reservoir; and a restricted pressure pump to pressurize the engine.

This system improves oil control, crankcase breathing and reduces wear through the use of the full-flow setup and a large oil supply. Although seldom seen on race cars in years past, this system is becoming more popular. The reason is the increasing cost of the competitive racing engine. Some racers are willing to use the heavier, more expensive and more complex dry-sump system (more things to break) in hopes of reducing long-term engine costs.

Dry-sump lubrication system adds weight, complexity and cost to off-road race car, but is best way to improve oil control and crankcase breathing. Large, external reservoir holds several *gallons* of oil, so overheated, dirty oil is never a problem.

IGNITION SYSTEM

Distributor—The stock distributor works fine off-road unless you are changing the carburetion. Most carburetors used off-road have no vacuum fittings. If you have changed to a different carburetor, you will have to install a centrifugal-advance distributor.

If you retain the stock Solex carburetor and intend to drive your VW to work, keep a vacuum-advance distributor. It will give smoother off-idle operation and better gas mileage.

For the serious off-roader, I recommend the Bosch 0 231 178 009 (VW part 211 905 205F) centrifugal-advance distributor. It produces almost 30° advance at just over 2500 rpm. This provides a *fast* enough advance curve for all-out acceleration, without coming in too soon.

Heavy-duty breaker points with a stiffer spring are not needed for the 009 distributor, as it uses the same point set as the high-RPM Porsche 911. However, the recreational off-road VW engine should *not* be revved much over 4000 rpm anyway, so point float won't be much of a problem. This distributor is available brand new at most VW parts outlets.

Dustproofing—To dustproof the distributor, pay attention to two areas: the joint between the cap and distributor housing, and the vents at the bottom of the housing.

Dustproofing the vents is easy. Bond a small piece of foam rubber over the vent holes at the base of the housing. Use weatherstrip adhesive or silicone sealer. Do not plug the holes with the adhesive or sealer; the distributor has to breathe. Otherwise, moisture in the air will condense inside the distributor.

Next, remove the cap and clean both mating surfaces. While the cap is off, use a half-width piece of duct tape to wrap the cap a couple of times. Run the tape tightly around the cap between the plug-wire towers and the hold-down-clamp lips. This has nothing to do with dust or water. If a rock or something breaks the distributor cap, the tape may hold the cap together so you can limp home.

There are many ways to keep dust from getting in under the distributor cap. These include sealing the outside with putty or silicone, gluing the cap on, stretching plastic across it with a hole in the middle for the shaft and rotor, and enclosing the whole distributor in rubber, plastic or foam. All of these methods work, but most are a pain if you have to get inside the cap.

Off-road shops sell a plastic seal that fits inside the cap and over the breaker plate. Although it is inexpensive and allows easy access to the distributor's innards, it won't seal as well as the following method.

I prefer to cut a piece of foam rubber about 12-in. long and 1-in. square. Wrap it around the distributor, where the cap meets the metal housing. Route it so the foam goes *under* the metal snaps that hold on the cap. Cut the

Vent holes in the base of 009 distributor must be covered, but not sealed. Foam and a pair of scissors do the job. Secure the foam to the distributor with weatherstrip adhesive.

Simple but very effective. Cap on left was damaged by a broken rotor in the middle of Baja. We replaced the rotor, but continued on with the broken cap. Notice that the tape on the right cap will not interfere with the mounting clips or plug-wire boots.

Many off-road parts outlets sell a boot that fits like a glove over the distributor cap—wires and all. After the cap and wires are installed, the boot is sealed to the distributor with duct tape. Photo by Ron Sessions.

Dustproofed: Fabricating a foam ring with a slightly smaller ID than the distributor OD makes an effective dust filter at the distributor body/cap connection.

foam so that the ends touch. Next, take the piece of foam off and cut both ends at matching angles and glue them together with 3M weatherstrip adhesive or equivalent. This will make the formed ring a little smaller than you originally cut it so it will be snug against the distributor-cap connection.

Now, pop open the snaps on the distributor and slide the foam ring in place, clip the snaps back on the cap and you're ready. In extremely dusty conditions, you can lightly oil this foam ring just like an air cleaner. If, in time, the ring of foam loses its elasticity and gets loose, a rubber band will hold it in place until you can make another one.

Waterproofing—The degree of ignition waterproofing you need depends on the driving conditions you expect. Generally, water is not much of a problem in the desert, except for an occasional stream or irrigation runoff. But those of you who go off-road where there are water hazards should pay close attention to waterproofing.

A set of quality ignition wires in good shape will handle the average rain and puddle. If you expect to encounter lots of moisture, spray the ignition wires with waterproofing ignition sealer.

The Volkswagen As Dinghy—If you have heard that Volkswagens will float and don't want to lose fire while giving it a try, you'll need to make watertight ignition-wire connections at the cap, plugs and coil. Some boat-supply shops sell plug connectors with a silicone-rubber boot that seals tightly on the plug insulator. You can also use connectors off a late-model Type 3 or '75-or-later Super Beetle.

Using silicone sealer, fill every crack at every coil- and plug-wire connection. Mold a large glob of silicone around the low-tension terminals at the coil and distributor. This should make the ignition waterproof except for the vent holes in the bottom of the distributor housing. If you sealed the cap housing this should be no problem. Even with the engine submerged, no water should get in unless you capsize or the cap is not fully sealed and air escapes!

This Baja Bug's engine never skipped a beat! The key to water-proofing the VW is to cover your bases. The one part of an ignition you don't protect will leave you sitting knee-deep in water. Photo by Mike Rehler.

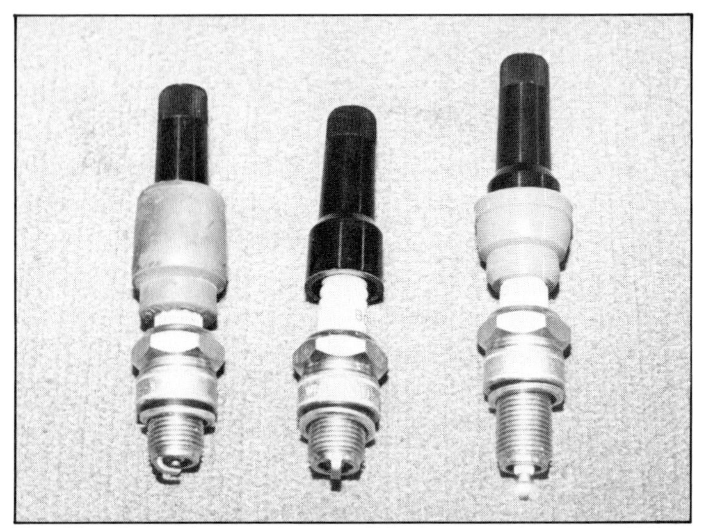

The thing that makes most cars quit in the water is steam off the engine that is trapped by the sheet metal over the cylinders. Plug connector on the center plug is probably what you've got on your VW. Connectors on the end plugs are the same, but with rubber ends that come on late-model Super Beetles. The black plug connectors don't have built-in radio resistors. Red ones do.

Coil Mount—The stock coil is fine and should be no problem provided it's the right voltage for the battery. With time, vibration may cause the sheet-metal fan shroud to fatigue and crack around the coil bracket. This problem can be prevented.

With a light piece of steel or aluminum, make an L-bracket about 2-1/2-in. long on one leg and 1-1/2-in. long on the other, page 21. Use a couple of sturdy sheet-metal screws and a hose clamp to mount the long end to the engine shroud and the shorter end to the top of the coil. If you don't take this precaution, center the coil in the stock bracket. This will lessen the chance of the shroud cracking.

GENERATOR/ALTERNATOR

These very sturdy units don't normally require any special off-road preparation. Generator brushes generally last 40,000—80,000 miles. Alternator brushes last even longer. Alternators or generators are kept clean by air forced through them by the engine cooling fan.

On the generator, some people use foam, a plastic cap or even duct tape over the top opening to keep debris out. Just as many don't. Never put anything over the bottom opening because this is where most of the dirt and dust is blown out. Most important is maintaining proper adjustment of the fan belt.

Belt Tension—Because fan-belt tension affects both engine cooling and electrical charging, it is important to check it often—at least every 5000 miles. If you're racing, check it before every outing.

When pushed with your thumb midway between the two pulleys, the belt should deflect about 1/2-in. Except on the late bus and Type-4 engines, adjustment is made by removing or adding washers between the halves of the generator pulley.

Adding washers deepens the V-groove and *reduces tension. Removing washers* makes the groove more shallow and *increases tension.* Extra washers should be stored between the large concave washer and outer pulley half so they'll be there for future belt adjustments.

Late bus and Type-4 alternator-belt tension is adjusted in the traditional manner—a retaining nut is loosened and the alternator is rotated toward or away from the crank pulley.

VALVE COVERS

Those fancy aluminum valve covers you see on most race engines are there for more than just looks. Most come with hardware so you can bolt the valve covers to the rocker shafts. This minimizes oil leakage and keeps the covers securely in place.

The stock bail-wire hold-downs have a habit of getting knocked off in the brush, releasing the valve cover and resulting in a disastrous loss of engine oil. If you'd rather avoid the cost of bolt-on covers, you can make the stock hold-downs more secure.

Bend two metal washers into a right angle. Weld the washers onto the stock covers, just above the hold-down notches. Install the covers and run wire or a cable-tie through the washers and around the hold-downs. This minimizes the chance of getting the hold-downs and covers knocked off.

ADDED HORSEPOWER

Getting more horsepower out of your VW engine obviously costs money. The price tag escalates the further you go. As the old saying goes, "Speed costs money—How fast do you want to spend?" The first 20 HP costs very little compared to that last 20 HP. And remember another hidden cost—the life expectancy of an engine decreases as its horsepower increases.

Your best horsepower buys are a big-bore kit, a bigger carburetor and suitable intake manifold, dual-port heads and a special exhaust system.

EXHAUST SYSTEMS

Choose an exhaust system that meets the needs of your car. Special matched-length 4-into-1 bolt-on exhaust systems are available for virtually every application, from street bug to single-seat competition buggies. Ex-

To keep the sheet metal from vibrating and cracking around a stock coil mount, add a second mount.

Remove this cover to check generator brushes. Put it back to keep debris out (rocks, sticks, brush, etc.)

tractor systems that use the familiar cone-shaped, open-megaphone stinger outlet will add about 5 HP to a 1500cc engine.

There are many special exhaust systems designed for street-legal sedans. The problem with all of them is the location of the collector and muffler(s), in the area below the rear bumper. This is not the spot to mount an expensive exhaust system if your plans include going off-road. If you insist on using this type of system, install a *sturdy* skid plate to protect the exhaust.

Most people associate speed with engine noise. "The louder it is, the more power you have on tap." Not true! The mufflers available for stinger and other types of off-road exhaust systems do not reduce performance appreciably.

A quiet-running car is a lot easier to live with on long outings, believe me. It will allow you to listen for telltale noises that could indicate a developing problem.

To get the best performance from a tuned exhaust, match the tubing diameter to the engine displacement: 1300cc-or-below, use 1-3/8-in. primary tubing; 1300—1835cc, use 1-1/2-in. primary tubing; anything larger than 1835cc, use to 1-5/8-in. primary tubing. As a rule of thumb, the exhaust tubing used in the headers should be about the same size as the

When you rev the VW engine, the generator belt appears to grow at the sides. To keep the belt on, some racers install "shoes" on the takeup side of the generator pulley. Photo by Ron Sessions.

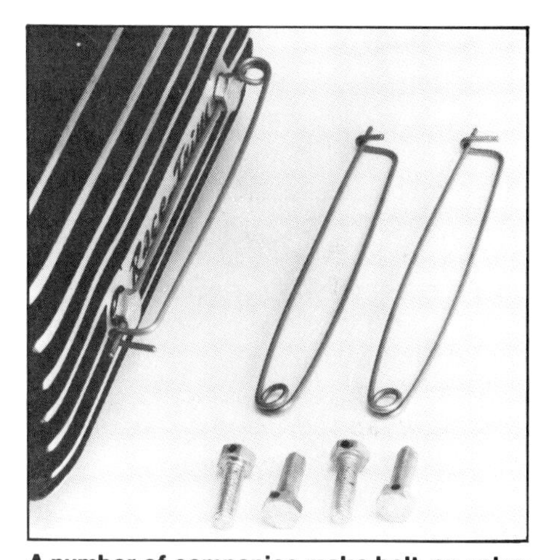
A number of companies make bolt-on valve covers. Race-Trim offers this handy safety pin to end the need for safety wiring or globs of silicone. Photo courtesy of Race-Trim.

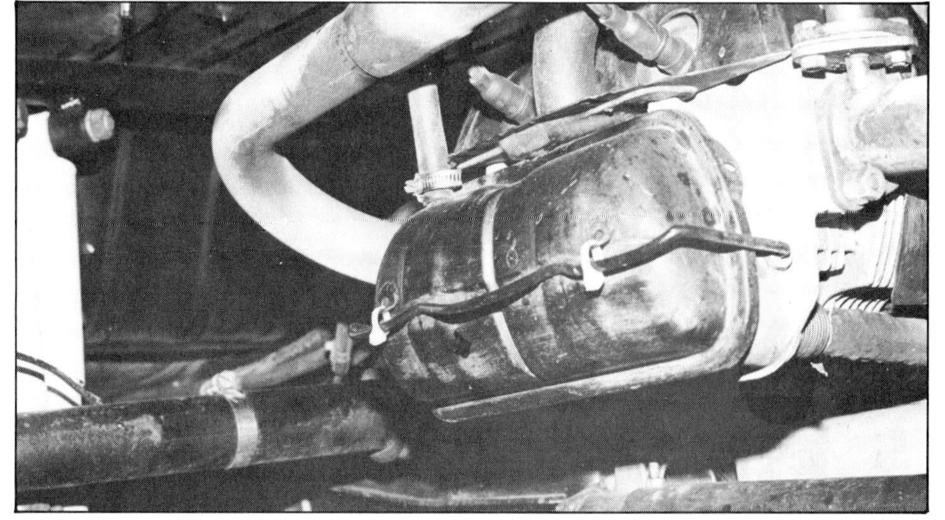
Here's a way to save the cost of bolt-on valve covers and still lessen the chance of getting the valve covers knocked loose. Weld two bent washers to the valve covers and use cable ties to secure the retaining clip. Simple, easy, and cheap.

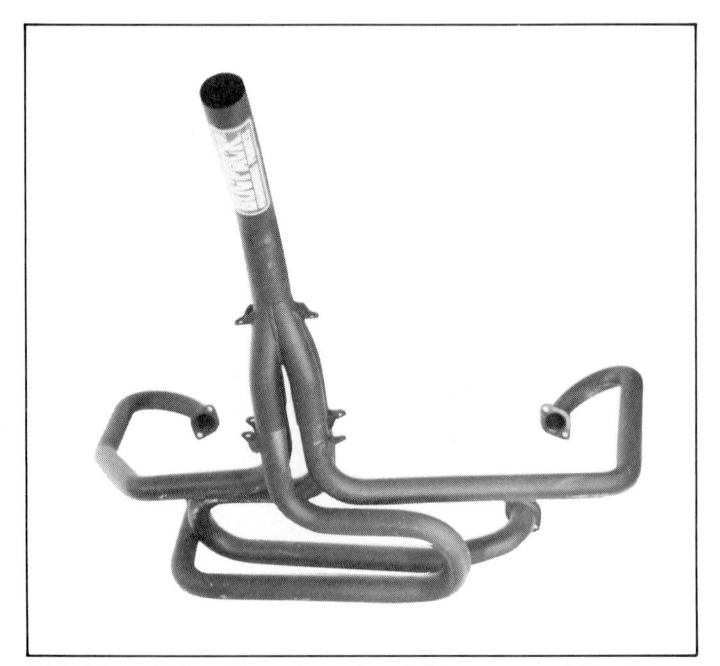

Tri-Mil-type Bobcat Extractor exhaust. My guess would be that at least 50% of the current off-road race cars use this system. Spring attachment at collector reduces breakage problem at slip flange. Photo courtesy of Bugpack.

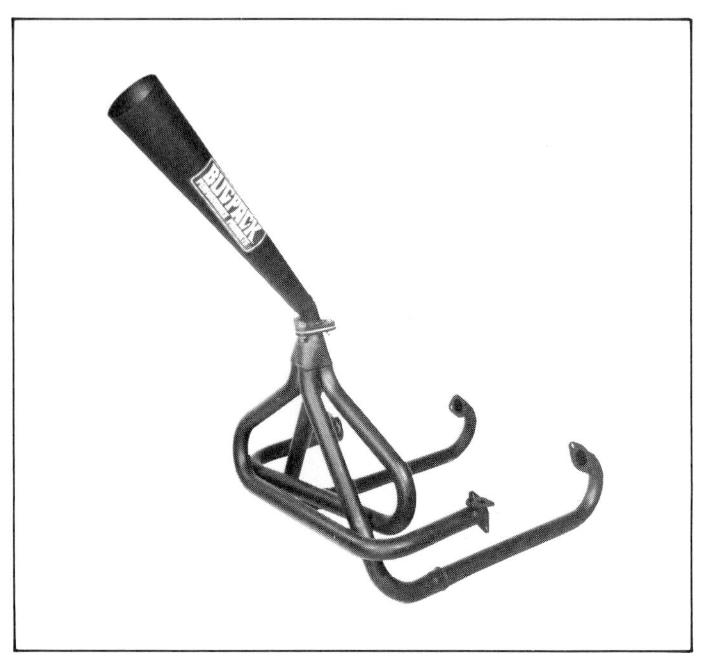

Competition stinger is designed with added ground clearance specifically for Baja Bugs. It fits all 1300—1600cc engines and comes with a bolt-on stinger flange. Muffler can be installed if necessary. Photo courtesy of Bugpack.

For a dual-purpose Baja Bug or recreational buggy, I highly recommend a muffled exhaust. With this Quiet Bobtail system, you can listen to the radio or carry on a conversation without resorting to a shouting match. Replacement heater boxes help warm the cockles in colder weather. Upswept pipes offer reasonable ground clearance. Photo courtesy of Bugpack.

Sometimes, mufflers are required when racing in certain noise-abatement areas. And some racers, such as owner of this Toyota-powered buggy, would rather do without din from open exhaust for hours on end.

exhaust-valve diameter. A collector and a compatible stinger can be used with any size primary tubing.

Because there are so many different exhaust systems for the off-road VW, make sure the one you want fits your car. If you can't try it out before laying your money down, make sure the store will take it back if it doesn't fit.

If it gets cold where you live, don't throw away the heater boxes. Many good exhaust systems are made to accommodate them. Finally, don't throw your stock system away. You may need to reinstall it for a muffler check if cited for an open system.

CARBURETORS

At first glance, a carburetor looks very complicated. Most people know very little about them. There have been entire books written about single brands of carburetors—how they work and how to keep them working. HPBooks publishes authoritative texts

Zenith 32NDIX: At desert off-road races, many winning cars use this popular 2-barrel carb. Photo courtesy of Race-Trim.

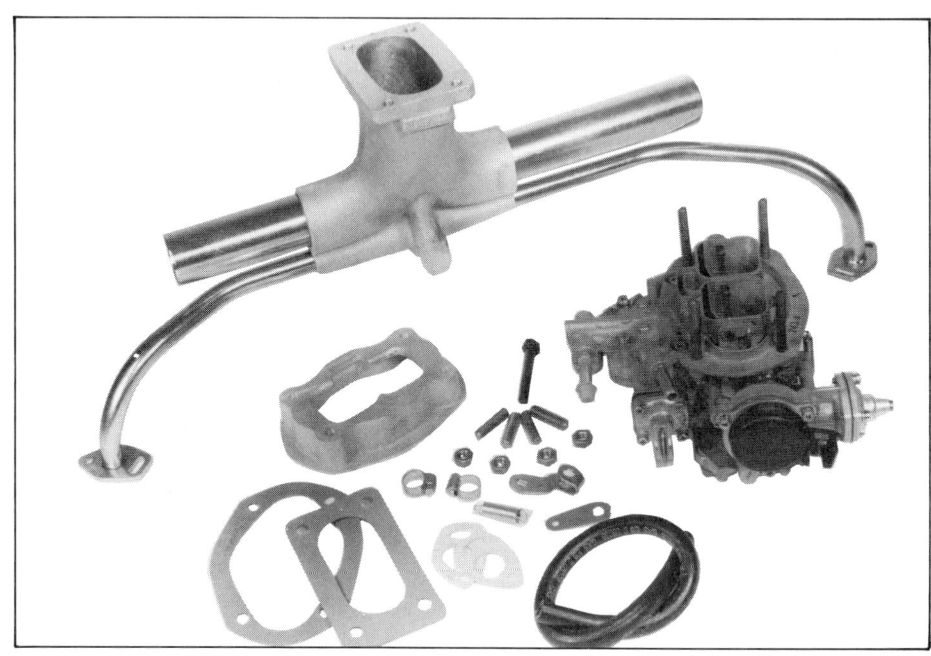

Holley 5200 manifold kit. This progressive 2-barrel carburetor is a great addition to a dual-purpose Baja Bug or recreational buggy with a 1600cc-or-larger engine. Cast plenum-type manifold can be adapted to single- or dual-port heads. Photo courtesy of Bugpack.

on Holley and Rochester carburetors.

Off-road driving imposes its own order of priorities on a carburetor. Dust contamination is a major problem. A carburetor used off-road must not be over-sensitive to dirt or dust. Even with the most effective air- and fuel-filtration system, some dust will get in.

Additionally, uneven terrain and traction characteristics require a carburetor that won't supply an excessively rich mixture under wide-open throttle, low-speed, high-load conditions. Severe vehicle attitudes require a carburetor with excellent fuel control—the float system should not be prone to flooding or fuel starvation.

For these and other reasons, many aftermarket carburetors that work well on a street-driven hotrod VW or quarter-mile machine won't work off-road. What follows is a discussion of the various carburetors used off-road, and my opinions on their suitability.

Solex One-Barrel Carburetors— For many off-roaders with little or no racing aspirations, or even those wanting to race in the limited-engine classes, the stock Solex carburetor is just the ticket. All Solex carbs are equipped with a vacuum fitting, so the stock vacuum-advance distributor can be retained.

The 40-HP 1200cc engines were originally equipped with the Solex 28 PICT-1 carburetor. For a little more power, it can be replaced with the Solex 30 PICT-1 carburetor originally used on '66 50-HP 1300s and '67 53-HP 1500s. The 30 PICT-1 is a simple, trouble-free carburetor that works fine on single-port engines up to 1835cc.

A similar 30 PICT-2 was used on '68—'69 1500s, and a 30 PICT-3 used on '70 models, but are more difficult to adapt because of their emission-control features. The limited-engine classes have rules that specify which carburetors are allowed and the maximum venturi size. Additionally, a restrictor plate between the carburetor and manifold is required in some restricted classes.

To prevent engine flooding, vent the float bowl to the atmosphere. Plug the original bowl vent into the air horn. Tap the float-bowl cover for 1/8-in. pipe thread. Screw a hose fitting into this hole and push a short piece of hose onto the fitting. Secure a piece of foam or a small motorcycle fuel filter to the end of this hose and secure the end so it is higher than the float bowl.

In '71, the Solex 34 PICT-3 was introduced on dual-port 1600cc engines. To accommodate the dual-port head, separate cast-aluminum ends were used. These were connected to the manifold's center section by high-temperature rubber boots.

The 34 PICT-3 will not fit earlier manifolds and vice-versa. Basically a larger version of the 28mm and 30mm PICTs, the 34 PICT-3 carburetor is an effective carburetor for larger-displacement dual-port engines, but is very intolerant of dirt. If you've got one, give it a try, but don't buy one for off-road use.

Holley/Weber 5200—Known in many circles as the Pinto carburetor, this Weber-designed, progressive 2-barrel is gaining popularity for use on VW engines. It features mechanical-secondary operation with linkage similar to Holley's performance 4-barrel carburetors. Its primary is jetted for good economy, and its secondary for performance. Both barrels are fed from the same float bowl.

The Holley/Weber 5200 can be used on 1600cc engines and larger. I recommend it as a good street/dirt compromise. It's seldom seen on race cars. A complete kit to switch over to the 5200, including carburetor, manifold, linkage and all attaching hardware costs over $200. One more thing. As this carburetor has a vacuum connection, it's a natural to use with the stock vacuum-advance distributor.

Weber Carburetors—Weber makes a lot of different carburetors for small-displacement engines like the VW. Parts catalogs are full of them. To many enthusiasts, "Weber" implies performance. Big-money desert and

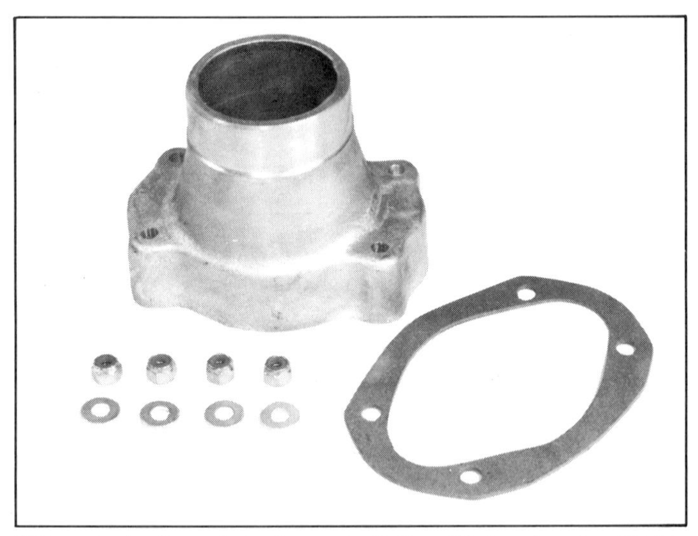

Kit to adapt off-road air cleaner to Holley 5200 carburetor. Photo courtesy of Bugpack.

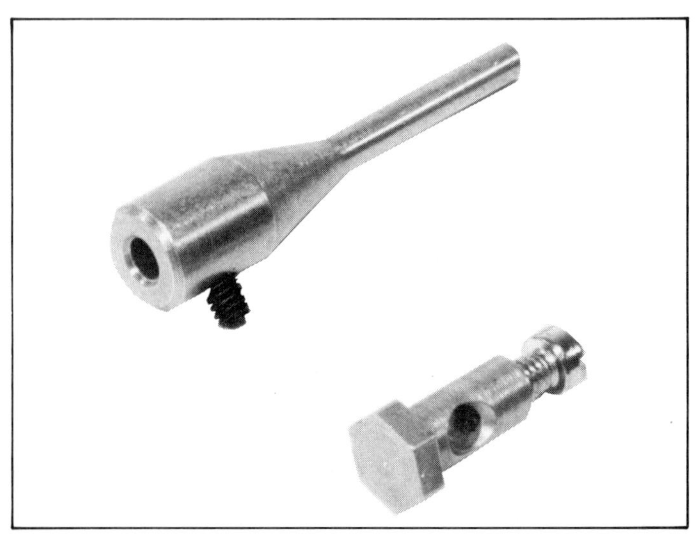

Kit to adapt aftermarket-carburetor throttle linkage to VW throttle cable. Effective cable length can be increased up to 1-1/2 in. Photo courtesy of Bugpack.

short-course or stadium racers consider the Weber an important engine component. But it is necessary to get these carburetors set up by someone who knows what he is doing.

Weber 2-barrels that can be adapted for use on VW engines include this bewildering array: 40IDF, 44IDF, 48IDF, 48IDA, 40MX, 44MX, 40DCN, 35DCNL, 38DCNL, 40DCNL, 40DCNF, 42DCNF, 34ICT, 36DFV and others. Often, Webers are used in dual-carb installations, compounding cost and tuning complexity.

Dual carbs are harder to set up. If everything isn't just right, throttle response and low-speed torque can suffer considerably.

Currently, most racers competing in the popular 1650cc-engine Class 10 use a single Weber 44IDF carb. Racers that use dual Webers in the VW engine in the Unlimited Classes I & II often use two Weber 48IDFs.

Because most Webers are non-progressive types used on individual-runner, ram-type manifolds, fuel standoff can be a problem.

Standoff is a phenomenon where fuel droplets are suspended over the carburetor inlet due to induction-system pulsations. Standoff leans out the air/fuel mixture at high rpm. Installation of a large airbox or air cleaner over the venturis is necessary to minimize problems with standoff. The design of this airbox can be critical in getting the ultimate performance from an engine.

Unlike most Weber carburetors used off-road, the 36DFV is a progressive 2-barrel. It's the design the Holley 5200 was based on. This Weber is a good compromise street/off-road carburetor.

Zenith 32NDIX—This is my favorite carburetor. If you check the top finishers in the VW unlimited-engine classes at any desert off-road race, you'll find a lot of Zenith 32NDIX carburetors. Why? The reason is not horsepower, although its two 24mm throttles do the same job as many larger carburetors. It is reliable, has good fuel control, and gives dependable performance over most terrain.

The Zenith float bowl should be vented to prevent flooding in the same manner as that described for the Solex. Speed Unlimited in Glendale, CA, is a major distributor for Zenith carburetors and components.

Like Webers, the Zenith carburetor is no bargain. Plan on spending the better part of $300 for the carburetor and a manifold. Also, because it has no vacuum connection, you must use a centrifugal-advance distributor.

INTAKE MANIFOLDS

There is a difference in single-carburetor intake manifolds. Basically, they fall into two groups—plenum and individual runner. When you swap for a 2-barrel carburetor, you must make a choice.

With the individual-runner manifold, one carburetor barrel feeds two cylinders. Its advantages are high mixture velocity through each venturi and greater low-speed torque.

With the plenum manifold, both barrels share a common plenum or air box beneath the carburetor. High-rpm performance is similar to that which would be obtained with a larger carburetor. The plenum manifold reduces carburetor tuning problems caused by excessive air/fuel standoff.

Without getting technical, here are my recommendations: With any carburetor other than a stock one, I prefer a plenum manifold. It meets the needs of around-town and off-road driving. If you're building a banzai race car with single or dual carbs and intend to run at or near the redline most of the time, an individual-runner manifold setup may be for you. It's as simple as that.

ROCKER-ARM RELIABILITY FIX

Most racers replace the stock rocker-shaft spring clips and washers with bolts and spacers. This ensures that the rockers don't come off the shaft at high rpm. Aftermarket rocker-shaft kits are available for this fix.

CAMSHAFT

Unless your plans include building a competition car or a hot street machine, stick with the stock camshaft. Why? An engine that pulls well at low rpm is an off-road necessity. With larger-OD off-road tires and rough terrain, you'll spend more time lugging the engine and "walking" over the rough spots than you will flying along with the engine at high rpm.

Most performance cams reduce

Single Weber 40IDF on individual-runner manifold. Photo by Ron Sessions.

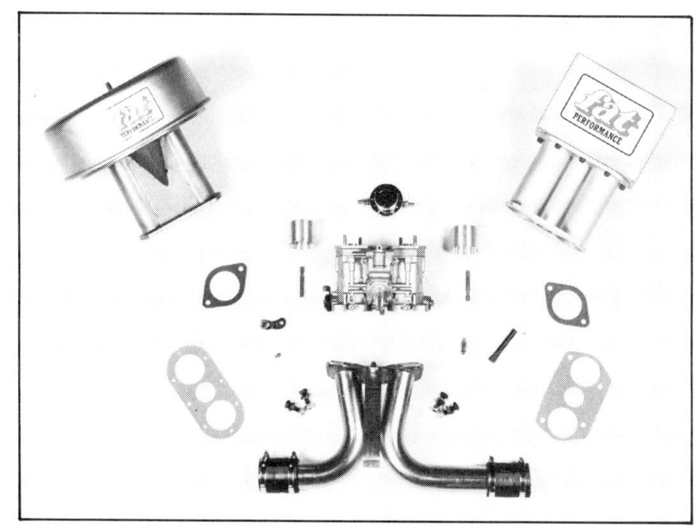
FAT Performance single-Weber kit includes 44—48 IDF carb, braced, high-flow manifold, fuel-pressure regulator, float-bowl vent hose, two-ball fuel inlet valve, and specific venturis and air bleeds. Airbox is specific to type of racing: short-course (left) or desert (right). Photo courtesy of FAT Performance.

Prevent float-bowl flooding in the rough stuff by venting float bowl to the atmosphere (arrow) or air-cleaner throat. Plug internal vent from bowl to air horn. Photo by Ron Sessions.

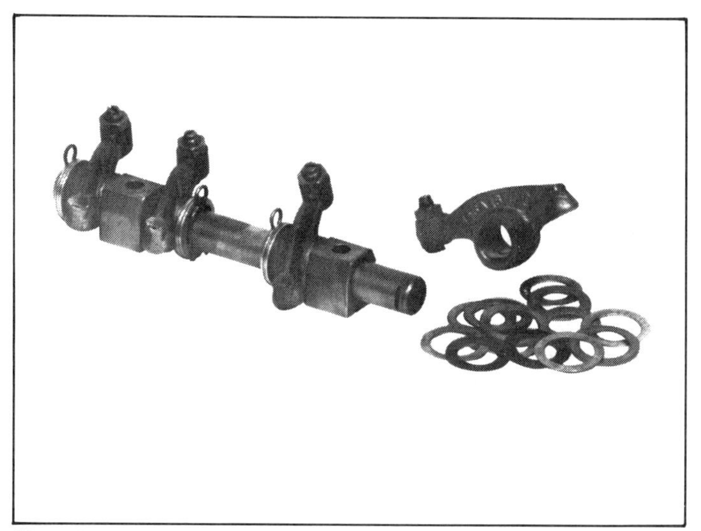
Claude's rocker-arm spacer kit eliminates stock-type spring washers that often fail under racing conditions. Photo courtesy of C.B. Enterprises.

low-rpm torque so you have to use 1st gear in the bad stuff. Worse yet, you may have to slip the clutch to keep moving and keep the engine running—bad for the clutch. A stock cam lets the engine chug on and keep pulling, even at low rpm.

When I'm building an off-road race engine, I use the Norris 407S cam. This 280°-duration, 0.447-in.-lift grind *requires the use of stock 1.1:1-ratio VW rockers*. The VZ-25 Engle or Sig Erson's 205 are other grinds to consider.

When installing a high-performance cam, use the lifters recommended by the cam manufacturer. Lifter and camshaft compatibility is very important.

Also, follow the cam maker's recommendations for shimming and/or changing the valve springs, valve-spring height and pushrod length. And, you must make sure there is adequate clearance between the valve guide and valve-spring retainer at full valve lift. There's a lot more to setting up a high-lift racing cam than just replacing the camshaft.

Because I've had good success with the Norris 407S and stock rockers, I've had little reason to change. Other camshaft and rocker combinations are consistant performers at the races. But, one thing you want to stay away from are aftermarket high-lift rocker arms. VW came out with a 1.25:1-ratio rocker arm recently that works great and is of high quality. Don't go any higher and expect all the related parts to wear evenly or be as reliable.

Common problems with high-lift rockers are: increased valve-guide wear, galling of the rocker shaft, a need to clearence valve guides, possible coil bind and very little margin of

Rebuilding it stock, full-race or anything in between, large mail-order companies have a wide range of kits to choose from. Photo courtesy of C.B. Enterprises.

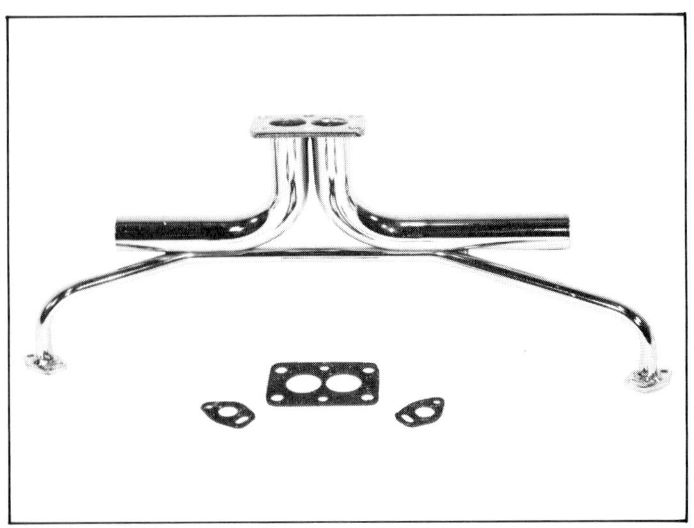

Individual-runner manifold for single 2-barrel Zenith, Holley Bug Spray and Carter carbs. Exhaust-heat-riser tubes provide quick warmup in cold weather. Photo courtesy of Bugpack.

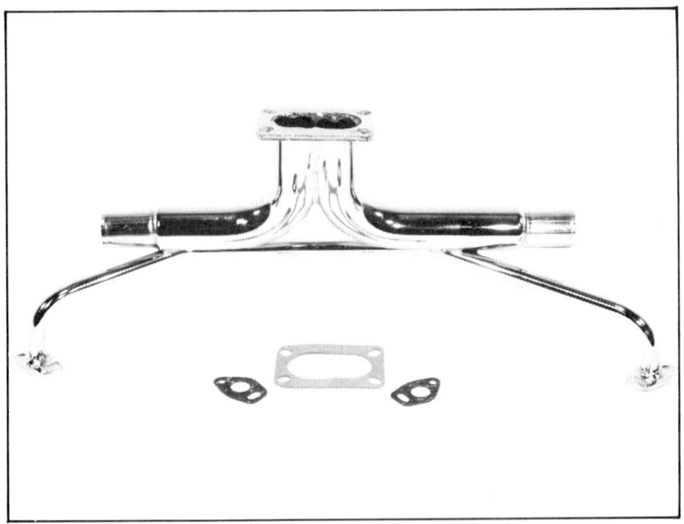

Plenum manifold for 2-barrel carburetors. Plenum reduces carburetor tuning problems caused by air/fuel standoff. Photo courtesy of Bugpack.

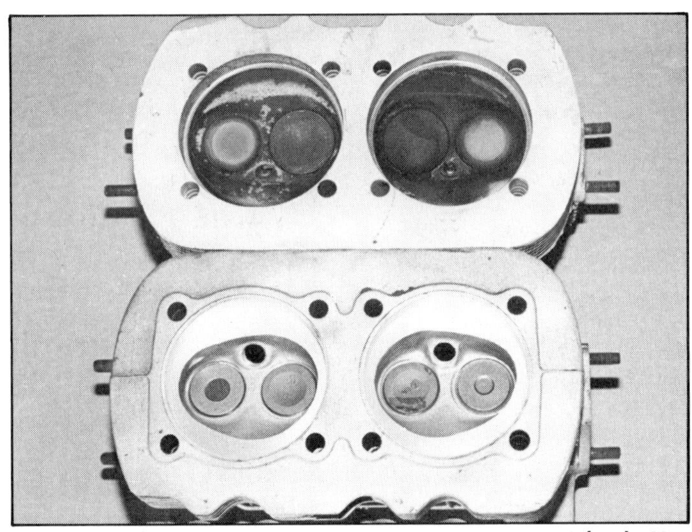

Stock head (bottom) and a big-valve competition head (top). Although it's dirty, the top head is worth about $600 more than the stocker.

error when shortening pushrods for proper geometry. High-lift rocker arms are often used with success on high-horsepower, short-course and drag engines. But these engines are taken apart often, inspected, repaired and reassembled. Great care is taken to get the valve geometry perfect. I think that rocker-arm reliability enjoyed with stock rockers is much more important for recreational use and desert racing than extra horsepower gained by switching to high-lift rockers.

MODIFIED CYLINDER HEADS

You will find a lot of so-called *VW racing heads* for $200 per set or less. Beware! A quality set of racing heads will cost $500 to $1000 or more, depending on who does them and what is done.

The problem I've found with cheap heads that are turned out in quantity is that they are often totally stock with two notable exceptions. Bigger valve seats and valves have been installed. And the valve-guide bosses have been ground away to give the big-port look buyers want.

Because the head modifier may not have used a flow bench to develop the port shape, he may not be aware that removing the guide boss can *reduce* the flow. At the same time, removing the guide boss takes away support so the guide can wiggle and work loose. This creates a high failure rate through the loss of valves and seats.

Further, some cheap heads do not have the correct springs, so the valves can float and break. The resultant engine damage is something you'll not appreciate. Just because a part is advertised as a *race* part does not mean that it is necessarily race quality.

Things to remember when buying modified heads: big ports and valves do not necessarily flow better than stock; valve-guide bosses are essential for reliability; and valve springs and retainers must match the camshaft.

Heads must be assembled very carefully to make sure there is adequate clearance between the valve guides and the retainers when the valves are wide open.

In racing classes using the restricted 1600cc engine, only the stock single-port head or its equivalent can be used.

When the VW case is bored out for 92mm barrels, the case is cut very thin on one side. It's not uncommon for a crack to start at the cylinder seating surfaces. To help prevent this problem, the thin half of the case has been reinforced by TIG-welding supports over the case webbing.

BIG-BORE KITS

Big-bore kits include barrels, pistons, piston rings and wrist pins. These kits are made in many sizes and quality varies greatly.

There are two kinds of big-bore kits. One type is known as the *slip-in* because the case and head ends of the cylinders pilot into the case and head without machining. This type has estremely thin walls at the case end. These are not generally used by serious off-roaders.

The other kind of big-bore cylinders are *oversize* at the head and case ends. Therefore the cylinder heads must be bored out to fit the cylinders. And the case must be split and disassembled so the cylinder openings can be bored to fit the cylinders.

If you are bewildered by the array of big-bore kits, I'm not suprised. Some off-road parts outlets are always trying to get more business by selling parts at cheaper prices. The manufacturers of big-bore kits want to keep their factories producing parts. So many of them will reduce the quality of the parts to get the cost down to what the seller says he will pay. The seller may accept a reduction in ring quality, leave out the rings or pins, or even accept lower piston quality.

As a result, it's not easy to compare big-bore kits, even though they may be made by the same manufacturer. Your best guide is the seller's reputation. Does he demand original-equipment quality? Will he stand behind the product? Will he sell only one piston and cylinder if that's what it takes to solve your problem? How has that company reacted to problems your friends may have had with engine parts they purchased from the company?

I use the N.P.R. 92mm big-bore kit to get 1835cc displacement with the stock-VW 69mm-stroke crankshaft. The N.P.R. kits are also sold under the names I.S.S. and H.Y.D. Another maker of big-bore cylinder kits is CIMA.

Before you decide on how big a kit you should install, give some thought to how you're going to use your car. If you regularly drive it around town and only occasionally take out in the boonies, stick with the 1600cc engine. The 1600's 85.5mm barrels are of the slip-in type and are easy to install, offering good economy, adequate power and good reliability.

If you use your car mostly off-road, the 1835cc engine works great. The 1835's 92mm barrels provide enough torque to let you roll along in the rough at a lower rpm, with no need to slip the clutch and less downshifting. The drawback to 92mm barrels is their size—bigger than the engine was designed for. The case must be bored to accept them. This creates a weak spot by the # 3 cylinder that sometimes cracks. Gusseting the case in that area can sometimes prevent the cracking problem (see photo).

VW sells new cases which have been redesigned to put more metal in this problem area. With the larger combustion chamber producing more power, the 1835cc-engine's barrel studs and nuts are more likely to loosen up than those of the 1600cc engine. To prevent this, retorque the heads periodically or when you see oil appearing where the barrel meets the case.

You can achieve the same type of horsepower increase without these potential problems by adding a counterweighted stroker crankshaft; 78mm for a 1791cc engine or 82mm for a 1883cc displacement (with 85.5 barrels) plus many other crankshaft and barrel/piston combinations. There are two problems in going to a stroker crankshaft and both of them are dollars. A stroker crank will cost more than four times as much as a set of barrels and pistons. Also, the new crank is not just a remove-and-replace procedure. As with installing 92mm barrels, it is a job for an experienced engine builder—more $.

Time and labor are expensive, whether it is yours or you are paying an experienced mechanic to build your engine. So, once you split your VW case, make sure that the important areas are checked and/or repaired before you put the engine back together. Don't get stuck with a flaw that may substantially shorten the life of your newly rebuilt engine. The following areas must be considered while you have the engine apart.

CASE SAVERS

Case savers are threaded steel inserts that replace the stock threads in the alloy case. These are the threads for the long cylinder studs. Expansion from overheating caused a lot of thread failures. When studs come loose, the cylinder-head seal fails and a compression leak develops.

In 1973, VW solved the problem by changing several things on the engine. These included a reduction of the stud diameter from 10mm to 8mm and the use of threaded steel inserts to give the 8mm studs a strong base. The genuine-VW 8mm studs are designed to give the exact expansion and contraction needed to protect the threads in the case and save the compression seal between the cylinders and the heads, even when the engine is overheated. This combination will ensure that you are giving your VW engine the best-possible chance for a long life.

Any time you disassemble an early case, have a machinist install a set of case savers and buy a genuine VW 8mm cylinder-stud kit (P/N 043 198 035). This will almost eliminate the possibility of the stud threads being stripped out of the case.

BARREL-SEATING SURFACES

Most VW engine builders lightly machine the head and case surfaces that mate with the cylinders (barrels). This is important to give the barrel a good, solid footing and a leak-free seal at the case connection. If the heads have never loosened and started to leak, you may be able to get by without machining the case. Even so, the cylinder-head-sealing surfaces should be flycut 0.005—0.010 in. to ensure a good seal. Remember, VWs don't use a head gasket.

ALIGN BORING

The VW engine pounds out its main bearings and bearing saddles to the point that the saddles eventually become out of round. Even a new set of bearings won't be held tight enough to support the crank and prevent excessive oil-pressure loss. To correct this, the case can be align bored 0.010, 0.020, 0.030 or 0.040-in. over and oversize bearings used. This is a universally accepted method of restoring a damaged case. A few engine builders still insist on buying a new case, instead of align boring a used one, and then have all of the expensive machine work done over again.

OTHER CONSIDERATIONS

Here are some other things to consider, depending on how you intend to use your engine.

Balancing—The higher you rev the VW engine, the more important it becomes to balance the internal parts. This a must on racing engines. The pistons, connecting rods, crankshaft, flywheel, clutch and pulley should all be balanced.

Dowels—The stock crank comes with four dowels to key the flywheel to the crank. If you are building an engine with a higher level of performance, add four more dowels to make a total of eight.

Machine Work—Cutting the case for big-bore cylinders or clearancing the case for a stroker crankshaft is all in a day's work for the experienced VW machinist with the correct tools. Stick with an experienced person or shop. I have used RIMCO for a lot of the machine work on my own engines.

Compression Ratio—With the lower octane ratings of today's gasoline, even the stock VW engine will sometimes have problems with pinging (detonation), especially when under load in higher gears. This problem is increased when the head is flycut to clean up the mating surface during a valve job, resulting in an increase in the compression ratio.

Alternatives include using high-octane racing gas, retarding the ignition advance or increasing the combustion-chamber size. At more than $3 a gallon, and if you can find it, racing gas is too expensive for the recreational off-roader to use on a regular basis. Retarding the ignition advance tends to raise engine operating temperature, decrease gas mileage and reduce performance in proportion to the amount the spark is retarded. So, increasing combustion-chamber size and thus, lowering the compression ratio, is the only option that makes any long-term sense.

This can be done in a couple of ways. One is to install shims or spacers between the barrel and the case. Although effective, it's not the best method because it lowers the combustion chamber down into the barrel. The other method is to have a machinist "strip cut" into the combustion chamber of the cylinder head. This is a routine job for a machinist familiar with VWs because a lot of the better mechanics will have the machinist step-cut the same number of thousandths out of the combustion chamber as he had to flycut off the mating surface of the head during cleanup. Ideally, you want to get compression ratio down between 7.5:1 and 7.0:1. A step cut of about 0.040 in. on a 1600cc engine's heads will bring compression ratio back in line and should stop the pinging.

When adding 92mm barrels and pistons to a 1300cc-or-larger engine, the resulting 1835cc engine will have a compression ratio of around 8.6 to 1. Much too high for today's gasoline. To reduce compression on an 1835cc engine, I suggest the following:

Dual-Port Head—Step-cut about 0.080-in. into the head for a total of not more than 0.135 in. Remember, the heads have been bored for 92mm pistons, so you can move your step cut out toward the barrel the same amount. With a deck clearance of 0.080 in., compression should be about 7.5:1.

Single-Port Head—Step-cut a total of 0.060-in. into the chamber, remembering to start the cut just inside of where the new, larger barrel will seat in the head. Set deck clearance at 0.110 in. to be fairly close to a 7.5:1 ratio. If pinging continues to be a problem, even with leaded premium and a slight decrease of spark advance, you can use a die-grinder and remove some metal from the chamber on the spark-plug side and/or increase deck clearance/height by adding barrel shims/spacers between the barrel and case.

Do-It-Yourself or Farm It Out—Even though the Volkswagen engine is relatively simple, tolerances and assembly procedures are very critical. Have the work done by an experienced Volkswagen mechanic.

Those parts you use, whether stock or special, are expensive enough without having to buy them twice. If you don't know a good mechanic whose work you can depend on, shop around for someone who's been in business for a number of years. Make sure he's familiar with what you want *before* you deliver your engine.

Also find out if you can supply the parts that you want him to install. Engine builders make money on parts because they buy them wholesale and sell them to you at retail. If they can't make the money on parts, they'll make it up somewhere else—labor perhaps. However, give it a try.

Don't bother supplying the engine builder with bearings, seals or other small items. Concentrate on saving money by shopping around for the bigger and more costly parts.

Type-4 desert-racing engine: To reduce weight, stock cooling tin was replaced with aftermarket fiberglass pieces.

Large-displacement Porsche six-cylinder engine drives through Porsche transaxle. Big horsepower costs big bucks. Some racers have gone to Hewland gearboxes to handle the power.

Pinto, Toyota, Rabbit, Honda, Renault, Mazda and other water-cooled engines are being tried in buggies with increasing frequency. Typical water-cooled setup shows midship radiator, long radiator hoses and electric fan.

Don't be suprised if the engine builder refuses to guarantee his work. He can't control how you use the engine, so don't expect any guarantee.

The old saying, "You get what you pay for," also applies to VW replacement parts and speed equipment. Cheap equipment will reveal itself the instant you start to push this little alloy engine harder.

Don't try to save money on machine work. Get it done right the first time by someone who is experienced with Volkswagens.

When you install 90mm-or-larger barrels and pistons in anything but the latest redesigned case, you're taking a gamble on cracking the case. Gusseting the rear webbing lessens the chance but does not eliminate it. Once a crack starts, it is almost always terminal for the case. I personally feel this is a good gamble because I've had good success with 92mm barrels and pistons, but other people would surely argue with me on this.

Building a Race Engine—If you intend to build a competitive race engine, I suggest that you read HP's book *"How To Hotrod Volkswagen Engines."* If you have the money to pay someone else to do it, get the job done by an engine builder who specializes in Volkswagen racing engines.

Talk to more than one engine builder and get some idea of what they charge and whether or not they'll work with you on parts. Race-engine builders are a strange bunch, so don't be too surprised if most of them tell you the other guys don't know what they're talking about. A competitive, full-race, big-bore off-road engine will cost $8000 or more including your engine as a core. It may be good for several races before it should be completely "gone through" or rebuilt.

OTHER ENGINES

VW Type 4—The Type-4 engine, so named because it originally appeared in the ill-fated VW 411/412, has been used in much greater numbers in the Porsche 914 and all 1972—83 air-cooled Type 2s. Because of its sturdier construction and due to its larger 1700cc, 1800cc and 2000cc displacements, a suitably modified Type-4 engine is capable of delivering a higher *reliable* output than the Type-1 engine.

The Type-4 engine is also heavier and more expensive than its Type-1 counterpart which, until recently, has limited its popularity. Helping to narrow the weight gap between the two engines to a mere 40 lb is an aftermarket kit that replaces most of the Type-4 cooling tin with fiberglass.

Most popular with engine builders is the Porsche 912 2000cc engine, with its freer-flowing head design. With large barrels, displacement of the 2-liter Type-4 engine can reliably be increased to 2600cc. Engine-building cost can run as high as $8000.

Porsche Six—Where class rules allow, some builders have gone to the Porsche 911 six-cylinder engine. These high-output engines are very expensive to build, easily double that of a full-race Type-4 engine, or three times as much as a full-race Type-1 engine—approaching $20,000 each! Additionally, six-cylinder Porsche power is hard on transaxles. Although successful in short-course stadium racing, these engines have generally not lived up to expectations in desert off-road racing.

Non-VW Engines—Off-road promoters have encouraged racers to try non-VW engines with recent rules changes. In addition to V8s in the unlimited classes, some racers are running Mazda rotaries, Renault V6s and other water-cooled engines. In Class 10, with its 1650cc-displacement limit, Toyota and Honda engines are popular.

chapter 2 TRANSAXLE & CLUTCH

If you drive like this, equipment takes a real beating. When a car lands on one rear wheel, the axle shaft, CV joints and transaxle are subjected to great stress. Photo courtesy of BFGoodrich.

TRANSAXLE

As the name implies, the VW transaxle combines the transmission and final-drive assembly in a single unit. Power is transmitted to the rear wheels through a pair of single- or double-joint axle shafts. Early transaxles, up to and including most '68 models, use a single-joint, *swing-axle* design. Some '68 and all '69-and-later types use double-joint axles with a *semi-trailing-arm* suspension.

Though the swing-axle and semi-trailing-arm systems are both, *independent rear suspensions,* the general public hangs the label *IRS,* or *independent rear suspension,* on the semi-trailing-arm suspension only. To eliminate confusion, I'll stick to the popular names: Swing-axle and IRS. For a thorough discussion of the design and handling characteristics of the IRS and swing-axle suspension, see page 67.

The late swing-axle transaxle and early IRS transaxle are nearly identical, except for the axle-shaft joint at the transaxle side gears. The reason suspension design is mentioned here is that the suspension determines what type of transaxle you'll use. You cannot use an IRS with a swing-axle transaxle or vice-versa. If you want the desirable handling characteristics the IRS offers, you must use an IRS transaxle.

SWING-AXLE TRANSAXLE

VW swing-axle transaxles fall into two major categories: *split-case* and *tunnel-case.* The split-case transaxle is similar to the VW engine crankcase. It's made in two halves. The split-case transaxle is used in the '61-and-earlier Type 1 (sedan) and '60-and-earlier Type 2 (bus/transporter).

1961-and-later Type-1 and all '60-and-later Type-2 transaxles have the tunnel case, which is basically one piece. The tunnel-type swing-axle transaxle is most often used because of its durability and good availability.

The '60—'67 Type-2 transaxle is unique in that it is used with a set of reduction gears at each rear wheel.

Swing-axle transaxle can be identified by the rigid axle tubes between transaxle case and wheel-bearing flanges. Tunnel-case transaxle, used in '61—'68 Beetles, has a one-piece case.

Early split-case swing-axle transaxle. Unless you're restoring an early Beetle, avoid these. Parts are hard to find and they are not durable.

Transaxle front cover (nosepiece) and shift rod (hockey stick). These parts can be replaced without disassembling the transaxle. And you don't have to be a transaxle specialist.

IRS transaxle drives rear wheels through an axle flange at each side. Each flange accepts a CV joint, which drives an axle shaft, or halfshaft.

For more details on this tunnel-type transaxle, see page 35.

The split-case transaxle has some major disadvantages. First, parts availability is poor. Also, it uses a small-OD pinion bearing that cannot handle high torque and shock loads.

The split-case transaxle's biggest drawback is its straight-cut, non-synchromesh first gear. Believe me, when you're out in the soft stuff and the engine can't pull in 2nd, you need to shift into 1st gear quickly without stopping. If you have to stop to shift into 1st, you'll probably get stuck. The transaxle cannot stand up to being repeatedly forced into 1st gear "on the roll" so don't try. Therefore, split-case transaxles should be avoided unless you're just using your car for transportation.

Before I get into the pros and cons of the various transaxle designs, a final word about the pre-'61 VW sedan. If you have one of these early Volkswagens, it may be wiser to restore it for street use. Good ones are getting rare—and *valuable.*

If that vintage VW is beyond restoring, replace the split-case transaxle with a tunnel-type transaxle. Simply replace the transaxle front cover, or *nosepiece,* and shift rod, or *hockey stick,* with corresponding parts from a '60—'67 bus transaxle. The bus shift rod has a countersink on the bottom for the shift-linkage setscrew. To make the conversion, a new countersink must be drilled on top of the shift rod.

There's another way to adapt the tunnel-type transaxle to a car originally equipped with the split-case unit. Weld a bus-transaxle front mount in the car—VW part 111 701 073C. Some solid-mount transaxle kits make this easy with a front-mount adapter that bolts right to the tunnel-type transaxle. See Solid-Mount Kits, page 32 for details.

IRS TRANSAXLE

Two types of IRS tunnel-case transaxles are used off-road: the Type 1 and the much "beefier" Type 2. The '68-and-later bus IRS transaxle is an stronger unit, with a heavy-duty final drive and, on '68—'73 models, super-low-ratio 5.38:1 ring and pinion. Even though it's about 50% more expensive than a sedan transaxle, the bus IRS transaxle is the one used on winning off-road cars.

Left, next to a sedan transaxle, is a '68-and-later IRS bus transaxle. The outside tells you the bus is the stronger unit.

Originally, most off-road products were designed for the swing-axle transaxle because it was more popular and plentiful. But as more IRS VWs are being converted to off-road use, more products are becoming available. Today, most of the internal parts—close-ratio gears, heavy-duty keys, washers, and so on—have been adapted to fit the IRS transaxle.

For all but the racer, the stock Type-1 IRS transaxle, with solid mounts, is strong enough. Many competitors successfully use stock components in their race cars.

IRS transaxles are a little more durable than the Type-1 swing-axle. Some of the strain and snap that are directly transmitted to the transaxle in the swing-axle unit is absorbed through the double *constant-velocity* (CV) joints of the IRS unit. Also, lateral forces, or side loads, can quickly destroy a swing-axle transaxle. On the IRS, some of these are transmitted to the torsion tube through the diagonal arms.

Trouble can occur when wheel travel is increased. Increased wheel travel is gained by lengthening the diagonal arms and spring plates and modifying the spring-plate stops. The severe angles—due to longer-than-stock wheel travel—and stresses placed on the CV joints cause them to fail. Race-car builders keep figuring out ways to get more wheel travel out of the IRS setup. As of 1982, racing buggies were being built with up to 17 inches of rear wheel travel. Every additional inch of wheel travel increases the angle the CV joints must handle. For more details, see "High-Angle Joints," page 38.

SWING AXLE vs. IRS

The Type-1 IRS-transaxle case is basically the same tunnel design as the Type-1 '61—'68 swing-axle transaxle, but differs greatly at the differential side gears and covers. This is because of the different type of axle shafts.

The biggest advantage of the swing-axle transaxle is its availability. It was installed in the early and mid-'60s Beetles, the cars most people modify for off-road. Eventually, the scrap-yard crusher will tip the scales in favor of the IRS cars. But right now there are more older swing-axle cars to choose from, and usually at better prices. If you're buying a rebuilt transaxle for your existing car, either type, swing-axle or IRS, will set you back a similar amount.

Whichever type you end up with, make sure the transaxle is in good working order. It's the hardest and most expensive item to work on. It's nearly impossible to fix *anything* in a transaxle without removing it from the vehicle and tearing it down.

SOLID-MOUNT KITS

Whichever mount you choose, you must change or improve them for off-road use. The Type-1 engine's sole support is the rubber-mounted transaxle. The engine literally hangs off the end of the transaxle. These bonded-rubber mounts work well for the street, but will come apart with normal off-road abuse.

With a broken front mount, it is difficult to shift gears because the shift linkage is misaligned. Driving a car with a broken front mount will break the transaxle nosepiece. Broken *saddle-type rear mounts*—the ones at the bellhousing—allow the engine to move around as the throttle is opened and closed.

Traditionally, racers have eliminated breakage by replacing the rubber mounts with solid metal mounts. Solid-mount kits are available through most Volkswagen off-road parts suppliers. Many kits come with three mounts: a saddle-type rear mount and clamping strap, a solid-steel or aluminum front mount, and a middle mount—also a strap—that clamps around the transaxle case at the nosepiece mounting bosses.

Middle mounts differ from kit to kit. On some kits, holes must be drilled in the frame horns to attach the middle mount. This weakens the frame horns. On others, the middle mount merely clamps around the horns. Still other middle mounts use a weld-on bottom section, or cradle.

Solid-mount kits are available to adapt different transaxles into the Beetle. For example, a front mount available from Chenowth allows the '61-and-later tunnel-type transaxle to be installed in a sedan originally equipped with the split-case transaxle. A Sway-A-Way kit adapts the '68-and-later bus IRS transaxle into the IRS sedan.

Before you install solid mounts, consider this. Solid mounts transmit engine and transmission vibrations and road noise to the chassis, and to the seat of your pants.

More recently, some racers have begun installing urethane mounts or rubber-lined, *tire carcass* mounts. Sway-A-Way's urethane mounts feature steel sleeves that eliminate the

Transaxle front-mount bracket. If damaged, new one can be welded in. It's available separately. Front bracket often cracks and breaks at welds with hard off-road use if middle mount isn't used.

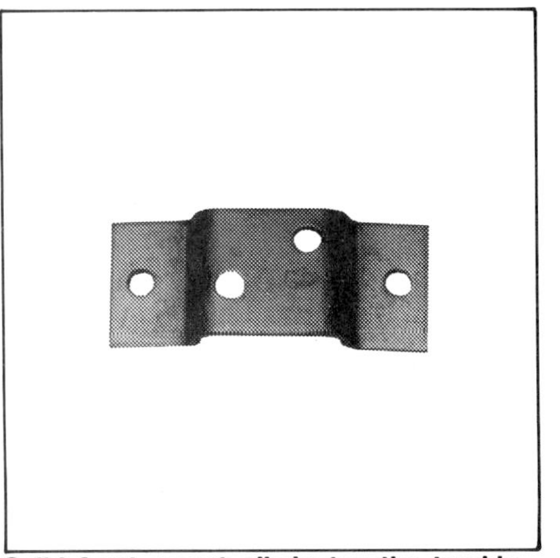

Solid front mount eliminates the troublesome bonded-rubber piece. Unfortunately, it also transmits drive-train noise and vibration to the driver and passenger. Photo courtesy of Chenowth.

Solid rear mount eliminates stock rubber saddle mounts. Strap clamps around top of bellhousing.

Precut front support for this late-bus IRS racing transaxle uses transaxle front-housing bolts for no-slop solid mount.

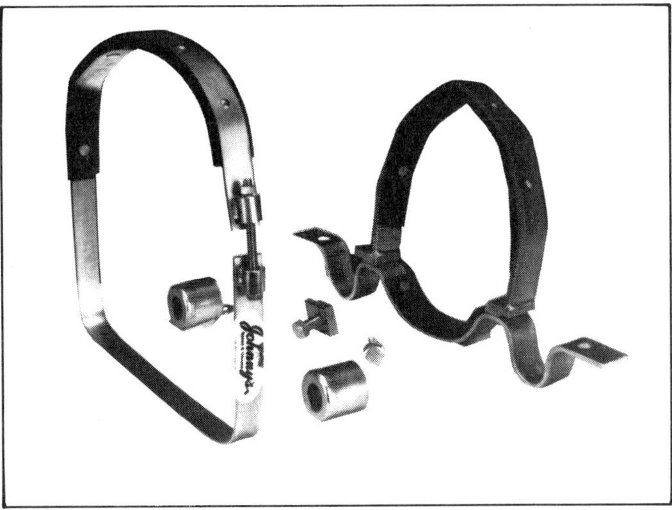

Transaxle mounts from Johnny's Speed & Chrome require no drilling or welding. Although not solid-type, these mounts secure transaxle to horns while damping noise and vibration. Photo courtesy of Johnny's Speed & Chrome.

Welded-in bottom half of middle transaxle mount, or cradle. A strap clamps down over the transaxle. This is for serious racers. See the photo at right.

Once the transaxle is dropped into the cradle, a strap is bolted in place on top. Photo by Ron Sessions.

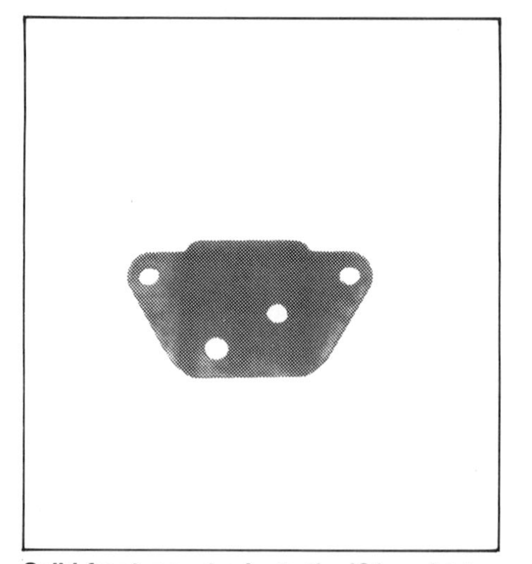

Solid front mount adapts the '61-and-later tunnel-type swing-axle transaxle to a '59-or-earlier sedan originally equipped with the split-case transaxle. Mount will not work with 1960-vintage sedan torsion tube. Photo courtesy of Chenowth.

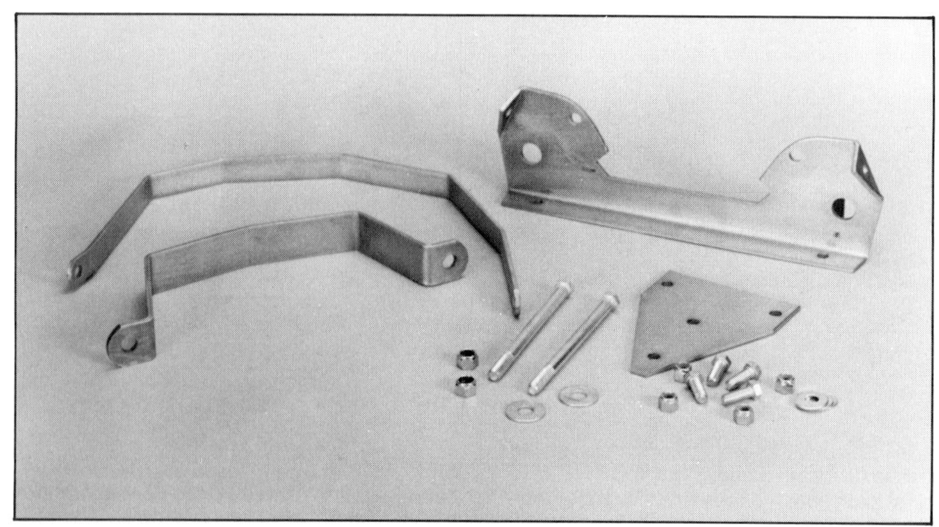

Solid-mount kit for the '68-and-later IRS-bus transaxle. Sway-A-Way kit adapts stronger bus transaxle into an IRS sedan. Photo courtesy of Sway-A-Way.

Transaxle-support kit from Crown features rubber-lined straps for middle and rear mounts. Photo courtesy of Crown Manufacturing.

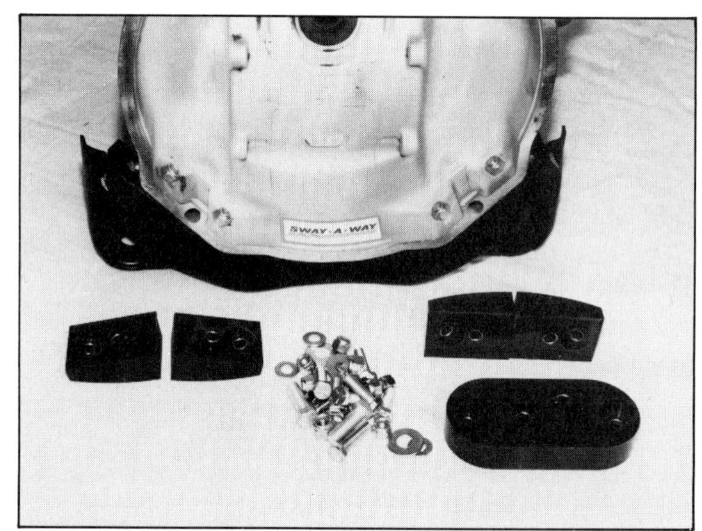

Sway-A-Way urethane mounts replace stock front and rear transaxle mounts. Urethane is many times stronger than rubber, yet damps vibrations effectively.

need for a steel backing plate. They fit in place of the stock mounts.

In the rubber-lined Crown mounts, the rubber becomes a sandwich between solid strap-type mounts and the transaxle case. Either of these mounts are excellent, especially for the casual off-road or street-driven car.

Installing Solid-Mount Kits—To install new transaxle mounts, solid or otherwise, remove the engine. Jack the transaxle off its old rubber mounts and install the solid mounts.

While you have the engine and transaxle out, inspect the clutch disc and pressure plate. Also check the clutch *release*—throw-out—bearing. If in doubt about its condition, replace it. It's cheap insurance. Also, this is a good time to reinforce the frame horns, torsion tubes and shock towers while you have easy access to them, page 75.

TO REBUILD OR NOT

A stock tunnel-type transaxle in good working order should last a long time, even if the engine is "tweaked" for higher output. A transaxle is expensive to tear into, so unless you plan to run a hot engine in a car set up for increased wheel travel, use the stock transaxle for awhile. This will give you enough information to judge whether changes are needed. You may decide a stock transaxle is fine.

If the transaxle is worn, have it repaired or buy an exchange transaxle. To rebuild one costs about the same as a rebuilt exchange transaxle, depending on how many parts need replacing. Sometimes, you'll get back an exchange transaxle anyway. You can also take a chance buying a used one—smaller price and bigger chance.

After living with the stock transaxle a while, you may not be satisfied with

Wider-spaced axle flanges on '67 and '68 swing-axle sedans and all swing-axle Type 3s (below) give more tire clearance in the fenderwell. Flanges move the wheel-mounting flange farther outboard.

its gearing. So this might be a perfect excuse to rebuild yours the way you want it. Late-type IRS transaxles have higher final-drive ratios than the swing-axle types.

By the time VW got around to using the IRS system, engine displacements were increasing with resultant increases in horsepower and, more importantly, torque. Due to the added horsepower and torque, VW could raise (numerically lower) transaxle gearing without hurting acceleration.

If your engine won't "pull" 4th gear with the big tires you're running, consider a set of close-ratio 3rd and 4th gears. You can also change the ring and pinion to a lower ratio and get a similar effect. But for a dual-purpose Baja Bug or recreational buggy, first gear will then be so low it will be useless for anything but pulling stumps.

While you're changing transaxle components, consider changing axle shafts too. In the case of a swing-axle transaxle in a sedan or Baja Bug, you can increase the rear track by swapping some parts. This allows for more effective wheel travel and adds clearance for bigger tires. Use the longer swing axles and wheel-bearing housings from a '67 or '68 Type 1 or any pre-'68 Type 3. These pieces bolt right up to the earlier swing-axle transaxles. Various aftermarket axles are available too. See "Heavy-Duty Parts," page 38.

Transaxle Economics—A fully reworked off-road race transaxle will cost nearly $2000. Serious racers have the transaxle completely gone through after *every* race at a price of about $500.

Many special parts are made to modify the VW transaxle for off-road competition. Special off-road parts include spider-gear assemblies, stronger side covers and cover supports, heavy-duty clutch-release shafts, hardened shift forks, close-ratio gears, hardened gear keys, forged axles and so on. But these parts won't be effective if assembly isn't done correctly.

Lots of folks can assemble a manual transmission without much difficulty. But how many do you know who can set up the preload, backlash and side gears on a final drive? It's almost a lost art. There are very few good racing-transaxle builders to choose from. You'll find them expensive and very busy.

Most off-road racers won't even attempt to rebuild a VW transaxle. Yet an uncanny number of shadetree tinkerers will give it a try. For these people, I have this advice.

Get a VW factory service manual and read the procedure until the information sinks in. Read it several times if necessary. Set aside plenty of time for the operation.

Try to get your hands on a Crown Trans-A-Jig. This is a jig for building the transmission part of the VW Type-1 transaxle—both '61-and-later swing-axle tunnel type and IRS type. Without the jig, it's virtually impossible to set the vital transmission clearances. For the final-drive part of the transaxle, you're still on your own.

BUS SWING-AXLE TRANSAXLE

For the low-budget off-roader who can't keep his swing-axle transaxle from breaking, a '60—'63-1/2-bus or Type-2 transaxle provides a good compromise. Racing use has proven it more durable than the Type-1 unit.

Its strength is derived from two large reduction gearboxes at the wheels. These auxiliary gearboxes take some strain off the gears in the main case. With more gears to share the torque multiplication, individual gear-tooth loads are reduced. So gears last longer.

TRANSAXLE GEARS & GEARS RATIOS

4th gear ratios available:

.82	1.14
.89	1.21
1.00	1.32
1.04	1.43

Ring & pinions available:

3.88
4.12 (Strongest)
4.38
4.62
4.86

Reduction ratios available:

1.26 (used ones are rare)
1.38
1.40
1.68

All these gears are available from, and will fit, the swing-axle, IRS and early-bus reduction transaxles.

Formula for finding final drive:
(4th gear) X (ring & pinion) X (reduction gear, if used)

Example: 1.32 X 4.12 X 1.38 = 7.51
or 1.32 X 4.12 X = 5.43

Which means: Every time the engine turns over 5.43 (7.51) times, the rear tires will turn over once.

A common stock ratio is: 5.78
1.32 X 4.37 = 5.78

There are 96 different final-drive ratios available, counting reduction gears, that range from a low gearing of 11.68 to 1, up to a high-gear ratio of 3.38 to 1.

1.43 X 4.86 X 1.68 = 11.68
.82 X 4.12 = 3.38

A '63-1/2-or-earlier reduction box with homemade adapter to lay the box back. This gives an automatic increase in wheel travel and tire clearance with the swing-axle suspension.

Stub axle and countergear from a bus-transaxle reduction box.

Another benefit of the bus transaxle is its lower (numerically higher) overall gearing. This gearing helps keep engine rpm up when using large off-road tires.

Still another bus-transaxle plus is the fact that its overall gear ratio can be changed without working on the transaxle itself. All you do is change two reduction gears at each wheel. In the '60—63-1/2 bus transaxle, four different reduction-gear ratios are available: 1.68:1, 1.40:1, 1.38:1 and 1.26:1. The 1.68:1 gearset is popular with off-roaders and is known as the "Alpine Gear" because it's a natural for mountain driving. Used examples are hard to find, but new ones can be special-ordered from Europe.

The '63-1/2—'67 bus transaxle is another story. Although it also uses reduction boxes, less-durable reduction-box bearings and lack of choice of gear ratios make it less desirable. On this box, 1.26:1 reduction gears were the only ones available.

Whichever ratio you choose, the stock '60—'63-1/2 bus transaxle will provide nearly as much strength as a modified Type-1 unit at a fraction of the cost. To adapt either of the '60—'67 bus transaxles to a sedan requires some work, as I will soon describe.

Also, adding a reduction-gear assembly at the outboard end of each axle increases *unsprung weight*—the weight that *moves* with the wheels. Add the weight of the braces and adapters needed to locate the reduction boxes and you've got an even bigger increase in unsprung weight. Unsprung weight slows suspension reaction time and detracts from handling.

Reduction-Gear Transaxle Swap—To modify the '60—'67 bus transaxle so it will bolt into a sedan, you must replace the transaxle *nosepiece* and *hockey stick,* with parts from a Type-1, tunnel-type swing-axle unit. This is simply a matter of unbolting one item and replacing it with another; no adjustments are necessary. The bus transaxle will then fit. But you will need some adapters to mate the reduction boxes to the spring plates.

Reduction-Box Mounting—Mounting the reduction boxes upright will give about three inches more rear ground clearance and will increase fenderwell clearance. You can buy adapters to mount the bus transaxle, axle tubes and reduction boxes to the sedan's rear-suspension torsion-bar housing and spring plates.

The problem with this setup is wheelhop under acceleration. Wheelhop occurs because the reaction torque of the reduction boxes causes the spring plates to flex in an oscillating motion.

John Johnson Racing Products offers a kit that permits using the boxes in the upright position with minimal wheelhop. The kit features a polyurethane-bushed strut to stabilize the lower end of each reduction box. The kit also includes axle overtubes to keep the axle tubes from bending.

Most people reduce rear wheelhop and increase the wheelbase of their car three inches or so by "laying down" the reduction boxes with adapters. Two versions of the laydown adapters are available: One positions the reduction boxes straight back in line with the spring plate. The other has a 10° drop. With the laydown adapters, use an axle overtube and brace to keep the axle tubes from bending. Laydown adapters are also available with four built-in shock mounts.

With all these adapters, you'll need to drill several holes in the spring plates. With the added leverage of the reduction boxes hanging off the spring plates, three bolts aren't enough. Most racers add at least two more bolts per spring plate. Also, when you use laydown reduction boxes, you must use special upper and lower shock mounts. See page 78 for details.

In case you're wondering, yes, you can use the bus stub axles and reduction boxes with a sedan gearbox, but the ring gear must be reversed, or *flopped.* If not reversed, the transaxle will have one forward speed and four reverse speeds.

Reduction-Box Transaxle Drive Characteristics—When you drive a car equipped with the bus reduction boxes, you'll notice some strange driving characteristics. Under acceleration, the rear suspension of a stock swing-axle or IRS car squats because of weight transfer. But with a bus reduction setup, the rear of the car rises and the suspension gets stiffer on acceleration.

The rear of the car now rises and the suspension stiffens because of two things: a change in suspension geometry and torque being applied to the reduction boxes. Without getting into a complex discussion of suspension theory, it's the old action-reaction game.

The dynamics of the revised rear-suspension geometry overcome rearward weight transfer under acceleration. The combination of moving the wheel center lower in relation to the spring-plate pivot and the car's center of gravity (CG), particularly with the upright-mounted boxes, and the tractive force of the tires now raises the car.

Additionally, the torque reaction of

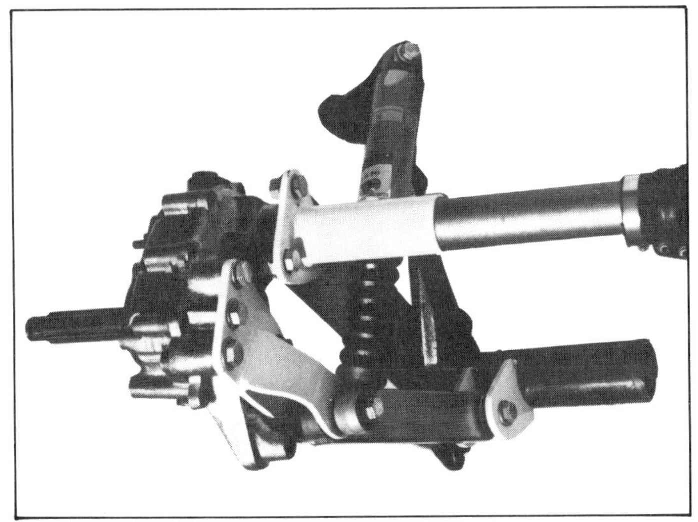

Johnson offers a kit to minimize wheelhop caused by axle-tube and spring-plate deflections. Wheel hop may occur whenever a reduction box is mounted upright. Note axle overtube and link rod. Photo courtesy of John Johnson.

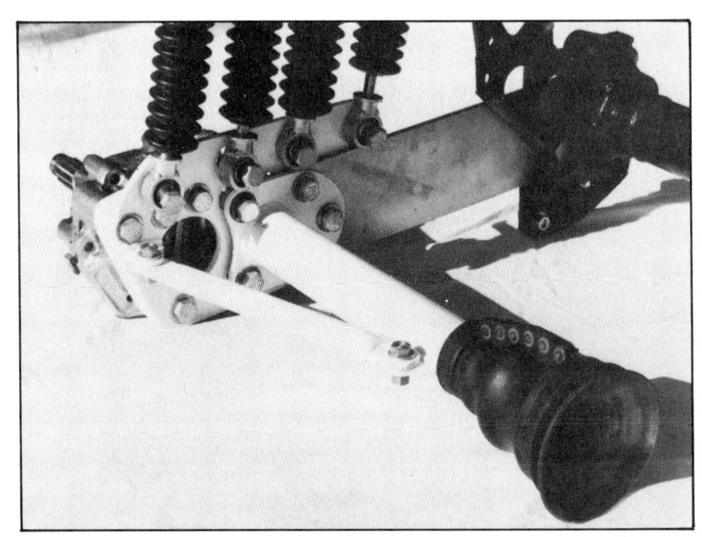

Adapter for laying back the reduction box includes four shock mounts. Axle overtube, brace and adapters without shock mounts are also available. Photo courtesy of John Johnson.

Metal may have to be removed from '66-and-earlier cases to make clearance for 200mm flywheels. Note marked areas. Flywheel can be used as a "milling machine" to clearance the bellhousing. With the engine loosely bolted up as close as possible, start the engine and gradually tighten up the bolts.

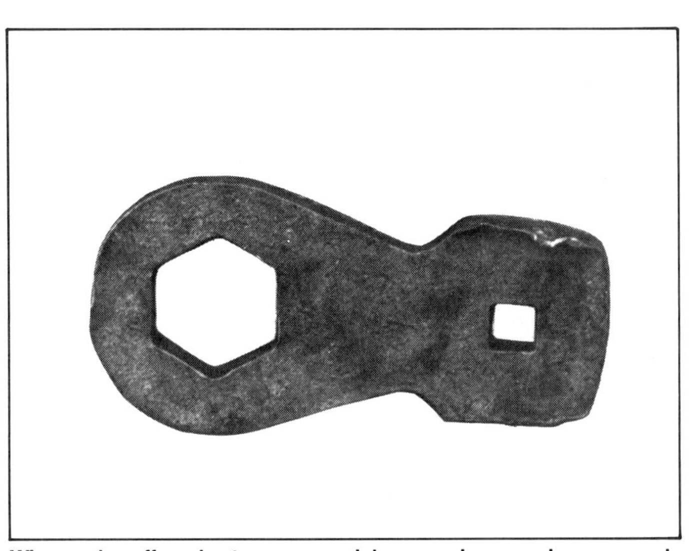

When racing off-road, a torque wrench is so much excess baggage and tightening to correct torque specifications an imprecise art. Most racers use this tool, a 5-lb sledge and a calibrated "feel" to remove and retighten rear axle nuts to the 200—220-ft-lb range.

the axle shafts at the reduction boxes—they turn in the opposite direction—tries to rotate the spring plates down and forward, thus raising the rear. Not only is the rear raised, it is held up as long as power is applied. This makes the rear suspension effectively stiffer.

Some drivers claim to be able to use the characteristics of the bus reduction box to their advantage. Here's how. Over rough terrain, careful use of the throttle can stiffen the rear suspension sufficiently to prevent bottoming out. If caught by suprise at too high a speed, brake before hitting the bumps, then apply the throttle to stiffen the suspension.

FLYWHEEL CLEARANCE

The tunnel-type transaxles, used in '66-and-earlier sedans and '65-and-earlier buses, were designed for the 180mm-OD flywheel. The '67-and-later sedans and '66-and-later buses used a 200mm-OD flywheel. When the larger flywheel is mated to one of the early transaxles, there may be interference between the flywheel and bellhousing. To accommodate the larger flywheel, some areas of the bellhousing must be cut or ground away.

Sometimes 200mm flywheels are inadvertently installed on earlier engines. This can happen when the engine is rebuilt or when the flywheel is replaced so the problem can show up anytime.

If you can't get the engine to slide into engagement with the transaxle, get it on as far as you can. Take a wrench and rotate the crankshaft at the pulley. Remove the engine.

The marks the flywheel made in the bellhousing show you where it has to be cut, or *relieved.* Sometimes only a little material has to be removed; other times it's a bigger job. Usually, material must be removed from the engine-stud-hole bosses. Sometimes material must also be cut from the lip just inside the bellhousing.

A hammer and a small sharp chisel can be used to remove metal from the boss. Be careful not to cut away too much at once. A grinder should be used to relieve the rest of the bellhousing, as this area is too weak and brittle to chisel.

The transaxle case is cast magnesium so don't create any sparks. **Magnesium chips and dust are extremely flammable.** Cut out or grind away some material, then wipe a little dirty oil on the areas you have relieved. Try installing the engine again. If it won't go, turn the crank again, remove the engine and cut away more metal where interference is indicated.

Early tunnel-type transaxles may also require relieving near the clutch arm for the larger flywheel. If you have one of these, compare your transaxle with a '67-or-later transaxle to see where to add clearance.

Many starter/flywheel combinations are possible. If you are switching from a 6- to a 12-volt system, install a 130-tooth, 200mm flywheel with a 12-volt starter. A smaller-ID starter bushing will be required in the transaxle bellhousing unless you use a *self-supporting starter.* Self-supporting starters were used in Porsche 912s and VWs equipped with automatic transmissions.

HEAVY-DUTY AND RACING PARTS

There is no such thing as a bulletproof VW transaxle. That is just salesman and catalog talk. Some racers can bolt a full-race engine to a stock transaxle and have it last many a mile. At the other extreme, there are people who can destroy a "bulletproof" race-transaxle with a stock engine in no time.

If you are hard on equipment—and you should know it if you are—have the transaxle professionally built, *with the best components.* If you're easy on equipment, you may get by with a much smaller investment in the transaxle.

Many of the heavy-duty transaxle parts replace parts that most people never break. But the more horsepower you feed into the transaxle, and the harder you drive, the more you'll need these heavy-duty parts.

IRS Shafts and Joints—Unlike the swing-axle suspension, there is angular movement between the outer end of each IRS axle shaft and the stub axle at the wheel. The amount of movement varies as the suspension goes through its travel. This requires a joint at the wheel and one at the transaxle. The IRS does not require an axle tube like that used with the swing axle.

Two basic types of drive-shaft, or axle-shaft joints are used in the auto industry: the *universal joint,* or *Hooke's coupling;* and the *constant-velocity (CV) joint.*

Usually referred to as a *U-joint,* the universal joint looks like two U's joined by an X-shaped *spider.* The spider is mounted in needle bearings at each yoke. One yoke drives the joint through the spider to the other yoke.

CV joints come in a number of different designs. The standard CV joint on IRS VWs utilizes steel balls between an inner and outer race. The balls are trapped in cages, much like those in a wheel bearing. Power is transmitted from one race to the other through the steel balls.

Other dissimilarities occur between the IRS and swing-axle suspension because of their differences in geometry and wheel-locating methods. The IRS axle is not used as a suspension-locating member as is the swing-axle's tube. Instead, the axle is free to move in and out of its joints.

The VW CV joints are designed for this longitudinal movement. In-and-out movement is necessary because the distance between the stub-axle flange and the transaxle side plate changes with rear-wheel travel. Not so with the swing axle.

The first thing off-road racers discovered with the Type-1 IRS was that the stock Type-1 CV joints didn't last long. After three races at the most, the cages would crack. For long-distance desert races, everyone carried spare axle shafts and CV joints for emergency repairs.

There was also a problem with the stock axle shafts. These shafts have shoulders that are held next to the CV-joint inner races—no problem for street operation. It's another matter when it comes to serious off-roading.

When a car lands on one wheel or hits a rock, the half-shaft shoulder overloads the CV joint. The high thrust load of the axle shaft slamming into the joint will crack or Brinell—dimple or indent—its races. This turns a CV joint into instant junk.

Constant-velocity (CV) joint transmits power via caged steel balls. This is a Porsche 930/935 Turbo CV joint. Off-road racers have adopted this joint because of its durability and ability to operate up to 24°. Photo by Ron Sessions.

IRS Axle Shafts—Early on, racers began machining off the axle-shaft shoulders to save the CV joints. But they discovered that this considerably weakens the stock axle shaft. Today, the remedy is to replace the stock shafts with heavy-duty aftermarket ones. Sway-A-Way makes an axle shaft with extra-long splines. This allows the axle shaft to "float" in the inner CV-joint race, preventing the axle shaft from hitting the joints.

Another feature of the Sway-A-Way shafts is that they twist, or wind up, like a torsion bar under high torque loads. This cushions or absorbs shock loads that would otherwise be transmitted to the joints and transaxle. Heavy-duty axle shafts are available from other manufacturers such as Henry's, Dura Blue and the Summers Brothers. Many types are offered in a variety of lengths for different vehicle widths and suspension travel.

High-Angle Joints—Not only does the shoulder on the stock axle shaft tear up the CV joints, Type-1 joints don't have enough angular travel to handle the increased suspension travel required of them on many off-road cars.

One way to increase travel has been to install modified Datsun Z-car or Spicer U-joints. These joints are more durable than the stock CV joints. The yokes must be clearanced to provide a larger angle of movement than they were originally designed for.

Another modification is required; splined axle shafts. Unlike the stock

IRS axle shaft with clearanced Datsun Z-car U-joints. Notice radiator hose slipped over shaft to protect splines against rock damage. Photo by Ron Sessions.

Rear suspension with over 10 inches of wheel travel uses Type-2 CV joints to handle the angles. Note safety-wired Allen bolts and protective sleeve over shaft. Note also braided-steel brake line and hydraulic clutch—all good stuff.

VW CV joints, U-joints are only capable of angular movement. They are not capable of in-and-out movement. Splined axles also add unsprung weight; an undesirable.

But as rear-suspension travel increased still further, the Z-car and Spicer-type U-joints also reached the limits of their angularity—clearanced or not—and lost favor. Filling the void of late has been the larger Type-2 CV joints from a '68-or-later bus.

While only slightly more expensive than the Type-1 CV joint, the bus CV joint is more durable and can operate continuously at 17°. It can handle 19° *intermittently.* The bus CV joint can be adapted easily to sedan IRS axle shafts if VW Thing (Type-181) flanges and stub axles are used.

The top of the heap is the 930/935 Porsche Turbo CV joints and axles. The Porsche CV joints cost about twice as much as the bus joints—and you will need four. To use these CV joints, Porsche flanges and specially fabricated axle shafts and stub axles must be used—an expensive proposition. If you can afford to go this route, you'll have joints capable of operating up to a 24° angle.

Special IRS Stub Axles—The stub axle is the short axle at the wheel hub. Early on, off-road racers found that the VW Thing Type-181 stub axles are stronger than the sedan units. Custom-forged stub axles are now available for almost any IRS setup. Among these, the larger-OD bus stub axles are favorites.

Special Swing-Axle Shafts—When it comes to the swing-axle transaxle, most racers merely shot-peen and polish the spade end of the stock-VW axle shafts. This process minimizes the tendency of the ends to gall and fatigue.

When buying used shafts, it's a good idea to clean them and have them *Magnafluxed*—magnetic-particle inspected. Late-style ('67—'68) fulcrum plates are best because they are rounded and have one or two grooves for lubrication. Those with one oil groove are for use in a transaxle in good condition. Those with two grooves are 0.00039-in. *oversize* and used with a worn axle shaft or side gears.

VW swing-axle shafts are manufactured within a tolerance range. Therefore, you must match the spade ends to the fulcrum plates and install the correct thrust washer. Don't fit these so tightly that there is no slop in the assembly. For off-road use, there should be enough clearance between each axle shaft and fulcrum plate to allow oil to pass to each side gear.

Heavy-duty swing-axle shafts are available from aftermarket sources. Although stock shafts are forged and surface-hardened from the factory, some aftermarket sources make axles out of a heat-treated, hi-tuff material.

This material is ideal for axle shafts, but the threads for the axle nut are a bit soft compared to stock VW shafts. So, it is important to use a *new,* clean axle nut. Gene Berg recommends using one from a Type 3 or one from '61—'62 Type 2. These are both forged nuts with flanges. Part number for the Type-3 nut is 311 501 221. Berg claims it is easier to torque.

Swing-Axle Overtubes—As mentioned earlier, swing-axle tubes can be strengthened at their outer ends. An axle *overtube,* slipped over the existing axle tube, provides this strength. Without it, the axle tube will bend under hard off-road use. For more information on axle overtubes, see page 80.

Special Hardened Rear-Axle Spacers—On the wheel end of each axle shaft, inside the bearing housing, there is a pair of steel spacers. These locate the wheel-bearing-inner race and brake-drum hub. On the IRS transaxle only, an additional spacer sleeve locates each stub-axle in its bearing housing.

With hard use and excessive axle-nut torque, these spacers will crush and/or crack. If you repeatedly need to retorque the axle nuts, chances are these spacers are at fault.

Extra-hard 4130 steel spacers from Sway-A-Way will cure the problem. For the swing-axle transaxle, Sway-A-Way's outer spacer is extra thick. This gives the additional clearance inside the brake drum required for wheel-stud conversions.

Whichever spacers you use, take care not to overtorque the axle nuts. Consider 220 ft-lb (swing-axle) or 250 ft-lb (IRS) as the limit.

Close-Ratio Gears—As mentioned earlier, large-diameter off-road tires seriously hamper performance, especially in 3rd and 4th gears. To counteract the problem, your options include changing the final-drive or transmission gearing. A larger displacement engine won't make things much better. If you're happy with the gearing in 1st and 2nd, don't change the

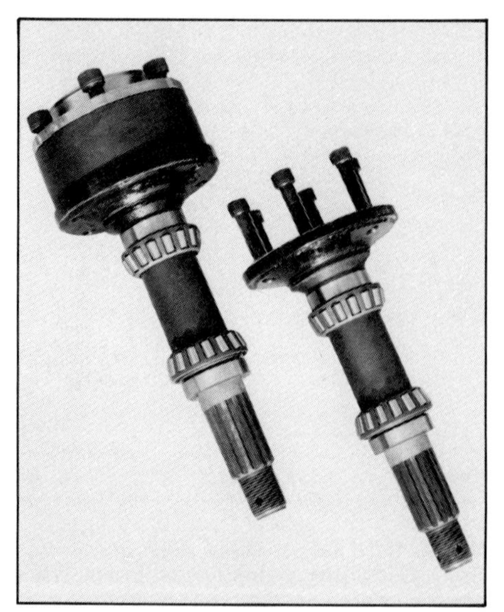

IRS stub axle joins outer CV joint to the wheel. Stub axles are large-OD Type-2 units. Photo by Ron Sessions.

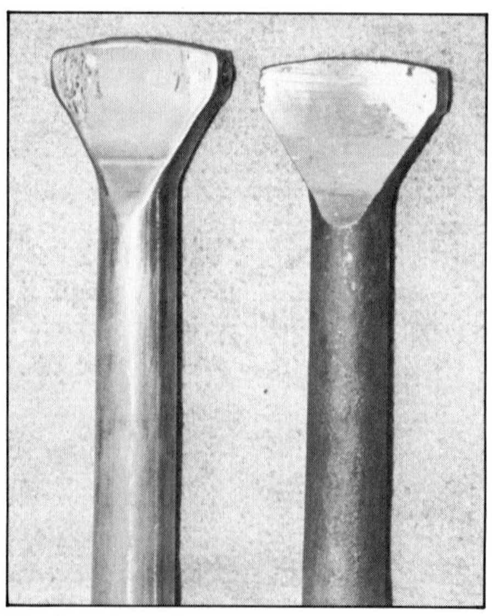

Stock swing axles are strong enough if you do a little work on their weakest area. Light grinding and polishing on the spade end will help prevent cracking in the area indicated by the rough spots on the shaft at left.

Tall rear tires can seriously hamper performance. Installing close-ratio gears for 3rd and 4th speeds can restore much of that performance. Photo courtesy of Crown Manufacturing.

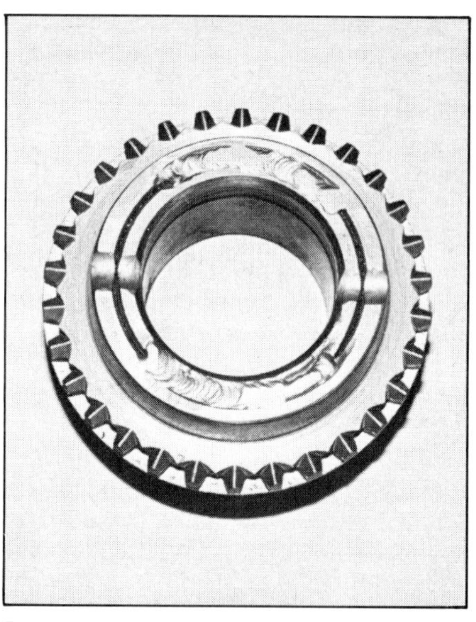

Pressed-in synchronizer hub is TIG-welded to gear to make sure it stays put. ID is honed after welding to restore shaft tolerances.

final-drive ratio. Changing the ring and pinion might make these gears too low. Most racers end up changing the gearing in 3rd and 4th.

In most VW transaxles, 3rd gear is usually 1.31:1 or 1.21:1, and 4th gear is an overdrive 0.89:1 or 0.82:1. Close-ratio gears available from aftermarket sources include 1.48:1, 1.58:1 and 1.71:1 for 3rd, and 1.04:1, 1.12:1, 1.14:1, 1.21:1, 1.31:1 and 1.43:1 for 4th. Judging by their popularity, the optimum gear ratios appear to be 1.58:1 for 3rd, and 1.31:1 for 4th, which is close to the stock "tall" 3rd gear.

If you do much pavement driving, remember: A close-ratio 4th will be like driving in 3rd gear all the time. That means lower top speed, more engine noise, more engine wear and lower gas mileage. It's a good trade-off only if your primary objective is good off-road performance.

TIG-Welded Gears—Both 3rd and 4th gears have a *synchronizer hub* pressed into them. The synchronizer matches gear speeds so gears can be engaged without clashing during gear changes. A common occurrence, especially when running a large engine, is the press fit comes apart and the transaxle binds up.

To prevent the possibility of this press fit coming apart, most off-road transaxle builders grind a slot and TIG-weld the two pieces together. Crown Manufacturing says that three or four 3/8-in.-long tack welds are sufficient. I've seen this done with a low-cost arc welder, but most people TIG-weld it to keep heat to a minimum. Local overheating can ruin the heat-treating of the gear and synchro. Also, have the welded synchro-hub ID honed to maintain proper shaft tolerances. Lap the thrust surfaces flat.

The '72-and-later bus transaxle comes stock with an electron-beam-welded 4th gear.

Steel Shift Forks—Some '69—'72 tunnel-type transaxles used a brass shifting fork for 1st and 2nd gear. These shift forks are soft and are not suitable for off-road.

Most transaxle builders replace the brass fork with a steel one. Therefore, if the transaxle has ever been rebuilt, it may have a steel fork.

It's been my experience that the stock VW forged steel forks are more durable than any aftermarket heavy-duty forks. If you feel better using cast aftermarket parts, go ahead, but I think you'll be money ahead using the less expensive, forged VW forks. Just make sure they are the correct width; 0.165 in. for transaxles built before mid-'72 and 0.205 in. for mid-'72 -and-later models.

Special Hardened Gear Spacers and Keys—Other items to be included in any transaxle used off-road are hardened gear spacers and keys. Although I've successfully raced cars that didn't have these installed, most good transmission builders install them.

The stock keys that hold 3rd and 4th gears on the transaxle mainshaft may shear off with off-road abuse. Considerably stronger than stock, the hardened aftermarket keys resist shearing.

Between 3rd and 4th gear on the pinion shaft of the '60—'67 tunnel-type transaxle are spring washers and a metal sleeve. These often fail in off-road use. Crown Manufacturing has a heat-treated spacer and beryllium-copper spring washers to cure this problem. Later-model transaxles don't need this improvement.

Heavy-Duty Side Covers—Under the

RACE-CAR GEAR-RATIO CHART

CLASS	RING & PINION	3RD GEAR	4TH GEAR
Open Classes—Desert Racing	4.86:1	1.43:1	1.05:1
Buggies—Barstow and Mint Courses	4.86:1	1.43:1	1.14:1
Restricted Classes—Desert Racing w/High-rpm Power Engine	5.38:1	1.58:1	1.32:1
Restricted Classes—Desert Racing w/Wide-Power-Band Engine	4.86:1	1.48:1	1.14:1
Class 9 and 1600cc Restricted Classes	4.86:1	1.58:1	1.32:1
Single-Seaters—Short Courses	5.43:1	1.43:1	1.05:1
Sedans—Short Courses	4.86:1	1.58:1	1.32:1

Popular gear ratios used in various off-road-racing classes. Data courtesy of F.A.T. Performance.

Stock 1st—2nd-gear shift fork (right) is brass on some models. It will bend under hard shifting. Crown offers a steel replacement (left) that resists bending. Photo courtesy of Crown Manufacturing.

transaxle side covers are the shims that adjust ring-and-pinion preload and backlash. The stock covers flex under load. This allows the ring gear to move away from the pinion gear, causing rapid wear or gear damage. This isn't much of a problem on a stock transaxle with a stock engine. But the more horsepower and the more wheel travel you add, the bigger this problem becomes.

Heavy-duty side covers are more rigid, to eliminate the flexing of the stock covers. HD side covers are available in both steel and cast-aluminum from aftermarket sources. The aluminum covers are lighter, for the weight-conscious racer, and slightly more expensive.

If you decide to use aftermarket HD side covers, buy them from a reputable source. Manufacturing tolerances can vary so widely that it can be nearly impossible to adjust ring-and-pinion preload and backlash.

In the case of an IRS setup, use the stock covers from a '75-and-later Beetle. They have proven to be very durable, and quality control is good, as usual. You can even use two left-side covers from the Super Beetle. The right side of the Super Beetle IRS transaxle is cast shut so it uses only one side cover. Use two of the same pieces on the right and left.

For fun buggies, a single HD cover on the *thrust side* should be sufficient. The thrust side of the transaxle is the side with the ring gear. On all but the swing-axle reduction-gear bus transaxle, the ring gear is on the left. On the reduction-gear transaxle, the ring gear is on the right side. Race cars need HD covers on both sides.

Heavy-Duty Axle-Tube Retainers— On swing-axle cars, you can solve two problems at one stroke. The stock side covers tend to flex and the stock axle-tube retainers leak. Add aluminum or steel heavy-duty retainers to keep the tube's outer flanges tight against the transaxle side covers. This solves both problems.

Crown Manufacturing's steel retainers come in one- and two-piece designs. To install the one-piece retainers, the axle shafts and tubes must be removed. The split design of the two-piece retainers enables them to be installed without removing the axle shafts from the transaxle. After removing the axle-tube-to-transaxle nuts, you can slip the split retainers into place.

Spider-Gear Kits—The stock-VW differential assembly uses two spider gears to drive the side gears. The axles connect to the side gears. In off-road use, drive-axle thrust destroys spider gears. This is especially common when a car lands on one rear wheel while accelerating in hard, rocky terrain.

Spider-gear kits have the load-bearing benefits of a four-spider differential, but at less cost. The kit consists of an extra shaft that supports two more spider gears. The new shaft

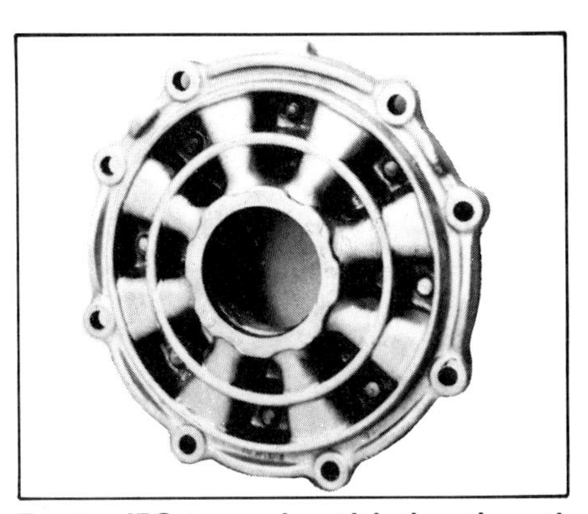

For the IRS transaxle, original-equipment side covers from the '75-and-later Beetle and Super Beetle make excellent choices. Photo courtesy of Phil's Inc.

incorporates a block with a hole in the middle so it can be mounted on the existing stock shaft. The stock carrier is retained, and no machining is required. Operation of a steering brake, if used, is not affected.

Although this may seem like a good idea, especially because of the low cost, it's my opinion that the kit creates a new problem as it solves the old one. The shaft carrying the two new gears isn't supported at its ends. Therefore the stock shaft must support four gears—twice what it was designed for. If you have a transaxle this far apart, do it right and install a *Super-Diff*.

Super-Diff—A Super-Diff is a specially machined, high-strength differential carrier that replaces the stock carrier. It incorporates four spider gears, thus doubling its load-bearing

For fun buggies, a single heavy-duty cover on the *thrust side* should be sufficient. On all rear-engined cars, except those using the swing-axle reduction-gear bus transaxle, the thrust side is on the left, as shown.

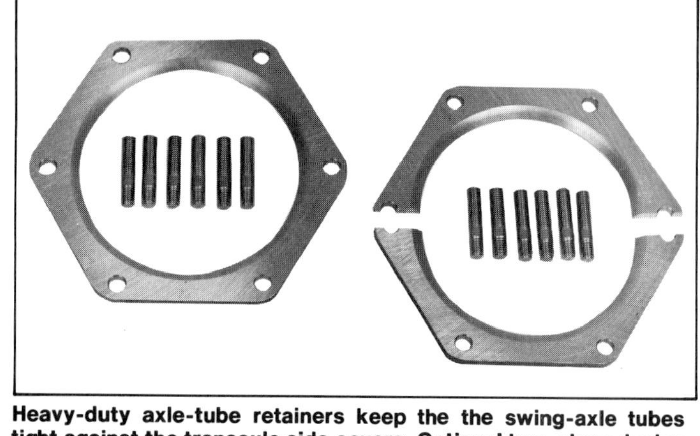

Heavy-duty axle-tube retainers keep the the swing-axle tubes tight against the transaxle side covers. Optional two-piece design allows installation without disassembling axle shafts from swing-axle transaxle. Photo courtesy of Crown Manufacturing.

Spider-gear kit doubles the number of spider gears in the VW differential—from two to four. Photo courtesy of Crown Manufacturing.

Racing Super-Diff (left) and a stock differential carrier (right). Features include a threaded collar (1) instead of the stock retaining ring; built-in oil scoop (2); and two sets of holes (3) for two spider-gear shafts instead of one. Scoop is drilled and opened up on one side. A smaller hole is used on the other side.

capability. Unlike the spider-gear kits, the Super-Diff carrier supports the extra spider-gear shaft at the ends.

The IRS Super-Diff manufactured by Crown features an oil reservoir between the bearing surfaces to aid lubrication, especially on steep terrain. On the Crown swing-axle Super-Diff, lubrication is aided by additional oil grooving and a cast-in scoop in the carrier. To accommodate the different direction of rotation of the sedan swing-axle transaxle and the bus reduction-gear transaxle, Crown center drills the special oil grooves. The customer finishes drilling the oil grooves according to the direction of ring-gear rotation.

A steering brake can be used with it. A Super-Diff is a great addition to any off-road transaxle.

All Super-Diffs are not alike. For the swing-axle transaxle, you have a choice of a Racing Super-Diff or a Snap-Ring Super-Diff. The major difference between the two is how the side-gear thrust collars are retained.

On the racing model, the ends of the carrier are threaded. These threaded ends accept collars, which screw in. They are then pinned for positive retention. This eliminates the stock method of retention that consists of a large retaining ring over each thrust collar.

Under the severe stresses of competition, the stock retaining ring can pop loose. This allows the collar to move out and the side gear move away from the spider gears. The result is a broken tranny.

On the other version of the Super-Diff, the stock snap-ring or retaining-ring setup is used. Unless your plans include racing, I suggest you stick with the Snap-Ring Super-Diff. Although weaker than the threaded-collar type, the Snap-Ring Super-Diff model is strong enough for recreational off-roading. It also offers one big advantage if you ever need it. It's repairable in the field.

With the Snap-Ring Super-Diff, you can replace a broken axle shaft without removing the engine, transaxle and all of the rear-suspension components. On the racing model, everything must come out to replace an axle shaft.

With large off-road tires, bending an axle at its outboard end is not uncommon. Sliding into a curb or hitting a large hole at an angle are common causes. Also, if you break an axle inside the transaxle and are miles from civilization, the Snap-Ring Super-Diff allows you to unbolt one side of the suspension, remove the side cover and axle tube, and replace the axle. All you need is another axle.

If you're building a high-performance, swing-axle off-road car and expect to tear into the transaxle now and then, use the Racing Super-Diff. Otherwise, stick with the Snap-Ring Super-Diff.

Limited-Slip Differential—An expensive item that has never caught on

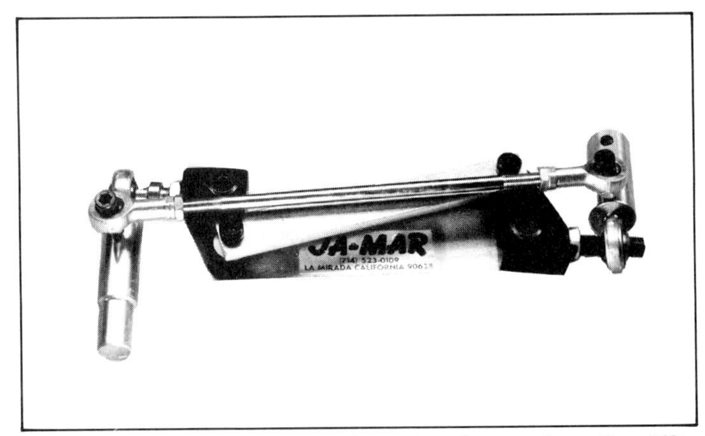

Single-seat-buggy driver's seat is centered over where the shift linkage would normally go. Ja-Mar side-shifter offsets linkage so the driver doesn't have shift lever between his knees. Photo courtesy of Phil's Inc.

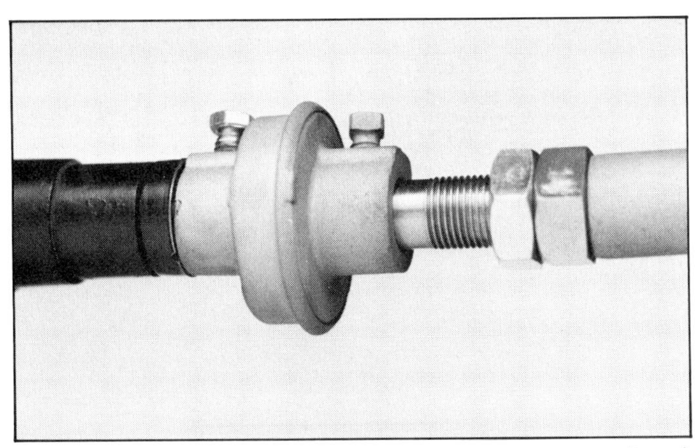

Shift-linkage problems? Unless you've had them, you can't appreciate this item. This Sway-A-Way piece provides complete linkage adjustment. Photo courtesy of Sway-A-Way.

200mm-diaphragm pressure plate (Luk 31141025) at left compared with 180mm coil-spring type. Speed-equipment companies offer special racing clutch plates for the VW, but I know this stock diaphragm pressure plate works great.

Left is stock clutch disc. At right is a ceramic or "puck" disc used in racing. Stock disc will fail if you get it too hot when slipped; ceramic disc just grabs better.

for off-road use is a limited-slip differential. Though it would seem that added traction in rain, mud and deep sand would be worth pursuing, the drawbacks of limited-slip far outweigh those advantages.

Off-roaders don't like limited-slip differentials because they can't be used with a steering brake. They also tend to make the car *understeer*, or "plow" in corners. Finally, there's that extra complexity—more things to break. Save your money.

Shift-Lever Mounting Box—When building a tube-frame buggy, you must provide a compartment for the shifter. Don't custom fabricate this part. Buy a prefabricated shift-lever mounting box and weld it to the car's frame. The shifter then bolts to this box. It is designed to use the stock-VW shifter nylon bushing.

You will need a shift-lever box when adapting the '68-and-later bus IRS transaxle into a sedan. Because the bus transaxle nosepiece sits higher in the car, the stock shift linkage must be routed on top of the tunnel.

Offset Shift Linkage—If you're building a single- or tandem-seat buggy, you'll have to offset the shifter to one side. Otherwise, you will be straddling the shift rod and operating the shifter between your knees.

The problem with offset linkage is joining the shifting mechanism to the transaxle. The push/pull and rotating motion required for shifting has to be translated from the rod at the shift lever to the transaxle.

Ja-Mar's Side Shifter solves this problem through a system of levers and rods. Although it looks complicated, all you have to do is make provisions for mounting the Side Shifter between the transaxle nose and shifter rod. Mounting is done by welding in a bracket and fitting the attachments to the shift rod and the rod entering the transaxle. Fine adjustments are made by lengthening or shortening a threaded rod.

Adjustable Shift Rod—In theory, a correctly set up VW transaxle needs no linkage adjustments. But off-road abuse can change the position of the shifter and transaxle relative to one another.

Sway-A-Way makes a shifter adjuster for the VW shift linkage. When welded into place, the kit makes it possible to adjust the length and rotation of the rod. Bent or worn shifting forks, bent shift linkage, bent frame horns, or even a bent frame can make this adjuster a handy item to sort out shifting problems.

When building a tube-frame buggy, the shifter adjuster can come in handy too. Just cut off the end of the shift rod just short of the desired length and weld in the adjuster. Adjustment is done by loosening a locknut and rotating a threaded rod.

IRS Bus Racing Transaxle—In the last few years, the '68-and-later bus IRS transaxle has almost completely

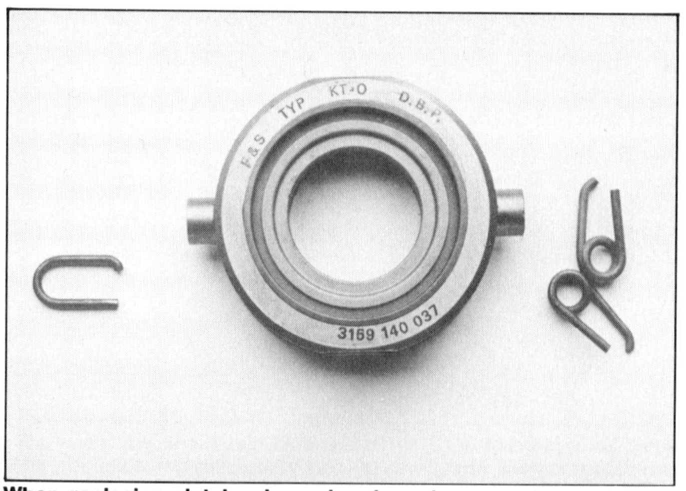

When replacing clutch-release bearing, always replace the old-type clips (left) with torsion clips (right). If you've ever pulled an engine merely to replace a release bearing that popped off, you'll appreciate these clips.

It's always a good idea to carry a spare clutch cable. Note all the other spares taped and cable-tied in place. Photo by Ron Sessions.

replaced the popular Type-1 IRS unit. The reason is horsepower. The 2180cc Type-1 VW race engine, once considered to be *the* hot engine, is now considered inadequate by many racers. Instead, the Type-4 engine is king. Whereas the Type-2 IRS transaxle is stronger than its Type-1 counterpart, it is by no means bulletproof when mated to the high-output Type-4 engine. Aftermarket beef-up kits are now available for the Type-2 IRS transaxle to handle this horsepower. Typically, the beef-up includes a Hewland gear pack and a high-strength spider-gear carrier similar to a Super-Diff.

CLUTCH

For street and pleasure off-road use, stick with the stock Fichtel & Sachs or Luk drive disc and pressure plate. The strongest is the 200mm disc and diaphragm-type pressure plate from the '63—'65 1500cc bus.

I personally like the diaphragm-type pressure plate, Luk 311141025, which offers over 900 lb of clamping force. Most short-course racers with high-output engines have gone to the Kennedy pressure plate. Kennedy modifies Fichtel & Sachs diaphragm-type pressure plates for clamping force ranging from 1800 or 2300 lb.

Diaphragm-type plates are lighter and require less pedal effort than coil-spring types. As an added bonus, the clamping force of a diaphragm spring *increases* as the clutch disc wears. They cost about twice as much as coil-spring pressure plates but are worth it.

All '67-and-later Type 1s use the 130-tooth, 200mm flywheel, clutch disc and pressure plate. To switch a

Neal hydraulic clutch uses a pulling slave cylinder which installs in place of the cable. Photo courtesy of Neal Products.

'66-and-earlier Type-1 engine over to the 200mm clutch, its 109 tooth, 180mm flywheel must be replaced with the 200mm type. When you do this, you must also relieve the bell-housing. See "Flywheel Clearance" on page 37 for more details.

Many racers mate the ceramic or puck-type disc with the 200mm Volkswagen pressure plate. This disc may cause objectionable chatter in a street-driven car with rubber engine mounts, but off-road it's virtually indestructible. It's affected little by oil and engages better as it gets hotter.

As far as the clutch *release,* or "throw-out," bearing is concerned, the stock ball-bearing type is fine. Use the newer-type torsion-spring retaining clips. The old C-clips have been known to come loose off-road. Never use a carbon-block release bearing.

Clutch-Release Shaft and Bushings—Although fine for recreational use, the clutch release shaft used in '71-and-earlier transaxles is a little weak. It can fail where the release fingers join the shaft. You can buy an aftermarket shaft that's been jig-assembled, heat-treated and TIG-welded, or you can reinforce the existing shaft.

The stock shaft comes only partly welded where the release fingers join the shaft. After removing the engine, you can 100% weld these fingers to the shaft without removing the shaft from the transaxle.

Reduce shaft-bushing wear by applying a few drops of light oil. If the bushings are worn out, install bronze aftermarket replacements.

Hydraulic Clutch—Is your off-road sedan breaking a clutch cable every time you turn around? Do you need a sanitary, bolt-in clutch-actuating system for your tube-frame buggy?

Ja-Mar and Neal Products offer hydraulic systems to operate the clutch. The master cylinder is similar to those used for brake systems. The *pulling* slave cylinder mounts to the transaxle side cover in place of the cable.

Free play adjusts in the normal manner, with a wing-nut on the end of the slave-cylinder rod. The pedal mounts conveniently to the floor, and can be bolted or welded down. Kits are available for both tube-frame buggies and sedans.

Morse Boat Cable—If you don't want the expense of a hydraulic-clutch system, but don't trust the stock cable when the going gets rough, there is a compromise solution. An 8-ft-long Morse boat cable, used with a John Johnson clutch-pedal assembly, can be substituted for the stock cable. The boat cable is roughly three times stronger than the stock cable.

Chapter 3 FRONT SUSPENSION & STEERING

The VW torsion-arm front suspension works so well off-road that you'd swear Dr. Porsche had raced the Baja before designing it. Photo by Tom Monroe.

FRONT SUSPENSION

Probably the most common VW-based off-road vehicle is the Baja Bug, or play buggy, with a 1500cc or 1600cc engine and swing-axle rear suspension. This combination has enough power to go where you want, but not enough to tear up the stock transaxle. The reason one car can hot foot it across the rough stuff and another can't, when both look the same, is suspension preparation. As good as the engine and transaxle are for off-road use, the Volkswagen suspension, both front and rear, is far better. You'd swear Dr. Porsche had designed the VW suspension for racing the Baja.

This chapter details modifying the Type-1 front suspension for off-road use. As you may already know, *Type 1* refers to the Beetle, Karmann Ghia and 181 Thing vehicles only. Although the more ungainly bus and Type 3 use front suspensions similar to the Type-1 design, they're not used with any frequency off-road.

MacPherson-Strut Suspension—Before getting into front-suspension modifications, there's one thing you should know. Not all VW Type-1 suspensions are created equal.

The MacPherson-strut front suspension originally used in the '71-and-later Super Beetle and Beetle convertible is not suitable for off-road use. Simply put, it's not durable.

This suspension uses coil springs rather than torsion bars. The springs are perched on top of what amounts to a combined shock absorber and long steering knuckle—a *strut.* Each strut is supported at its bottom by a single ball joint and transverse control arm. The control arm is located by a combination strut-rod/stabilizer bar. A sheet-metal tower built into the body structure and a bearing assembly locate each strut at the top.

VW made good use of this suspension design to maximize luggage capacity and shorten turning diameter in the Super Beetle. However, do *not* take the Super Beetle off-road, even for short distances. The solitary ball joints and insert-type shock absorbers of the strut-type suspension are below par. And a sharp impact may cause the strut rod locating the lower arm to break, allowing the lower arm to bend and fold back.

Torsion-Arm Suspension—Other than allowing for a high degree of wheel travel, what makes the VW torsion-arm suspension so adaptable to off-road use is its durability. Much of this is directly attributable to the suspension geometry. The wheel moves back, up and away from an obstacle. This reduces the shock loading on the VW suspension components.

The suspension geometry on most other cars allows only for vertical travel and a small amount of front-wheel *compliance*—deflection at rubber-bushed joints. This small amount of compliance does not allow any substantial fore/aft movement to reduce shock loads.

The Type-1 torsion-arm front suspension falls into two major groups: pre-'66 kingpin (also known as *link-pin*) and '66-and-later ball-joint type. The major difference is in how the torsion-arms attach to the steering knuckles—either by kingpins and link-pins, or by ball joints. Both front-suspension assemblies are removable but won't interchange unless you weld on a whole new *front clip*—the front part of the floorpan.

State-of-the-art coil-over front torsion-arm setup from The Wright Place has 15 in. of wheel travel. Torsion arms are 4-in. longer than stock.

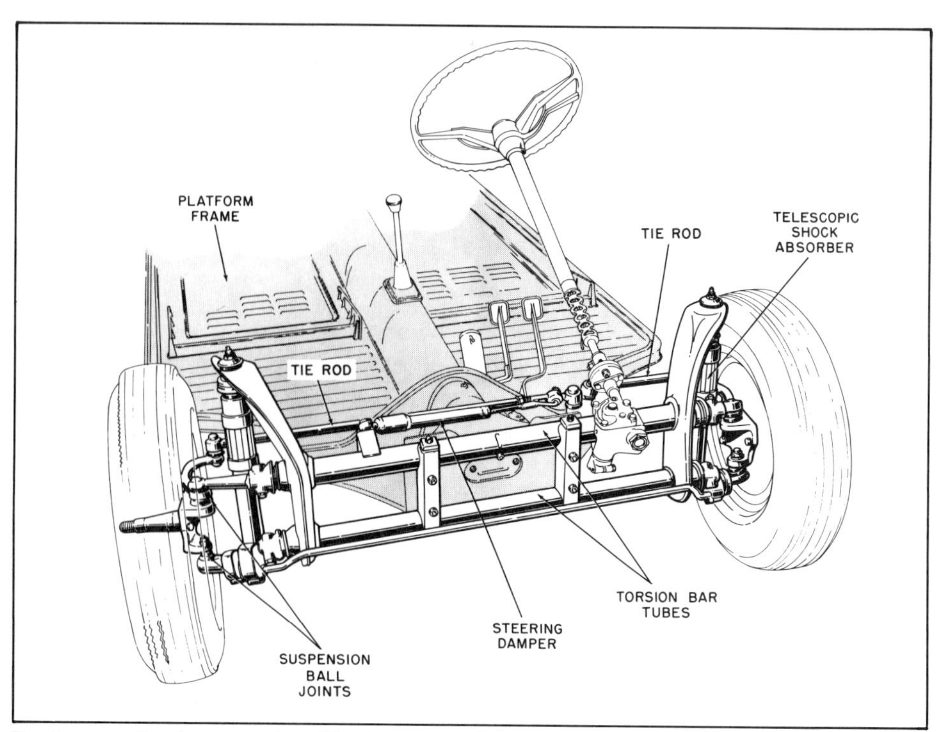

Torsion-arm front suspension. Stout construction, trailing-arm geometry and lots of wheel travel make it a natural for off-roading. Ball-joint front end shown; kingpin front end is similar. Photo courtesy of Volkswagen of America.

If your off-road car has the ball-joint front end, skip ahead to page 62 for modification details. Only a small portion of the information under "Kingpin Front End" applies to your vehicle.

VW calls the entire torsion-arm front-suspension unit the *front axle* in their factory manuals. Although it's not an axle in the true sense of the word, that is how the German word translates into English.

The VW torsion-arm front suspension consists of two transversely mounted tubes with stacks of torsional-steel leaves inside. Collectively, each stack of leaves makes one *torsion bar.* At the outboard ends of each torsion bar, a pair of *torsion arms*—trailing arms—are fitted with hefty setscrews.

Each torsion arm fits into a torsion tube, supported by needle bearings. When the wheel hits a bump, the torsion arms rotate up, twisting the torsion bars in their tubes and producing torsional spring force. Wheel movement and the action of the torsion bars are damped by a shock absorber at each wheel.

KINGPIN/LINK-PIN-TYPE FRONT END

The kingpin front end is the one most frequently used on both play and race cars. Why? Because it's been around longer and is still more plentiful in junk yards than the ball-joint front end. And it's far more durable. Because more people use these units, more special parts are made for them.

Many modifications can be made with the front-suspension assembly installed. If you intend to make extensive modifications, I recommend removing it. The front-suspension assembly simply unbolts from the car. On a sedan, there are six bolts—four to the forward end of the floorpan, and two to the body beneath the fuel tank. Disconnect a couple of brake lines, unbolt the steering shaft at its flange and out it comes.

With the front-suspension assembly out of the car, it's much easier to work on. And you won't have the fuel tank to worry about if you do some welding. You'll also have good access to the *frame head,* or front clip. This clip can be strengthened to prevent problems that may develop due to the higher loads being put into the front suspension.

Let's start with the easiest and least expensive modifications.

Front Stabilizer Bar—The front stabilizer bar was first used on '60 Type-1 models. It is a torsion-bar spring, connected to each lower torsion arm. For street use, the front bar aids directional stability and reduces the tendency of the car to *oversteer,* or spin out, by increasing front *roll stiffness.* Roll stiffness is the resistance to a car rolling or leaning over while it's cornering.

A front stabilizer bar is of no use in an off-road car. Big rear tires can cause excess *understeer*—the tendency of the front end to "plow" or push out in turns. Removing the front bar helps restore the car's suspension balance. Therefore, remove it and set it aside. It can be used to beef up the tie rods.

Tie Rods—On a stock VW front end, the steering box is offset to one side. This requires a long and a short tie rod. The long tie rod bends very easily when the car is taken off-road. If a tie rod is bent, the front tires will be *toed out*—spread apart at the front. You won't go very far in soft terrain with the front tires trying to plow up the countryside. So, you'll have to fix it right there, wherever "right there" is.

This is not too bad if you have a tie-rod-end puller, a "pickle fork" or two ball-peen hammers to break the tie-rod ends loose. If not, you can crawl under the car and try to bend back the tie rod.

Sometimes this isn't easy because not only are you trying to bend the rod, you are also trying to rotate the

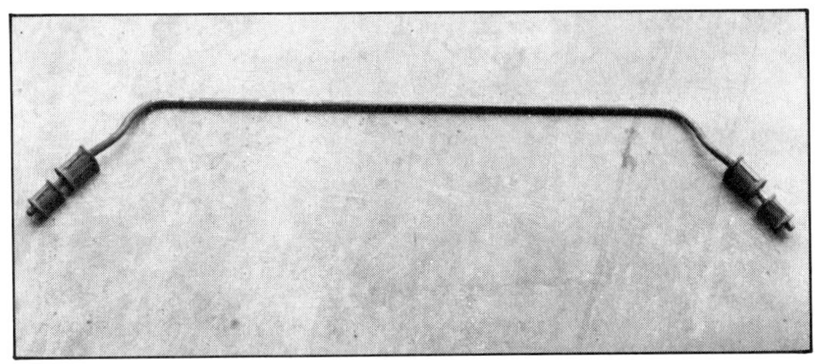
Stock front stabilizer bar is of no use in the dirt. Remove it.

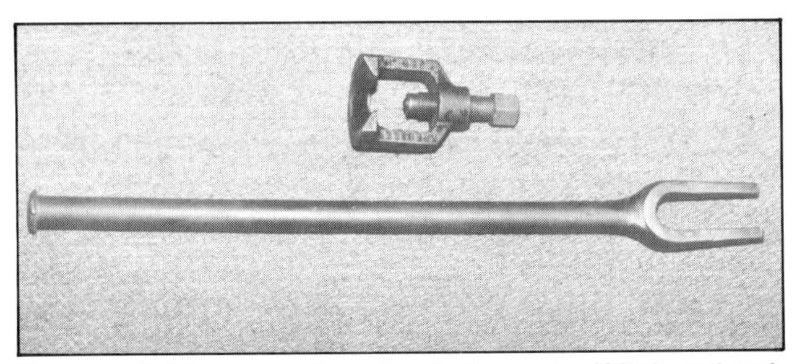
Tools needed to remove tie-rod ends. Expensive VW tie-rod-end puller above. Bottom tool is inexpensive "pickle fork."

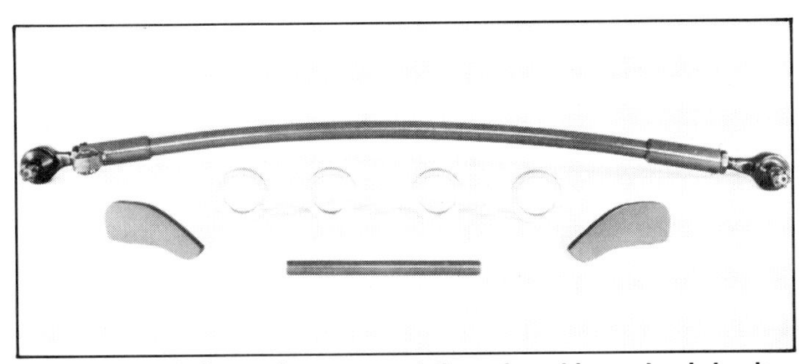
Crown Beef-A-Steer. Long curved tie rod cushions shock loads. Short steel rod is internal stiffener for short tie rod. Gussets are welded on to brace the steering arms. Aluminum thrust washers replace stock rubber seals between torsion tubes and arms. Photo courtesy of Crown Manufacturing.

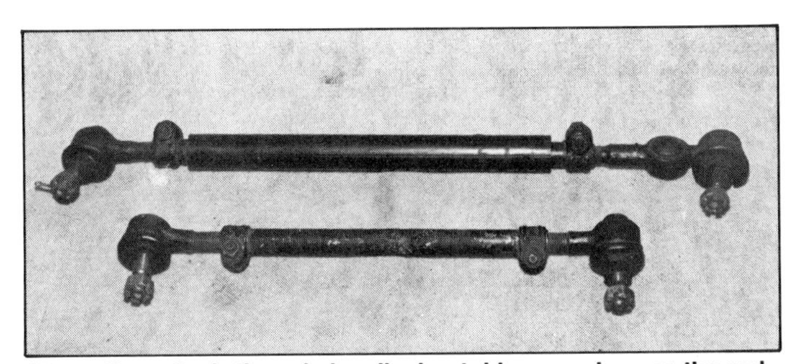
Strengthen stock tie rods by slipping tubing snugly over the rod. Make sure it won't interfere with the tie-rod-end clamps at the ends. Holes should then be drilled through the outer tube and the tube welded to the rod through these holes.

tire as well. To make it even more difficult, the tie rod will tend to rotate around the tie-rod ends as you work on it. Whichever way you do it, you'll end up with grease up to your elbows and a semi-straight tie rod that's now much weaker than before. So, it's best to prevent this problem by modifying or replacing the stock tie rods.

There are two schools of thought here. Some off-roaders want flexible tie rods so they'll bounce back into shape; others want them stiffened so they won't bend in the first place. When the tie rods are stiffened, shock loads that might have bent a tie rod now telegraph into the steering arms, gear and mountings, so you'd better reinforce these as well.

A flexible spring-steel tie rod to replace the stock long tie rod is available from Crown Manufacturing. Although this 3/4-in.-OD rod is certainly up to the task, it's not used often in race cars. When traveling at high speeds over rough terrain, the rod will flex enough to let the front-wheel toe change in and out. This makes the steering a little unpredictable and can give you a thrill now and then.

There are a number of ways to stiffen tie rods. The most inexpensive method is to reinforce your stock rods. Remove the tie rod from the car. Unscrew one tie-rod end and its clamp, counting the number of turns so you don't lose the toe-in adjustment.

Take the tie rod to a metal yard and find a piece of 0.050-in.-wall tubing that will fit snugly over the tie rod. Cut the tubing so it covers all the rod except the ends where the clamps are.

Drill two 3/8-in.-OD holes all the way through, about 4 in. from the ends. Slide the tube over the rod, center it and puddle-weld through the holes to secure the tube to the rod.

Now take that stabilizer bar you removed and cut a piece about 4-in. shorter than the tie rod. Slide it inside the tie rod. The 1/2-in.-OD piece of bar will rattle around a bit, but adds stiffness to the rod. Install the tie-rod end, screwing it in the same number of turns it took to get it off. Put the tie rod back on the car. If you lost the adjustment, set toe-in at 1/4 in.

Although the short tie rod is naturally stiffer because of its length, it should also be reinforced for rough off-road use. No one makes a flexible short rod, so that option is out. Reinforce the short tie rod in the same manner as the long one.

If you have more money to spend and want stiffer tie rods, you can go to 4130 chrome-moly steel or 2024 T3 aircraft-grade aluminum tubing. Most racers use 1-in.-OD tubing with 3/16-in. wall thickness. The tubing is tapped 11/16-24 left- and right-hand threads for Ford 3/4-ton pickup or Econoline tie-rod ends. You can buy these already made up from the Wright Place. The Ford ends use jam nuts instead of clamps.

Tie-Rod Ends—Because of the severe angles and the pounding encountered off-road, stock tie-rod ends don't last long. If your plans don't include racing and you want to retain the stock VW tie rods, you'll still need stronger ends.

In '68, Volkswagen started using stronger tie-rod ends with larger-OD ball studs. To use these ends on earlier models, the steering arm on the earlier spindle has to be taper reamed to fit. Also, the earlier pitman arm has to be taper reamed or replaced with a later pitman arm.

To save the cost of buying new VW tie-rod ends, some off-roaders get theirs at the junkyard. Their reasoning is that used street-driven tie-rod ends are like new for off-road operation. If you decide to go this route, check the rubber boot over the rod end. If it's cracked, torn or missing, go on to the next one. Damaged boots allow dirt and grit to work into the joint, greatly accelerating wear.

Most racers use custom tie rods machined to accept heavy-duty Ford or Chevy ends. These hefty ends are ac-

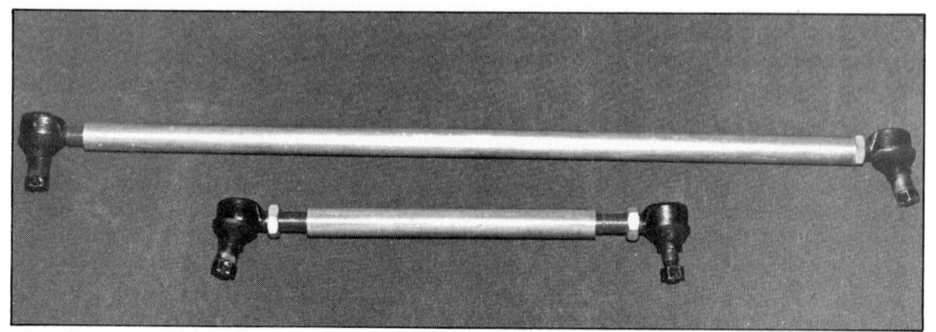

Aircraft-grade 2024-T3 aluminum tie rods from The Wright Place are racing-caliber stuff. Ends are threaded to accept Ford 3/4-ton pickup-truck tie-rod ends.

Three tie-rod ends. Left is Ford-truck end used with custom tie rods. Notice big stud and fine threads. Center is '68-and-later VW end with larger-OD ball stud. Right is '67-and-earlier end.

BILSTEIN OFF-ROAD SHOCK-ABSORBER CHART

Part No.	Ext. Length (in.)	Comp. Length (in.)	Travel (in.)	Valving Reb./Comp. (lb)
B46-032	15.25	10.24	5.00	255/108
B46-034	16.22	10.71	5.51	360/160
B46-040	15.25	10.24	5.00	330/130
B46-620	16.12	10.31	5.81	185/75
B46-628	16.34	10.63	5.71	287/86
B46-638	22.05	13.62	8.43	178/71
B46-033	15.43	10.43	5.00	330/130
B46-349	22.05	13.82	8.23	275/78
B46-350	31.10	19.37	11.73	306/80
B46-351	31.10	19.37	11.73	275/78
B46-359	22.05	13.82	8.23	178/71
B46-360	25.83	15.79	10.04	275/78
B46-361	25.83	15.79	10.04	178/71
B46-930	16.22	10.87	5.35	360/160
B46-1361	25.83	15.79	10.04	191/79
B46-1362	25.83	15.79	10.04	70/71
B46-1348	22.05	13.82	8.23	75/75
B46-1351	31.10	19.37	11.73	271/98
B46-1352	31.10	19.37	11.73	191/79

There's more to off-road shock-absorber selection than short-travel and long-travel. Use this chart to determine the basic lengths and valving of gas-pressure shocks. Chart courtesy of Bilstein of America, Inc.

tually less expensive than the stock VW ends. A clamp-on bracket can be used to accommodate single or dual steering dampers. As with the late-VW ends, the steering arms and pitman arm must be taper reamed to 7° included angle or 1-1/2-in. taper per foot to accept the Ford or Chevy ends.

Front Shocks—The first thing you'll notice when you go off-road with a stock VW front end and stock double-tube shock absorbers is that the shocks will progressively lose their damping ability the harder you run.

A shock absorber damps by channeling fluid through various valves to convert mechanical energy into heat. Off-road, the stock VW double-tube shocks have a hard time dissipating this heat. The hotter they get, the less they can damp. Also, rapid movement of the piston creates a pressure rise inside the shock, which causes air bubbles—*aeration.* The more the fluid aerates, the less the shock valves can damp movement.

If your recreational off-road car has a stock-travel front suspension, I heartily recommend using high-pressure-gas, single-tube shock absorbers. They'll be a little stiffer on the pavement, but worth it off-road. About 250—350 psi of high-pressure nitrogen gas keeps the fluid from aerating spontaneously.

Many of the shock-absorber manufacturers have developed "racing" shocks for use in SCORE and High Desert Racing Association Class-11 competition. These reasonably priced shocks are stock-length, stock-travel, with stiffer valving. In my opinion, they're just the thing for the recreational off-roader.

KYB, Bilstein, Fox Shox/Phoenix Phactory, Rough Country, and others, offer good quality off-road shocks for these classes. I've enjoyed good service with the KYB Gas-A-Just shocks, but you'll find many people who swear by other brands.

Racing Shocks—When it comes to racing, shock selection becomes more scientific. I recommend contacting the shock-absorber manufacturer or distributor of your choice and getting his advice. Information you should have handy for the decision-making process is the car's front-end weight, front-wheel travel, torsion-arm length and torsion-bar spring rate. Heavy cars with stiff springs and

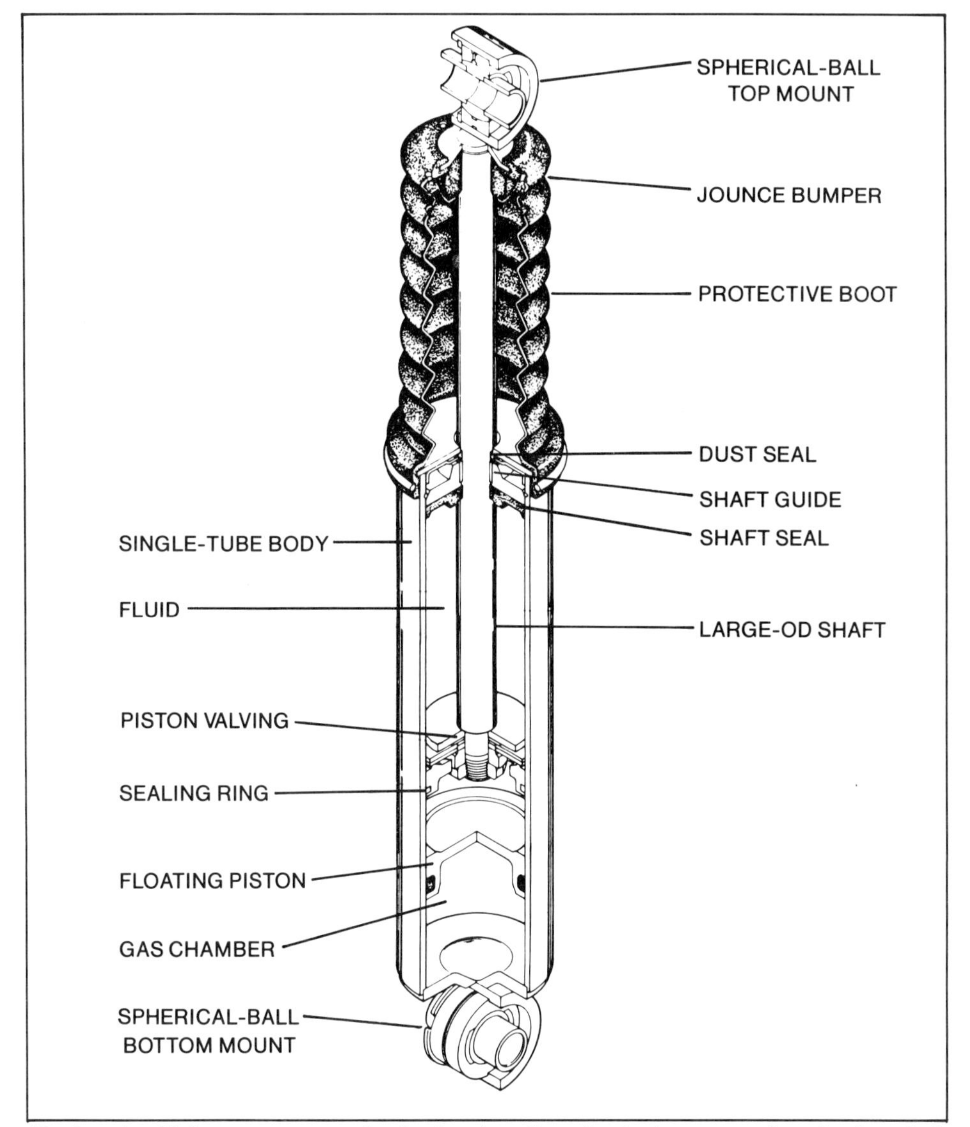

Single-tube, high-pressure-gas shock absorber. Gas shocks have superior wheel control under off-road-racing conditions. Drawing courtesy of KYB of America.

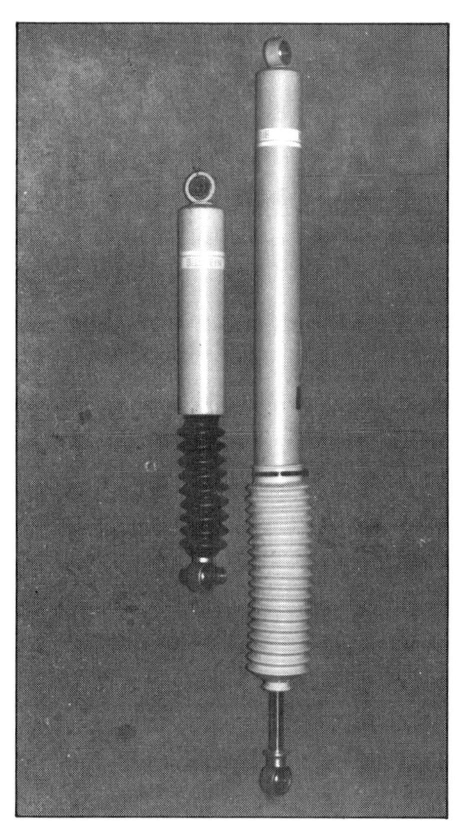

Most people think in terms of two different lengths of shocks for the VW front end—long and short. There are many variations in between.

KYB GAS-A-JUST
Off-Road Shock-Absorber Chart

	Ext. Length	Comp. Length	Travel	Valving Reb./Comp.
KOR 611	15.59 in.	10.67 in.	4.92 in.	175/90 lb
KOR 612	15.59	10.67	4.92	300/175
KOR 621	22.01	14.13	7.88	275/145
KOR 623	22.01	14.13	7.88	200/100
KOR 631	17.13	12.60	4.53	175/90

Before running out and buying a set of off-road shocks, measure your car's suspension travel. Getting the "right" valving is more difficult. But basically, heavy cars with high-rate torsion bars need stiff valving. Light cars with low-rate bars need milder valving. The longer the suspension travel, the milder the valving can be. Chart courtesy of KYB of America.

short- or medium-travel front ends need stiff shock valving. Light cars with stock springs and medium- or long-travel front ends need milder shock valving. There's a big gray area between.

To lessen the chance of shock damage, I recommend using shocks with about 1-in. more travel than you need—about 1/2 in. at *jounce* (suspension compressed) and 1/2 in. at *rebound* (suspension extended) is good. If a suspension stop is bent, the extra-travel shock will have some travel left before bottoming or topping out. Otherwise, when the shock is fully extended, the chances of the shaft bending or breaking are greatly increased.

Off-road shocks traditionally have very stiff rebound valving to prevent the car from becoming airborne as the suspension unwinds from full jounce. Don't go too stiff on rebound. If rebound valving is too stiff, the car may "ratchet down"—not rebound fully after each bump—over washboard surfaces. Thus, the car gradually loses jounce travel and ground clearance.

Cooling Fins—As mentioned earlier, shock absorbers turn mechanical energy into heat. Touch a shock that's just run an off-road race and you'll be reminded of the time you discovered how hot Mom's kitchen stove could get. Shock oil can get as hot as 200F (95C).

In the early days of off-road racing, there weren't that many good shocks available. High-pressure-gas shocks weren't yet adapted to off-road. Multiple-shock installations were few and far between. So, off-road racers often asked shocks to do more than they were designed to do.

To help dissipate heat, aluminum cooling fins were used. The cooling fins slip over the outer tube. They provide an added heat sink by increasing the surface area exposed to the air. Today, cooling fins are seldom seen because of the advances in shock-absorber technology.

Reservoir Shocks—One of these advances is reservoir shocks. A separate reservoir adds fluid capacity to the shock and isolates the fluid from the heat caused by piston movement. The reservoir is mounted "piggyback" on the outside of the shock tube, or remotely with a connecting tube to some part of the vehicle, such as a frame tube or roll cage. To aid fluid cooling, the remote reservoir can be located in the airstream.

Extra Shocks—Under hard off-road conditions, heavier vehicles such as sedans and Baja Bugs need higher-rate springs to control suspension movement. To damp this higher spring rate, use a shock with stiffer valving or multiple shocks at each wheel. With multiple shocks, valving can be reduced, thus reducing heat buildup in each shock.

Shock Mounts—Whenever you use a shock with a rubber or urethane grommet at either end, add a large metal washer under each mounting nut to keep the shock on its mount. If you're using *heim joints*—spherical ball rod ends—at the mounts rather than grommets, don't worry about the washer.

Top Shock-Mount Fix—On the stock kingpin front end, the upper shock-mount bolt threads into a weldnut inside the top of each shock tower. If you look at a cross-section of this bolt, you'll see that the threaded portion has a smaller diameter than the unthreaded or shank portion.

The weakest part of this bolt is where the threads end and the shank portion begins. This point between the weldnut and shock bushing also has the highest loads. Under routine off-road operation, the bolt will often break at this point. Also, rust can freeze up the bolt so it will break off when you're trying to remove it. By drilling a hole through each shock tower, you can prevent these problems.

By using a longer bolt and nut, the loads are more evenly distributed. If you're going to drill the weldnut, it's easier to do it when there's not a broken bolt inside. Use a good-quality 15/32-in. drill bit—or 1/2-in. bit if that's as close as you have—and a 1/2-in. electric drill to cut through the hard weldnut material. If the bolt is broken off inside, the drilling operation will take awhile. Use light oil to

In the early days of off-roading, cooling fins were used to help dissipate heat. Some people still use them. Photo by Tom Monroe.

Air-suspension system from Giese Racing: Large air shock takes place of conventional torsion leaves and hydraulic-type racing shocks.

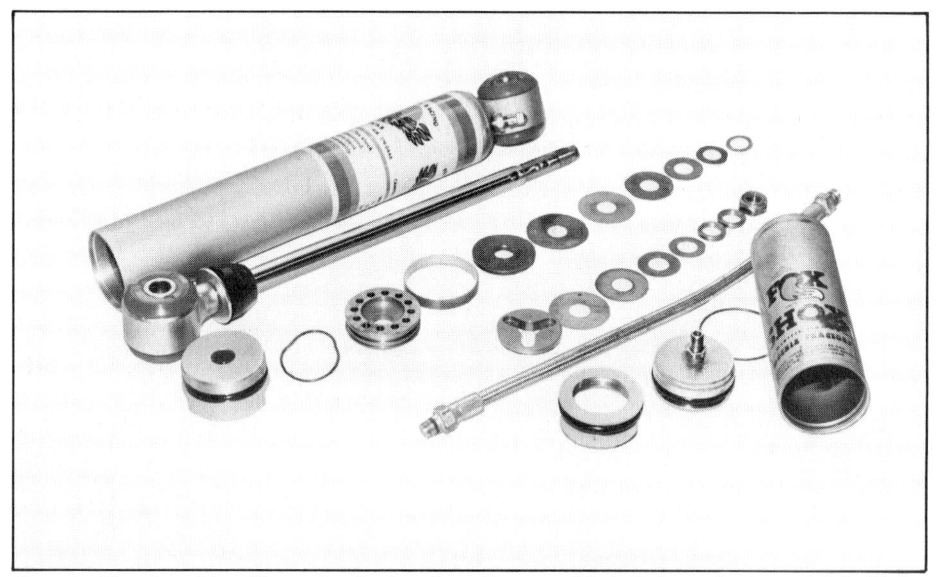

Phoenix Phactory/Fox Shox shock absorber disassembled. Unique feature of this high-pressure-gas shock is that it is easily disassembled. Valving, shock oil and worn parts can be changed or replaced. The Phoenix Phactory people recommend changing the 5W oil every 3—4 months on a race car, and once a year on a street car. Photo by Ron Sessions.

Heavy cars need more damping to control higher-rate springs. Dual shocks can prevent overworking a single shock.

Stock shock tower drilled to accept extra long bolt. Bolt presses against the body to support the tower. Note that the spot-welded seam has been rewelded. Washer under bolt head prevents rubber bushing from working loose.

WARNING
Being a fireman by profession, I have seen what gasoline can do to a person who is careless with it. When welding on the front end, remember there is a gas tank right above your head. Even though the tank may look empty, highly explosive vapors remain inside.

The tank is only held on by four bolts. Don't take a chance; remove the tank and set it out of the area—out of the room if possible. Besides being safer, the front end as a whole is a lot easier to work on with the gas tank removed. Another good habit to get into is to weld in a well-ventilated area.

lubricate the drill tip. Make sure the bit is sharp and the motor strong. Now you can run a 15/32- or 1/2-in.-OD bolt through the mount and the shock is secure.

Reinforcing the Shock Tower—Before bolting up the shock, remember that the shock tower is nothing more than two stamped pieces of metal sandwiched together and spot-welded at the seams. With a good off-road shock and some rough roads, the shock tower will bend and bind the shock. Sometimes the spot-welded seam will fail completely and the tower will split. The standard off-road fix is to run a weld bead around this seam to keep it from splitting and to gusset the tower to prevent bending.

Sedan and Baja-Bug Procedure—To reinforce the shock towers, there are some fairly expensive kits you can buy that include several precut metal pieces to be welded on. But here's a cheap way to do the same job on a sedan or Baja Bug.

Directly behind the hole you drilled in the shock tower is a strong box-shaped inner panel built into the body. You can't drill through it because the gas tank is on the other side. But you can use it to support the shock tower and limit its bending.

The distance between the inner panel and the back side of the shock tower varies a little with each car. But it is usually large enough to get a nut between the panels and onto the shock bolt. If not, use a pry bar or blunt punch through the bolt hole to make room for the nut. When you have enough room for the nut, find a 15/32- or 1/2-in.-OD bolt that will go through the shock and shock tower and butt against the inner panel.

Adjust the length of the bolt so it goes through the nut and comes out to push firmly against the body. Do this by shimming under the bolt head with a washer or two or grinding away the threaded end of the bolt. Strong inner panels will help support the top of the tower.

Now and then, after hard use, check that the end of the bolt is not making a hole in the body. If it is, tack-weld a 3x3-in. piece of 1/16-in. steel plate over the inner panel where the bolt bears against it. Then readjust the shock bolt to compensate for the added material. Remember that the stock gas tank is on the other side of that panel.

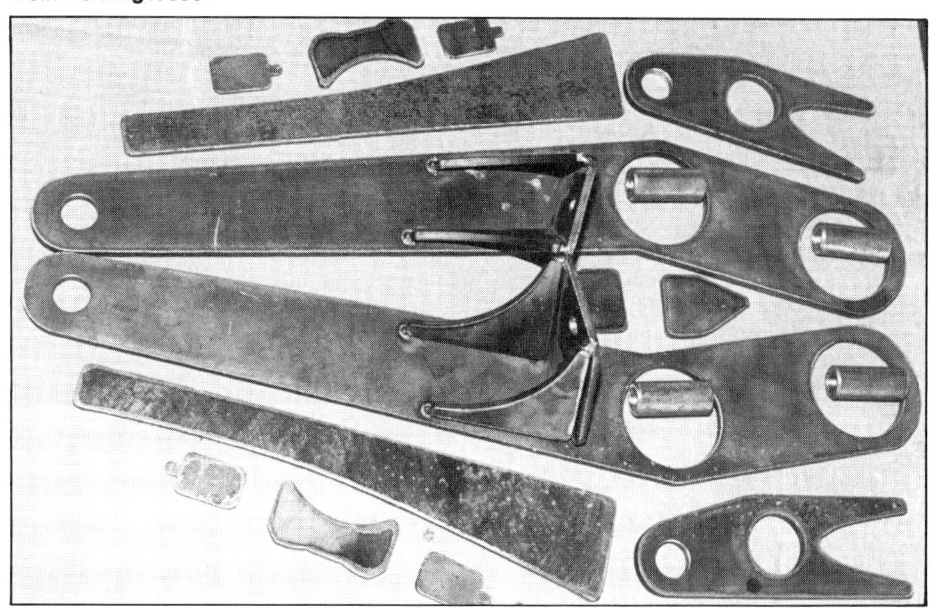

JT Machine in Lakeside, CA makes these front-end beef-up kits. They manufacture shorter-travel models to beef up the stock shock tower. This long-travel-shock kit adds about 4 in. and has a jounce stop above the torsion arms.

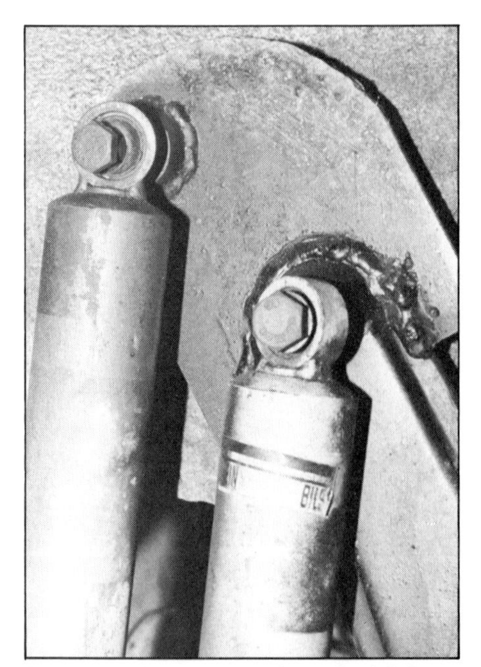

Top mount for extra shock can be fabricated a number of different ways. This is not one of them. Don't try this method unless you reinforce the stock shock tower.

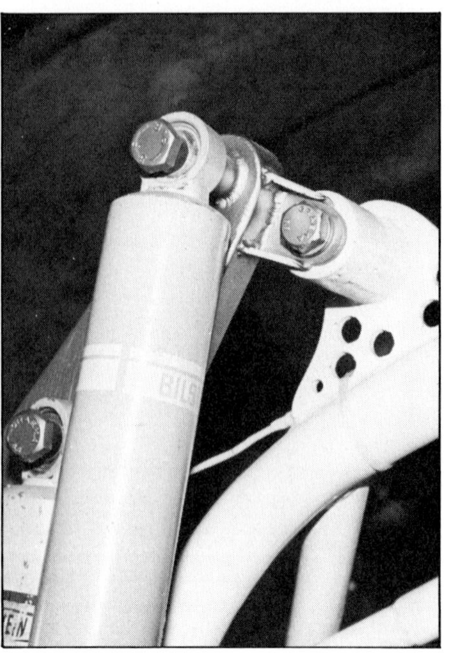

Race-car shock mount is attached to roll-cage supports.

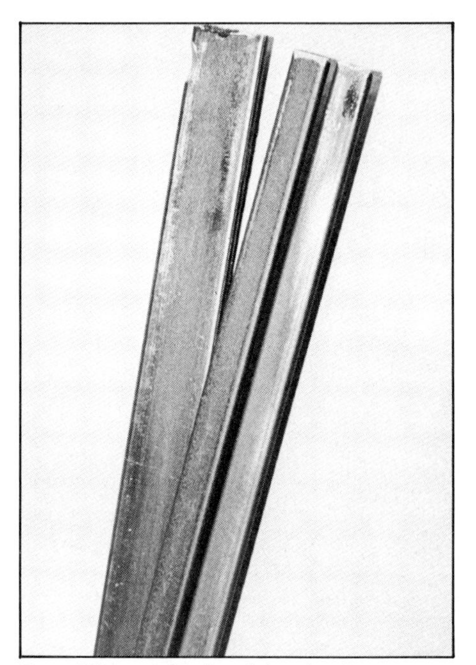

Two different kinds of torsion leaves are used inside each torsion tube of a kingpin front end. There are four full-width leaves (left) and four split leaves (right).

Buggy Procedure—In a buggy, bracing the shock tower is a simple matter because the body is not in the way. Simply brace from the side of the tower to the torsion tube.

Front Torsion Bars—As mentioned previously, each of the two tubes that runs across the front of the car houses a set of eight torsion leaves. Each set of spring leaves functions as a torsion bar. The four outside leaves in each tube are 1/2 the width of the other leaves, allowing for more progressive springing.

The bars run the entire length of the tubes and are anchored in the middle by a square hole and a setscrew. Torsion arms fit over each end and are secured to each bundle of leaves by a similar setscrew. As the torsion arms pivot with vertical wheel travel, the torsion bars twist and the spring force in the bars increases.

More spring force or a stiffer suspension can be had by making the torsion bars harder to twist. You do this by pre-twisting or preloading the existing bars, or by installing *higher-rate,* or stiffer, leaves available in the aftermarket. For more information on preloading the front bars, see the section on "Cutting and Rotating Torsion-Bar Tubes."

Spring rate refers to the torque required to twist a torsion bar one degree, expressed as so many inch-pounds per degree of twist. This translates directly to *wheel rate*—pounds per inch of wheel travel. A higher-rate spring will allow less vertical wheel movement than a lower-rate spring when a car hits the same bump.

High-Rate Torsion Leaves—One easy way to stiffen the stock front end is to replace the four split leaves in each tube with two full-width ones. The full-width leaves have a slightly higher spring rate than the split ones. The cheapest place to find extra torsion leaves is in a junk yard. Look for a VW with front-end damage. Most yard keepers don't like to part-out a good front end.

If you want the suspension stiffer still, get some high-rate torsion leaves. These are available from aftermarket sources such as Sway-A-Way. Because Sway-A-Way's leaves are all full-width and slightly thicker than stock, they use one less leaf per stack. The thicker leaves increase spring rate by about 20%, but do not increase stack thickness.

High-rate springs are usually needed in Baja Bugs with stock or limited wheel travel. Generally, the more travel the suspension has, the lower the spring rate needed.

Cutting and Rotating Torsion-Bar Tubes—If you don't want to shell out much money for special bars but would like a stiffer front suspension and more ground clearance, there's a cheaper alternative. It only takes a little more work. This alternative is pre-twisting or preloading the front bars by cutting the torsion tubes and rewelding them.

If the center anchor tilts the torsion arms down against their lower stops, it will take a stronger force to start the arms moving toward their upper stops. I recommend this fix for most sedan and Baja Bug owners who don't want to go to the cost of high-rate torsion bars and a custom, long-travel front end.

If you race the car, the preloaded bars will sag after three or four races. This is because rotating the tubes forces the bars to twist more every time the suspension moves from stop to stop. Also, because the bars are preloaded against the bottom stops, I recommend replacing the stock rubber stops with tough urethane stops available in the aftermarket.

To do the job right and make a clean cut, you need a pipe cutter large enough to fit around the torsion tube. Ideally, the cutter should have four blades because the other torsion tube will prevent the cutter from traveling all the way around.

If you can find only a one-bladed

Cutting and rotating the front end. With one cut already made, the second cut is started on the top tube. Original position has been marked (arrow) on the left cut.

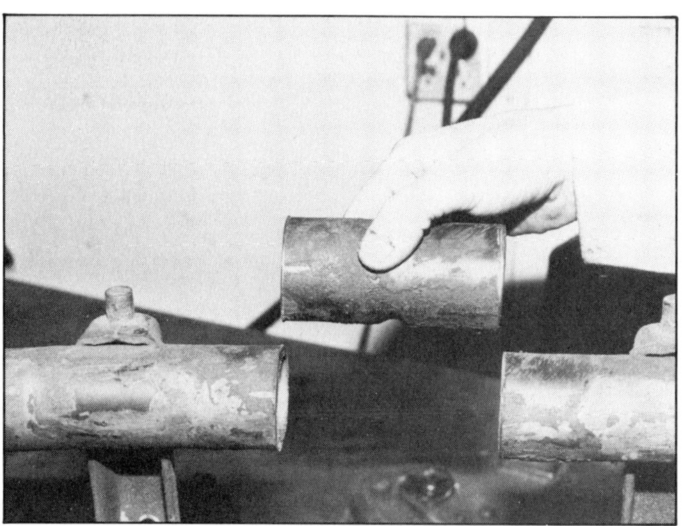

After second cut is made, section is free to rotate. Do one tube at a time, so the other tube keeps the front end together.

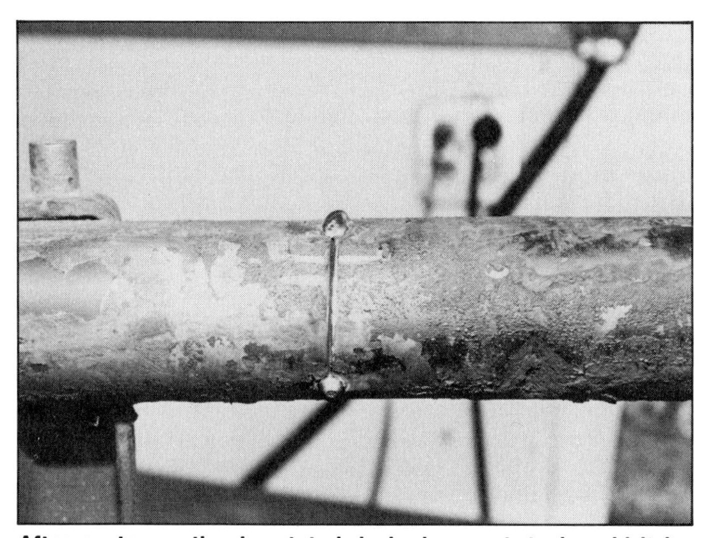

After center section is rotated desired amount, tack-weld it in place. For a stiffer front suspension and more ground clearance, rotate the setscrew up. For a softer ride and low front end—*California look*—rotate the setscrew down.

Completed cut and rotated torsion tubes. Weld them fully when you're done.

cutter, use a hacksaw and carefully saw through what you can't reach with the cutter. I've heard of people doing this whole job with only a hacksaw. But if you can locate a cutter, use it.

The cut-and-rotate procedure must be done with the suspension removed. If you are making several modifications to the front suspension, I recommend disassembling it. But, if the suspension is in good shape and you're not replacing torsion leaves, you can do the job without disassembling the suspension.

Disassembled Method—Remove and disassemble the front suspension. Take the torsion leaves out of the tubes, marking which side faces up. Torsional steel takes a set. If you install used leaves so they bend in the opposite direction, they'll fail sooner.

I recommend cutting one tube at a time so the other tube will hold the torsion-tube assembly together while you work. *Scribe a straight line lengthwise on the tube across where you'll make the cut.* Now cut the section with the middle setscrew out of the top torsion tube. Make two cuts about 1-1/2 in. from each side of the locking nut on the setscrew.

After the section is free, rotate it 1/8—1/4-in. upward, using your reference marks as a guide. It doesn't seem like much, but 1/8 in. will make a big difference. A full 1/4-in. rotation will make the stock-shocked suspension a little stiff for trips back and forth to work, so you had better enjoy feeling every crack in the road.

Make sure the section is rotated so the setscrew is being moved *upward.* Otherwise, you'll have less ground clearance and less travel than stock. Tack-weld the section and make sure it is lined up straight. Then weld it solidly in place. Repeat this process on the bottom tube.

When reassembling the front suspension, you may have trouble get-

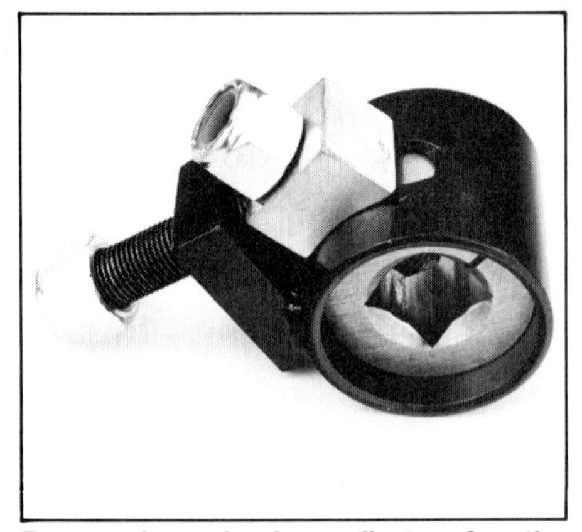

Front-end torsion-bar adjuster for the kingpin suspension by Sway-A-Way. Photo courtesy of Sway-A-Way.

Wright torsion-bar adjuster for their custom wide front end. As with the Sway-A-Way unit, preload and ground clearance is adjusted by loosening the locknuts and rotating the center anchor. Photo courtesy of The Wright Place.

ting the torsion leaves as a group back through the square hole at the center of each tube. If so, wrap the leaves with a heavy rubber band or an old pushrod-tube seal about 1 in. from the end. This will hold them together and keep them straight as they go in.

Assembled Method—To cut and rotate the torsion tubes with the front suspension assembled, you need an assistant. Simply mark and cut one tube at a time with the torsion bars still inside. Grab the cut-out section with a long pipe wrench and twist it up 1/8—1/4 in. Hold it there while your assistant tack-welds it.

Torsion-Bar Adjusters—Aftermarket torsion-bar adjusters allow you to set the angle of the center anchor, thus setting preload and ride height. The adjustable feature is ideal for racers wanting to change preload for different course conditions.

Many front-torsion-bar adjusters available today are more suitable for the street than off-road. These were designed primarily to lower street cars for the "California Look," and are not strong enough for off-road pounding.

Sway-A-Way offers a race-proven front-torsion-bar adjuster. You have to cut the center section out of each torsion tube using the process just described. But instead of rotating the piece you cut, you weld in Sway-A-Way's adjuster. The adjusting screw sets the angle of the anchor and a locknut keeps it there. When properly installed, this adjuster allows about 40° of torsion-bar adjustment. This translates into more than 3 in. of travel adjustment—1-1/2-in. upward and 1-1/2-in. downward.

The Wright Place also makes a front-torsion-bar adjuster for racing, but it only fits their custom-fabricated front end. Although it can be installed after the fact by cutting the tubes as previously described, most racers buy the front-end unit with the adjuster already installed.

Torsion-Tube Widening Kits—Torsion-tube widening kits can be used to widen the front track so it matches that of the rear. A typical kit consists of two tubular-steel sections. These are welded into the center of the stock tubes after the tubes are cut in half. Sections are available in lengths of 2 in. up to about 7 in. Depending on the design of the widening kit, the stock torsion leaves are cut in half, or the last third of the leaves are cut off and a second shortened set used for the other side.

The stock center anchor is cut out and replaced with dual anchors—one for each stack of torsion leaves. Sway-A-Way makes an adjustable center section so preload and ground clearance can be adjusted. Either stock or high-rate springs can be used. Remember that the shorter the spring, the more quickly it winds up and acts like a high-rate spring. Using shortened, high-rate springs may make the front suspension too stiff on all but the heaviest Baja Bugs.

Strengthening The Sedan Front Clip—On a VW sedan or Baja Bug, off-road use can seriously weaken the forward end of the floorpan—the part the torsion-tube assembly bolts to. This part is known as the *frame head* or *front clip.* It's replaceable in the case of a collision or if you want to change from a kingpin to a ball-joint front end or vice versa. There's an aftermarket kit that allows you to weld a ball-joint front clip onto a Super Beetle.

This piece is the sole support for the front suspension, yet it is only spot-welded as it comes from the factory. While the front suspension is unbolted, it's a good time to weld the seams. Get a flashlight, stick your head in there and look around. You'll find lots of spot-welded joints. Get a welder and *weld around the edges of all the spot-welded flanges.*

Before starting, remember that welding sheet metal is a lot easier if you clean off the grease, dirt and undercoating. An acetylene or propane torch works well to burn off undercoating. Don't overlook cleaning the bottom of the clip.

Don't slop any weld in the torsion-tube cradle while welding around those edges. Four long front-suspension mounting bolts thread into lugs welded into the clip. With hard use, these lugs will break loose. You can redo the stock welds to reinforce them, but be careful not to burn holes in the surrounding metal. If the lugs get damaged, you can drill them out and use longer bolts with nuts. The stiffer you make the suspension, the more trouble you'll have with the clip because it's the only support for the front suspension.

Front-End Supports—An easy way to strengthen the front suspension of any sedan or Baja Bug is to bolt on prefabricated upper and lower front-end supports. Without the supports, the torsion tubes can bend backward if an object is struck hard enough. The supports are made out of the same tubing used for roll cages.

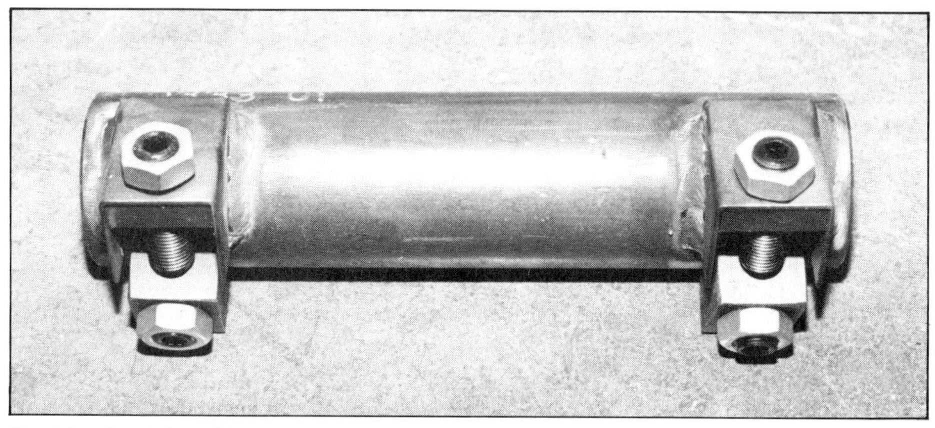

Most torsion-tube widening kits consist of two sections of tubular steel. The sections are spliced into the middle of the tubes. This is Sway-A-Way's adjuster for a widened front end. Each setscrew adjusts one set of leaves.

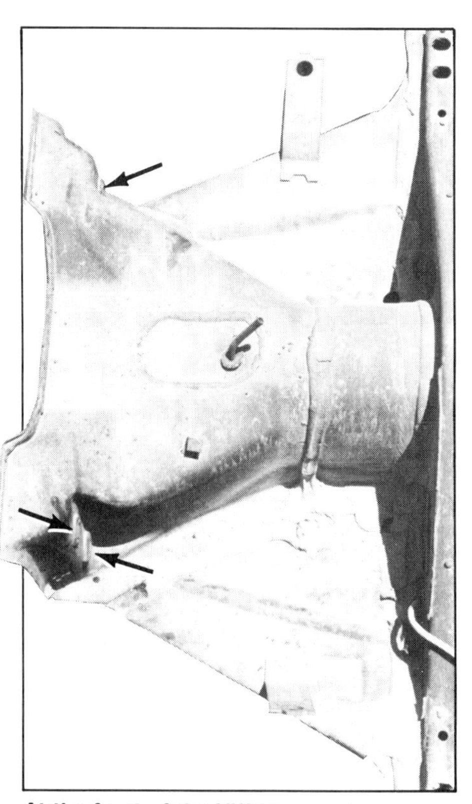

At the front of the VW floorpan is a separate, welded-on section called a *front clip*. This section supports the front suspension. Two welded lugs (arrows) on both sides of the clip should be checked occasionally for cracks around the welds. Under the layers of putty are spot-welded seams that should be rewelded.

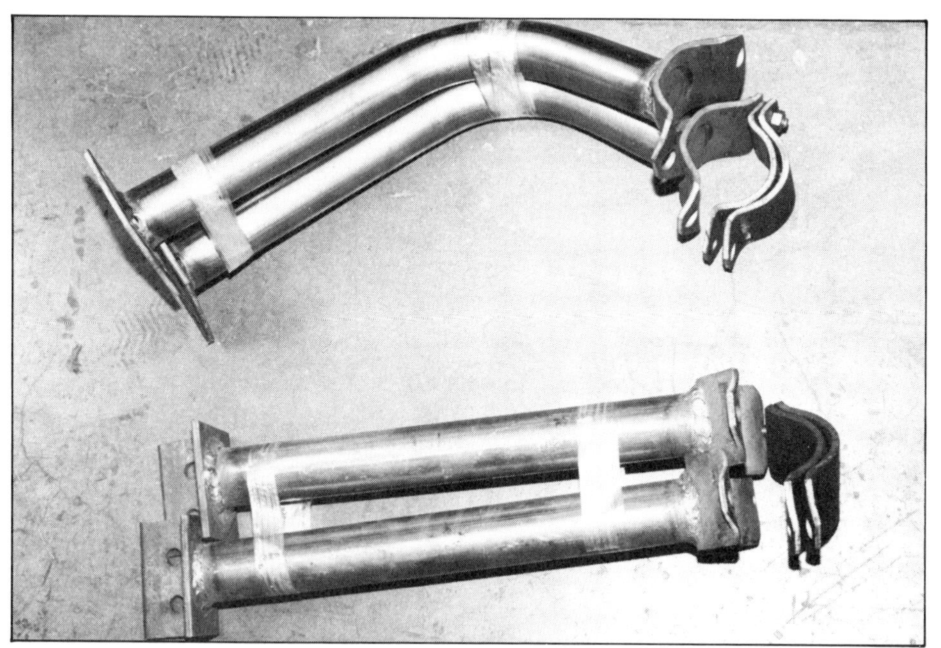

Front-end supports strengthen the front torsion tubes. Stock gas tank must be removed to use upper supports.

The supports install between the fire wall or floorpan and the torsion tubes, tying into the car's structure. They come with clamps for the torsion tubes, and bolt flanges for the body mounts. Front-end supports are available from most off-road buggy-supply houses.

Front-Suspension Stop—With stiffer torsion bars, increased front-wheel travel, front-end supports, and some good off-road shocks, the front suspension can be used harder. This will reveal another weak spot.

The stop between the torsion arms, which is a rubber cone or snubber on a metal arm, limits wheel travel in both directions. It stops downward travel—rebound or droop—of the upper arm and limits upward travel—jounce or bump—of the lower arm. Even with the torsion tubes twisted and high-rate torsion bars installed, severe use will bend the stop *up*.

With stock suspension, small bumps cause the front end to bottom, limiting speed over a given terrain. But with a stiffer suspension, there is less wheel movement over small bumps, and the car can cover ground much faster. A car traveling faster has a lot more momentum, so when you hit that bigger bump and bottom the suspension, the stop is apt to bend upward.

With the stop bent upward, the downward travel of the upper torsion arm is limited, which reduces ground clearance on the front end. If the stop

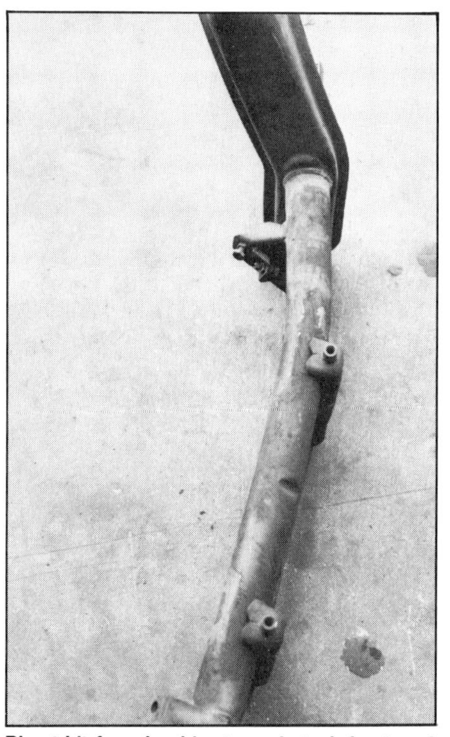

Direct hit from boulder turned stock front end into so much junk. Front-end supports might have prevented bent tubes. If this was a high-dollar racing front end, it could be straightened in a large press.

Panzer torsion-tube brackets were originally used to mount a front-suspension assembly onto a tube-frame buggy. Here, a wise Baja-Bug racer is using a pair of them to strengthen the torsion tubes.

Gusset on the stop between the torsion arms prevents the stop from being bent upward when the suspension bottoms out hard.

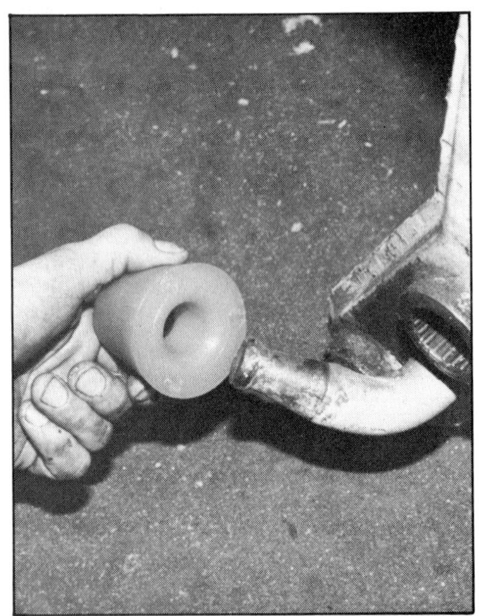

Soft rubber of stock VW snubber will not stand up to off-road abuse. If you're keeping the stop between the arms, knock off the stock snubbers and slip on tough urethane replacements. These are by Sway-A-Way.

Are dirt and crud causing premature torsion-arm bushing wear? Homespun fix on this recreational buggy helps keep crud out. Photo by Ron Sessions.

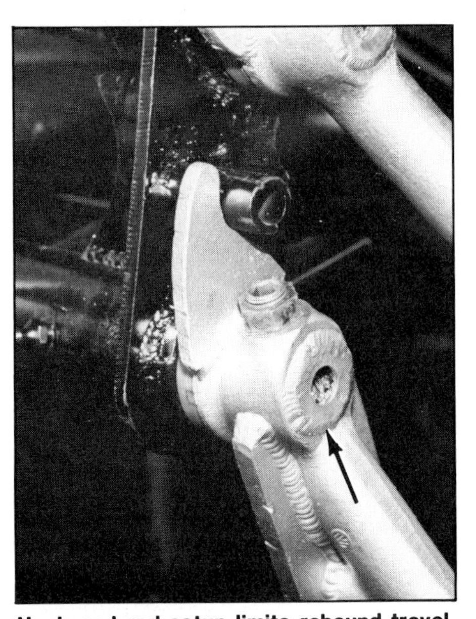

Hook-and-rod setup limits rebound travel. Washer (arrow) welded over torsion arm prevents torsion leaves from working loose.

the stop is bent far enough, jounce travel will be restricted by the shock, causing it to bottom internally. This will destroy a shock. If you bend the stop on one side, as can happen when a car lands on one front wheel, that side of the suspension will sag.

Reinforce The Stop—The solution is to reinforce the stop by welding on some *gussets,* or corner braces. Fabricate two gussets for each stop by cutting two triangular pieces from approximately 1/8-in.-thick steel plate, 3-1/2x3-1/2x3-1/2 in. Next, cut 1 in. off *one* of the points of each triangle. The pieces will look like a triangle with a flat top: 3-1/2-in. base, 2-1/4-in. sides and a 1-1/2-in. top. See the picture for details.

Put a reference mark on the stop flush with the end of the rubber snubber so the gussets won't interfere with its installation. A piece of masking tape wrapped around the stop works well. Knock the snubber loose with a hammer and slide it off.

Place one gusset with its long side next to the seam on the shock tower and its small side on the front of the stop. In this position, it should not interfere with where the rubber snubber fits—see your mark. Hold the gusset in place, and weld it to the shock-tower seam.

Next, press the opposite end of the gusset against the stop and weld where they meet. Now take a hammer and bend the corners of the gusset around the stop and finish welding. In-

Seat-belt strap limits rebound travel. Strap eyes are bolted to the upper and lower shock mounts. Photo by Tom Monroe.

Bolt and jam nuts on upper arm makes an adjustable rebound stop. Stop bottoms against lower-torsion-arm pivot. Photo by Tom Monroe.

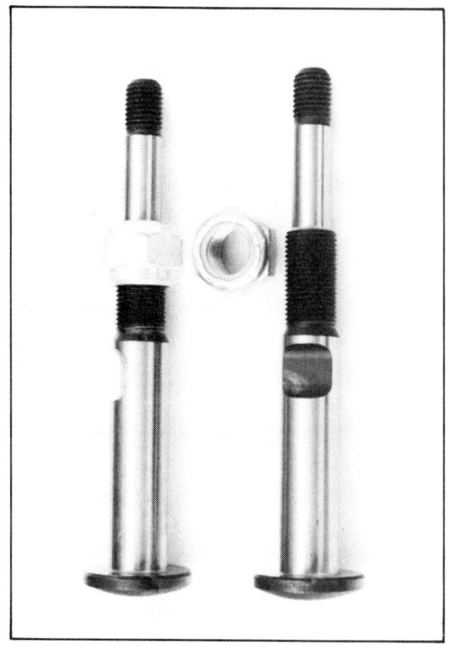

Sway-A-Way shock-mount link pins installed in lower torsion arms make perfect lower mounts for long-travel shocks. Photo courtesy of Sway-A-Way.

stall the second gusset at the rear of the stop and do the same. This job can be done without dismantling the front suspension.

Heavy-Duty Snubber—One of the most effective items for controlling suspension travel without transferring excess loads into the front end is a heavy-duty urethane snubber. The stock rubber unit is useless with stiffer front ends and will split under hard use.

Even heavy-duty snubbers can get cut and nicked from severe contact with the torsion arm. But this is no problem. The snubber can be rotated slightly so the damaged section faces away from the arm. Heavy-duty snubbers are available from most off-road buggy-supply houses.

Moving The Stop—The biggest change in the Volkswagen kingpin suspension occurred when racers removed the stop between the torsion arms and placed it above the upper torsion arm. With no bottom stop and high-rate torsion bars, the torsion arms droop much lower. This increases ground clearance and increases total wheel travel from about 7-3/4 in. to approximately 10-1/2 in.

Without the bottom stop, it's no longer necessary to heavily preload the front bars. The longer travel of the front suspension allows it to absorb bumps without being overly stiff.

To accommodate a long-travel front end, long-travel shocks must be used. This makes the stock upper mounts useless because they must be much higher with the longer shocks. Specially fabricated upper shock mounts can be welded onto the front part of the car's roll cage, or the shock towers can be raised. Custom-fabricated, long-travel torsion-tube assemblies or kits are available.

To help prevent shock-absorber damage, it's necessary to keep the shocks from acting as rebound stops. There are a number of ways to do this. One method of limiting downward travel is with a hook-and-rod arrangement on the torsion-tube uprights, page 56. Another is by attaching a steel cable or seat-belt-type strap slightly shorter than the shock's extended limit between its upper and lower mounts, shown above. Still another method utilizes an adjustable bolt-and-locknut-type metal stop on the upper arm. This stop bears against the pivot point of the lower arm, also shown above.

Link-Pin Bushings—The stock link-pin bushings in the torsion-arm link or carrier wear out very quickly in the dirt. Stock bushings are a soft material that provides long service life on the street, but gets beaten out in hard off-road use.

When these bushings get sloppy enough to need replacement, install tougher aluminum-bronze bushings available in the aftermarket. Link-pin bushings can be pressed in and out, or driven in and out with a special driver. Aluminum-bronze replacements are considerably more expensive than stock, but well worth it for serious off-roading. They rarely have to be replaced, except in a race car.

Hardened Link Pins—The stock VW link pin is perhaps the weakest link in the front suspension. Bottom the suspension by hitting a big rock or crashing into a big hole, and the case-hardened link pin will probably be the first part that breaks.

Through-hardened link pins made from high-tensile 4130 steel are available from aftermarket sources. Those made by Wright and Sway-A-Way also feature fine-threaded ends and locknuts that effectively clamp the carrier and torsion arm together and ease adjustments. Both stock and aftermarket pins are held in place by a torsion-arm clamp bolt.

Also available are special link pins with a threaded extension to act as a lower mount for an extra shock. These save you the cost of fabricating the mount separately or using a custom lower torsion arm with built-in dual lower mounts. Note that the link-pin shock mounts shown above are made of high-tensile 4130 steel.

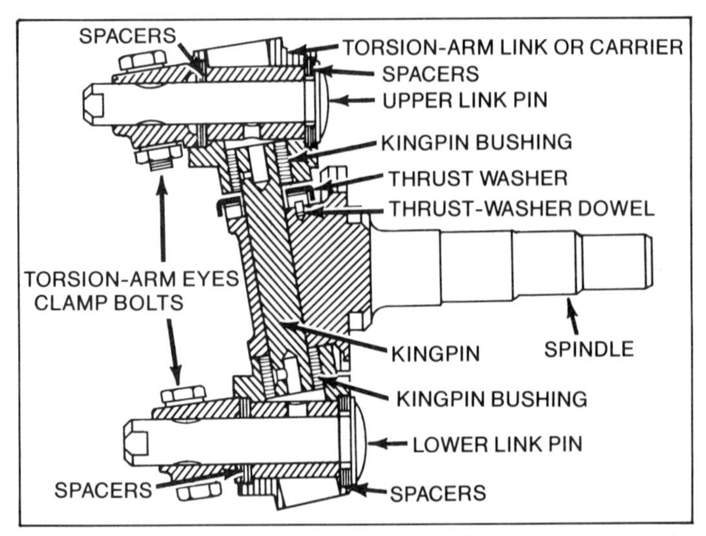

Cross-section of link-pin, kingpin and spindle assembly on stock VW front end.

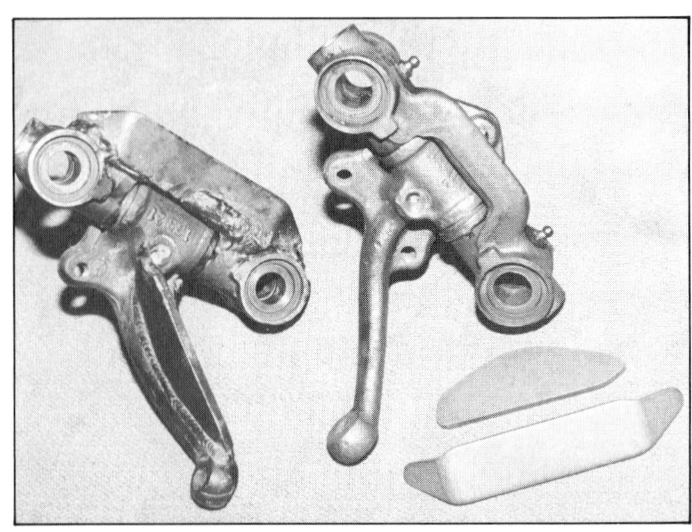

Gussets reinforce the stock VW spindle and carrier. These are available through most off-road parts outlets.

Kingpin Bushings—Stock kingpin bushings will last a long time if kept lubed, especially on a play car. Harder aluminum-bronze replacements are available in the aftermarket, but seldom needed. Whichever type of bushing you use, it must be pressed into place, then reamed to fit the kingpin.

Hardened Kingpins—Although stock kingpins rarely break, even under racing conditions, they can bend. As with the link pins, through-hardened, high-tensile 4130-steel kingpins are available from aftermarket sources. It's claimed that the high-strength kingpins strengthen the link itself, because the pins resist bending loads.

Worn spindles can be saved by using oversize kingpins. As with the stock pins, special pins are available in 0.001-in. oversizes up to 0.005 in. After reaming the bushings to fit, new pins are pressed into place. Break in the new pins and bushings after lubricating.

Torsion-Arm Link or Carrier—The front-suspension upright link, or *torsion-arm carrier,* supports the kingpin on which the spindle pivots. The carrier can be beefed up by welding a full-length 3/16- or 1/4-in.-thick steel gusset on the back of the carrier, just inboard of the grease fittings.

Make a template by tracing the back side of the carrier onto a piece of cardboard. Using the template as a guide, cut the gusset out with a torch or bandsaw. When welding the carrier, be sure it's assembled with the kingpin in place or it will warp.

Precut and bent gussets are available from aftermarket sources for a few dollars each. If you have a race car and more money than time, stress-relieved carriers with chrome-moly 4130 gussets are also available.

Torsion Arms—Stock VW torsion arms are forged. For recreational use, they'll last for years and years. Many racers use the stock arms with success too, *Magnafluxing*—magnetic-particle inspecting—them after every race. In severe use, the stock arms may *brinnel,* or develop indentations, where they ride in the torsion-tube needle bearings. Also, the stock arms may crack and break at the end of the machined section that fits into the tubes. By sleeving this section with chrome-moly steel, brinneling is reduced and the crack-prone area is reinforced.

VW off-road front-end specialists, such as The Wright Place, can sleeve stock arms. Other racing variations on the stock torsion arms include welding in extra shock mounts and full gussets.

For the racer, extra-long, tubular chrome-moly torsion arms are available. These arms are 1—4-in. longer than stock and fabricated with single or dual lower shock mounts. Tubular construction reduces unsprung weight. However, the longer arms are tough on torsion springs, and spring replacement is required every 2—3 races with the stock VW front end.

Spindles—Stock VW spindles last a long time in a recreational buggy or street-driven Baja Bug seeing minimal off-road abuse. Go racing with the stock spindles and they'll bend and break in short order. Next to the stock link pins, the spindles, which are integral with the steering arms, are the weakest links in the kingpin front end. The left spindle is particularly vulnerable because it is drilled for the speedometer drive.

First the bending problem. Now that you've reinforced the stock tie rods or replaced them with heavy-duty pieces, the steering arm is seeing greatly increased loads. Unless it is reinforced, the steering arm will bend. To strengthen the steering arms, weld on steel gussets.

Using the same material you reinforced the carrier with, cut and weld a piece to fit into the curved side of the steering-arm portion of the spindle. Trim it to line up with an imaginary line drawn from the base of the arm to the outside edge of the tie-rod hole. See the photo on this page as an example.

When welding in the gusset, be careful to keep heat to a minimum. If using a *wire feed*—MIG—or arc-welder, only weld about an inch at a time and then let the spindle cool. Weld another inch, let the spindle cool, and so on. Never cool by quenching in water, as it will make the spindle brittle. It will break rather than bend. When a spindle breaks, your car may lose a wheel or steering control—or both.

Some people like to weld the thrust washer, or spacer, to the spindle to increase the load-bearing area. The advantage of this is questionable because even if heliarced at a low temperature,

The Wright Place sells these reinforced VW carriers. Note beefy 3/16-in.-thick gussets running the full length of the carrier.

Stock-VW torsion arms are forged steel and will last many years in a recreational off-road car. Here, stock lower arm has been modified to use long-travel shock.

Urethane torsion-arm bushings are designed to replace stock Micarta bushings or needle bearings at ends of torsion tubes. Photo courtesy of Sway-A-Way.

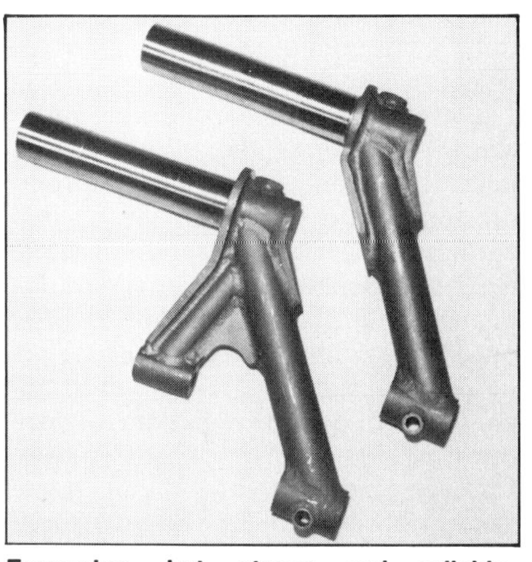

Expensive, but strong and reliable. Custom-made, tubular chrome-moly-steel torsion arms are from The Wright Place. Photo courtesy of The Wright Place.

The Wright Place's spindle/link-pin combination for kingpin suspension provides required strength for racing. Photo courtesy of The Wright Place.

the spindle should be heat-treated again.

Left-Spindle Fix—The speedometer-cable hole through the center of the left spindle makes this spindle weaker than the right one. This hole should be filled and the spindle strengthened. Otherwise, in rough use, the spindle will probably break at the machined taper for the wheel bearing.

Get a Grade-5 or -8 1/2-in. bolt over 6-in. long. Cut off the head and enough of the shaft to leave 6 in. Don't worry about the threads at the end. Now take the bolt and the spindle to someone with a lathe. Put the bolt in the lathe with the threaded end out and turn it down to 21/64-in. OD, 5-1/8-in. back from its threaded end. See the sketch on page 60.

While it's turning in the lathe, use a file or emery cloth to smooth any rough or sharp surfaces. Both diameters on the bolt are now just a shade bigger than the hole and counterbore in the spindle. Now put the bolt in the freezer, and the spindle in the oven at 300F (150C).

After about 30 minutes, remove both items and drive the bolt into the spindle. If the bolt is very hard to get in, remove it with a punch from the other end. You can take a little more off the bolt yourself. Use a 1/2-in. electric drill and some emery cloth—sort of a poor man's lathe.

After you've driven the bolt all the way into the spindle, lightly *tack-weld* its large end in place. Do this because any other method of welding will create enough heat to change the metallurgy of the spindle.

Racing Spindles—If you want to go fast, but are short on time and long on money, your best performance buy is probably a set of racing spindles. Plan on spending a few hundred dollars. Most aftermarket racing spindles start out as stock VW forgings. Those offered by The Wright Place come with the steering arm already gusseted, speedometer-cable hole in the left spindle filled and the spindle OD increased for oversize wheel bearings.

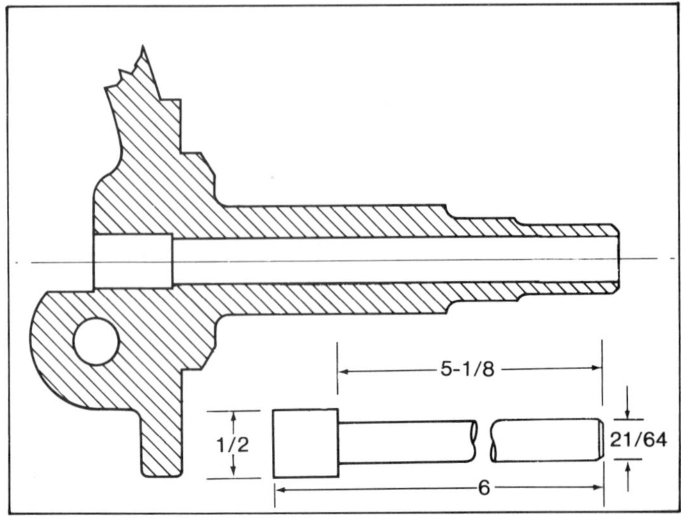

Strengthen the left spindle by machining a 1/2-in. bolt like this and pressing it into the speedometer-cable hole. Check spindle ID before turning bolt on lathe. Dimensions are in inches.

Three spindles popular with off-roaders. Bottom right is a stock-VW spindle. Bottom left is a '61—'63 Porsche spindle that fits the VW suspension. Top is a Wright-modified VW spindle with gusseted steering arm. Both the Porsche and Wright units have gusset-reinforced stock carriers.

The Wright Place offers these aluminum clamp nuts as replacements for the stock nuts and locking plate. Wheel-bearing adjustments are easily made with a small Allen wrench and adjustable wrench.

The steering arm can be taper-reamed to fit Ford, Chevrolet or Porsche tie-rod ends.

Years ago, racers were using '61—'63 Porsche 356 spindles—part 644 341 651 00 left, 644 341 652 00 right. Though they offer a larger-OD spindle and steering arm than the stock VW arm, they're not as strong as the aftermarket racing spindles available today.

To use Porsche spindles, a larger-ID Porsche inner wheel bearing and bearing spacer must be used. And you can fix the left spindle by filling the speedometer-cable hole as with the stock VW spindles. Today, Porsche units are hard to find and *very expensive*. My advice is leave them for the sports-car crowd and restorers.

Wheel Bearings—While I'm on the subject of spindles, there's something you should know about VW front-wheel bearings. The ball-bearing type used on all '65-and-earlier sedans with the kingpin front end should not be counted on to last with hard off-road use. Tapered-roller bearings were introduced on the ball-joint front end in '66. These are the bearings to use off-road. The tapered rollers offer almost twice as much bearing area as the old ball bearings. Therefore, the tapered rollers are far more durable.

Take out the outer ball-bearing races and press in new ones that match the roller bearings. The old races can be split and driven out. To install the new races, heat the drum to 200F (95C) in an oven, and chill the races in a freezer, then drive in the races with a proper-size bushing driver. Before heating the drum, make sure you get every last trace of grease out or the kitchen will get very smoky.

Spindle-Clamp Nuts—To hold the front brake drum onto the kingpin spindle, VW uses two large nuts and a lock plate with bending tabs. Without the special wrenches VW recommends, adjusting the front wheel bearings is not the easy job it should be. This is particularly true, considering how often the front-wheel bearings need adjustment with off-road use. Also, the lock plate should be used only once.

When running off-road, you should check wheel bearings for excessive play every few hundred miles or so. It only takes a moment to wiggle the wheels to check.

The Wright Place sells aluminum clamps to replace the nuts and lock plate. The clamp screws onto the spindle just like a nut, until the desired amount of bearing play is reached. Then it's locked in place by tightening a small Allen bolt that clamps tightly onto the spindle. These adjusters can be used over and over again. I consider them one of the better off-road buys.

Camber Adjustment—When reassembling the kingpin suspension, camber is adjusted with eight 0.020-in. spacers on each link pin. As pictured in the sectional drawing, page 58, the spacer washers fit over the link pin against the inboard and outboard ends of the carrier. Those spacers placed between the torsion arms and carrier affect front-wheel camber. Leftover spacers are placed between the carrier and link-pin head to keep link-pin play to a minimum.

The chart, page 61, tells you where to put the spacers to adjust camber based on a measurement of torsion-arm offset. Although the adjustment is really not that critical for off-road use, why not have it right? If you position the link-pin spacers according to the chart, you'll set front-wheel camber to the VW-specified 1/2° negative setting. Regardless of the camber setting, it's important to install all eight spacers on each of the link pins so the suspension will work correctly.

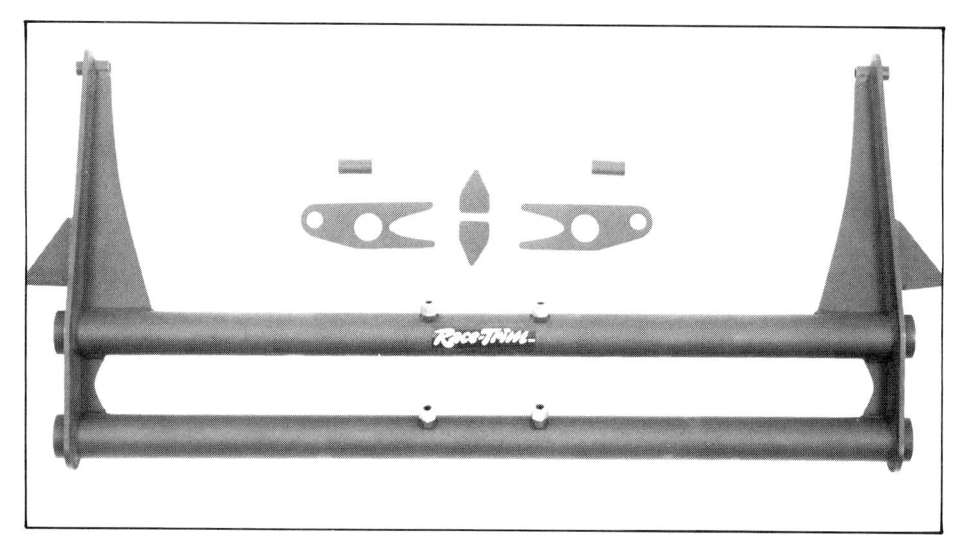

One of many of the lower-priced long-travel torsion-tube assemblies. This one is 5-in. wider than stock and has 4-in.-taller shock towers. Photo courtesy of Race-Trim.

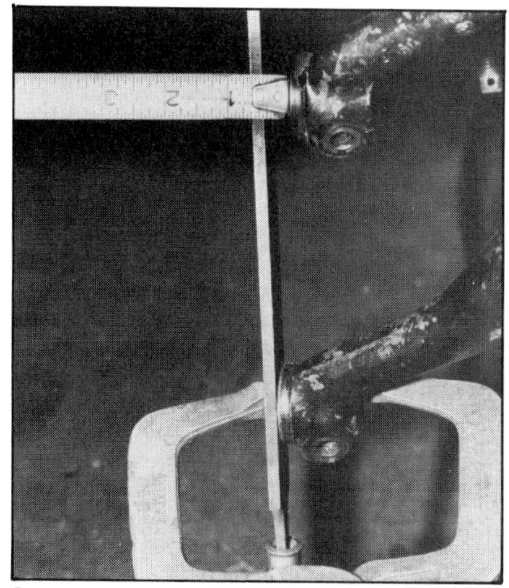

Simple and accurate way of measuring offset. Clamp a straightedge—in this case, a file—onto the lower torsion arm. Measure offset with a ruler. Judging by this photo, lower arm is bent.

1960–'65 CAMBER ADJUSTMENT CHART

Torsion Arm Offset (mm)	Upper Torsion Arm: Inner Spacers	Upper Torsion Arm: Outer Spacers	Lower Torsion Arm: Inner Spacers	Lower Torsion Arm: Outer Spacers
	Number of Spacers			
5	1	7	5	3
5.5	2	6	5	3
6	2	6	4	4
6.5	3	5	4	4
7	3	5	4	4
7.5	4	4	3	5
8	4	4	2	6
8.5	5	3	2	6
9	5	3	1	7

Position of spacers on either side of the torsion-arm link controls front-wheel camber. To obtain the VW-specified 1/2° negative camber with a given torsion-arm offset, place the spacers as per the chart.

Long-travel suspension from V-Enterprises uses third bundle of torsion leaves to control 4-in.-longer-than-stock torsion arms. Extra leaves are operated by arm riding between bottom shock mounts.

All the measurements on the chart are in millimeters (mm). Because most of us don't have a millimeter ruler handy, here is a ballpark conversion to inches: one inch equals 25.4mm. With a straightedge clamped to one arm, measure the offset to the other arm in inches. If the offset measures 5/16 in., using the conversion, 5/16 in. equals 7.9375mm. Round this off to 8mm and look it up on the chart. Position the spacers accordingly.

If the offset is less than 5mm or more than 9mm, look for a bent torsion arm. You can generally see a bent one by lining up the one in question against a good one. Put the machined parts of the arms on a flat surface and the rough-surfaced arms hanging off the edge. Line up the different parts of the arms and the bent torsion arm should be apparent. Usually, the lower arm is the one that gets bent.

Never use less than one spacer between the arm and carrier, or wear on the torsion-arm link will be greatly accelerated.

How Far Do You Want To Go?— There's still more that can be done to the Volkswagen front suspension, but step back for a moment and assess your car's needs. I assume most people reading this book don't intend to build an all-out race car. That's why I've concentrated on explaining how to get the most performance out of your off-road dollar.

At about the $200 or $300 mark, you begin to pay substantially more for that extra performance. Of course, how much money you spend and how you spend it is up to you. I hope I've given you enough information to help make your decision easier.

Full-Race Suspensions— If your aspirations are full-race with a budget to match, you'll want to invest in a long-travel suspension. Several companies offer adjustable torsion-tube units up to 10-in. wider than stock, with shock towers to mount two shocks per wheel. To complement the long-travel torsion-tube units, torsion arms 1—4-in. longer than stock are also available. However, the two sets of torsion leaves in the stock torsion tubes cannot handle the increased twist caused by the travel of arms more than 1-in.-longer than stock.

Solutions to the travel problem are many and varied. V-Enterprises gets up to 15 in. of travel by adding a third bundle of torsion leaves between the existing torsion tubes that act on a lever riding between the shock mounts on the lower torsion arm. The Wright Place replaces the leaves with adjustable "coil-over" springs that fit over the shocks, also netting 15 in. of travel. And Giese Racing does away with the torsion tubes and shocks entirely, replacing it with air suspension. The Giese system uses one large air shock per wheel and pressure is adjustable for vehicle ride height.

Ball-joint front end. Welding in a simple gusset helps keep the shock tower from flexing and breaking.

VW Thing ball joints are pressed in from above, making them stronger. You can use them in a sedan front end if you also use Thing torsion arms and steering-knuckle assembly. Photo by Ron Sessions.

Short-course racer is toughing it out with a ball-joint front end. Curiously enough, upper mount for second shock is run off stock shock tower—not the way to do it. Photo by Ron Sessions.

BALL-JOINT FRONT END

VW phased out the kingpin front end in the Type-1 sedan after the '65 model year. In '66, ball joints replaced the kingpins, link pins and carriers retaining the spindles to the torsion arms. A ball joint is a swivel joint that can rotate both with wheel travel and steering angle.

The ball-joint front end retains the basic front-suspension configuration, with torsion tubes, leaves and arms. But except for the steering gear and tie rods, none of the parts are interchangeable with the kingpin front end.

Shock Absorber As Suspension Stop—For off-roaders, the major shortcoming of the ball-joint front end is its lack of a built-in rebound, or lower stop. On bump or jounce, the travel is limited, or snubbed by a hefty rubber bumper built into the top of the shock. On rebound, the wheel stops when the torsion bars unwind and "relax." Because there is no lower stop against which to preload the torsion bars, this stock suspension should not be cut and rotated without limiting rebound travel.

The work of the shock is important here as it slows the momentum of the rebounding suspension. If the shock is worn out, it no longer damps rebound forces effectively. Consequently, the suspension may try to force the shock to extend farther than it was designed to go. This is the characteristic that causes the most trouble with the ball-joint suspension off-road. Any attempts to preload the torsion bars without making several other modifications will cause the torsion arms to try to exceed their intended downward travel. This eventually destroys the shocks and the ball joints.

Simple Fixes—The simplest way to improve a ball-joint front end is to get some good off-road shocks and reinforce the shock towers to handle the increased forces. As with the kingpin front end, the shock towers are spot-welded. Reweld these seams with an arc- or wire-feed welder.

The weakest part of the shock tower is at the base where it intersects the top torsion tube. On heavy cars like Baja Bugs, the tower may bend and crack there. Reinforce it by welding a couple of small gussets onto each tower. Use triangular-shape gussets, as big as the area will allow. Brace from the top of the tube to the inside of the tower.

Tie Rods and Ends—Another easy fix applies to '66 Beetles only. It's important that you replace the long tie rod with one from another year car. The long, adjusting tie-rod end on the '66 model is too weak for any off-roading, even recreational use. For more details on beefing up tie rods and ends, see page 46.

Other Fixes—Because this front end has not seen the off-road use the kingpin type has, there is not yet a universally accepted method for setting it up. The ball-joint front end is seldom raced. Because the angular travel of the ball joints is restricted, wheel travel is, too. Maximum wheel travel is necessary for optimum off-road performance.

Most of the ball-joint front ends you see in the dirt are on pre-run vehicles or play cars. Everyone I've talked to who does use a ball-joint front end off-road seems to have a different way of setting it up. If it works for him he's sure his is the way to go.

Ball-Joint Problems—On the '66-and-later VW sedan, the ball joints are pressed into the steering knuckle from opposite directions. The upper ball joint is pressed in from the bottom and the lower ball joint is pressed in from the top. When the wheel hits a hard bump, the force tries to unseat the lower ball joint and tries to push the upper ball joint into its seat.

Gimme That Thing—One ball-joint front end was designed with a little more heavy-duty use in mind. You can use the torsion arms and steering knuckle assembly from a Volkswagen Thing—Type 181. Both ball joints are pressed in from the bottom.

This unit is a lot stronger under full jounce, but weaker on a suspension that tries to rebound farther than the ball joint will permit. With lots of wear from off-road use, or any attempt to exceed the limits of angular travel in the ball joints, the ball joint will loosen and be forced out.

Oversize Ball Joints—Whichever

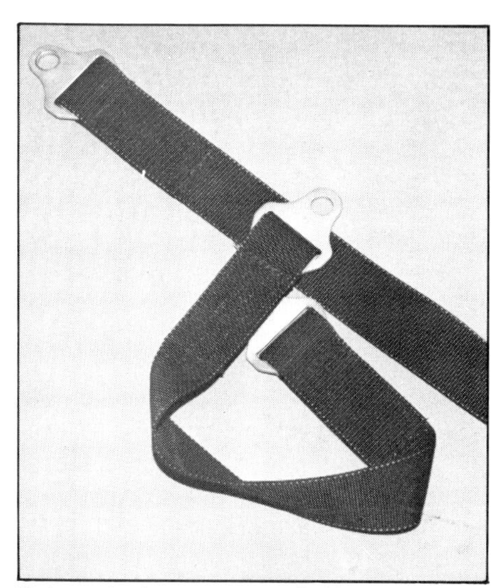

Rebound straps are available from most safety-equipment people. They bolt to the upper and lower shock mounts. The straps should stop the shocks 1/2-in. short of their extended length to keep them from being damaged.

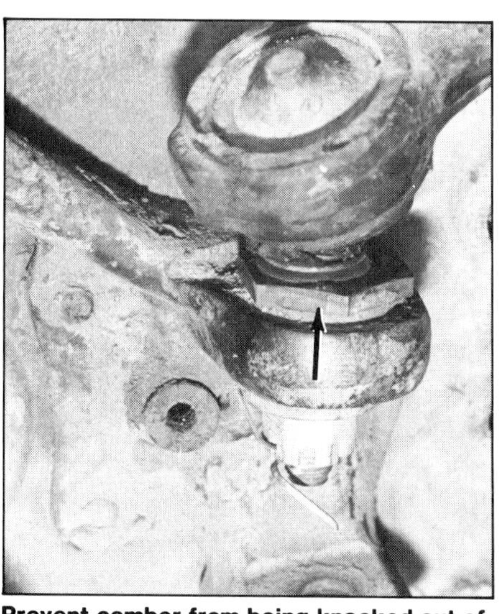

Prevent camber from being knocked out of adjustment by tack-welding the large adjustment nut (arrow) on the upper ball joint to the arm.

Torsion-bar adjuster for ball-joint front end. Replaces stock center anchor. Allows fine adjustments to ride height and stiffness. Photo courtesy of Sway-A-Way.

way you end up going, inspect the ball joints frequently. They are usually the first component to fail. Every time a ball joint is pressed in or out, the fit in the torsion arm gets a little sloppier than before. Volkswagen makes oversized replacement ball joints to cure a loose press fit. The torsion arm does not have to be machined to accept the larger OD joint.

Preloading the Torsion Leaves—You can get a little more ground clearance and a stiffer suspension by preloading the front torsion leaves. The principle and the procedure of cutting and rotating the torsion tubes is the same as that explained on page 52 for the kingpin front end. The only major difference is that the ball-joint front end uses ten leaves in each tube and the kingpin type uses eight.

I've seen some off-roaders stop the modification process after preloading the front suspension, and simply install an extra shock at each front wheel or just carry a couple of spare shocks. They explain that the shock will limit the suspension's rebound travel, sacrificing itself to keep the ball joints from breaking. When the shock finally breaks after being repeatedly overextended, a spare shock is installed and the car keeps going. Dual front shocks might just provide enough rebound damping to prevent being overextended.

Limiting Rebound Travel—Most people don't like to destroy expensive off-road shocks, so they limit the droop of the torsion arms by another method. A simple but effective travel limiter can be fashioned out of nylon strap, cut just a bit shorter than the shock's extended length. The strap is fastened to the upper and lower shock mounts.

Simpson, Deist and many of the safety-equipment manufacturers can fabricate these limiters out of seat-belt material for your car's needs. I've also seen steel cable used. Jack the front wheels off the ground so the shocks fully extend. Measure the distance between the upper and lower shock mounts. Subtract about 1/2-in. from this and you've got the eye-to-eye length needed for your limiter.

A neat way to limit travel is to fabricate a hook and rod. Drill a hole through the shock tower between the torsion tubes from the outboard side. As shown on page 56, weld in a 1/2-in.-OD rod or bolt so it sticks out over the lower torsion arm.

Next, fabricate a hook out of about 1/4-in. steel plate and weld it to the top side of the lower torsion arm. Check hook location by raising or extending the suspension to just short of full rebound. If done correctly, the hook will engage the rod just before the shock is fully extended.

High-Rate Torsion Leaves—Another good way to stiffen the ball-joint suspension is to install a set of high-rate torsion leaves. Those available from Sway-A-Way increase spring-rate about 20%. This makes the front suspension harder to compress without preloading it. It rebounds and stops at the same point as the stock springs. For more basics on spring rates see page 52.

If you're using high-rate torsion leaves with a ball-joint front end, the torsion arms and steering knuckles from a Volkswagen Thing are a big advantage. With each bump, both upper and lower ball joints are pushed *into* their seats, not one in and one out as in the sedan ball-joint setup.

Camber-Adjuster Problems—Camber adjustment is done at the upper ball joint with an eccentric bushing between the ball-joint stud and steering knuckle. To adjust camber, you rotate the eccentric bushing. In street use, the tightness of the ball-stud nut on the steering arm will keep the bushing from working loose. But under the severe pounding of off-road driving, the bushing can work loose, taking the camber adjustment with it.

If the bushing loosens up, the ball joint may come out of the arm. To prevent this, put a single tack-weld on the camber adjuster at the steering arm. Be sure to have the front end all together and the camber adjusted. Camber should be adjusted only with the proper front-end alignment equipment. Should it ever become necessary to change the camber setting, you can break the tack-weld with a hammer and cold chisel.

As with the kingpin front-end, precise adjustment is not all that critical in the dirt. If you are using your car for daily street use, don't tack-weld the adjuster.

Other Modifications—For information on beefing up tie rods, tie-rod ends, multiple-shock-absorber installations, strengthening the front clip, torsion arms, and spindles on the ball-joint suspension, see the appropriate heading in the kingpin section.

STEERING

The VW steering linkage is a little different than most cars because it has no idler arm. To accommodate left- or right-hand steering, the tie rods are of unequal length and both attach directly to the pitman arm.

The stock Type-1 steering box is a sturdy little unit. It will last for years on a street-driven Baja Bug or buggy under light recreational use. Under the severe conditions encountered in off-road racing, the stock box will last one race, maybe two or three at the most, before coming apart.

Of the stock boxes used with the torsion-arm front suspension, the '60-and-earlier unit is strongest. It's a worm-and-sector design. A worm gear on the steering column drives a gear rack or sector on the steering shaft. Although no longer available from VW, you can occasionally find one in a junk yard. If it's seen only street use, all it may need is a set of bearings, a few seals and an adjustment.

Beginning in '61, VW switched to a worm-and-roller box. With this box, the sector is replaced with a roller gear. The roller gear is mounted on a pin so the teeth can rotate as they mesh, reducing friction. New Brazilian-made TRW worm-and-roller boxes are available at reasonable prices. If the current box has slop that won't adjust out and you don't plan to race the car, get one of these new boxes.

Both units have a fairly decent steering ratio for off-road use—14:1 for the worm-and-sector and 15:1 for the worm-and-roller. Though neither of these ratios is what you might call quick, they work out to about 2-1/2—2-7/8 turns lock to lock, which is a good compromise between directional stability and twitchiness off-road.

Living With the Stock Box—In

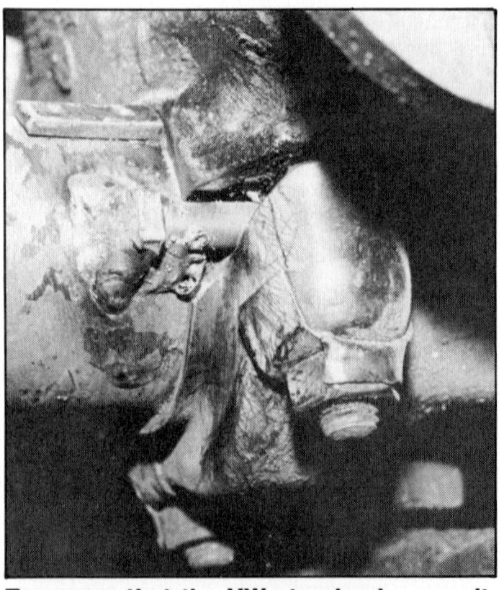

To ensure that the VW steering box won't rotate on the torsion tube on hard bumps, weld a bolt or similar piece of metal on the tube so it "keys" the box through the gap at the clamp.

rough use, the pitman arm transmits shock loads into the steering shaft. The steering shaft then bangs into the play-adjustment mechanism and cast-magnesium steering-box cover. The adjuster pounds into the cover and the end play of the steering shaft becomes excessive.

A few years ago, Gil George of FUNCO devised a way to make the '60-and-earlier worm-and-sector box live longer. He placed a Torrington-type needle-bearing and shim or stack of washers between the steering-box housing and pitman arm. If you are making this fix, leave about 0.010—0.020-in. clearance so the steering doesn't bind. When the car hits a bump big enough to force the pitman arm up, the load is distributed into the housing, not into the sector shaft and steering-box cover.

The '61-and-later box has more than one problem. With pitman-arm thrust transferred into the housing rather than the steering shaft, the next weakest point will break. With severe shock loading, the roller shaft will break at the roll pin where the teeth rotate.

Steering-Wheel Diameter—Often overlooked, steering-wheel diameter has a direct effect on steering effort. The larger the wheel, the lower the effort. The first thing many people do when they modify a sedan is throw away the stock 15-in. VW steering wheel. They often replace it with a dished, padded, chrome-spoked wheel of smaller diameter that "looks good." When steering-wheel diameter is reduced, you lose leverage so steering effort is increased. If space is not a factor, I recommend going to a larger-than-stock, 16- or 17-in. wheel as used in sprint cars.

Steering-Box-Mount Fix—The VW steering box is kept from rotating on the top torsion tube by a lug or small pin. This lug or pin fits in the gap between the steering box and its mounting clamp. In hard off-road use, it may not be strong enough. Use a bolt, *Woodruff key* or fabricate a piece of metal to fit in the gap between the steering-box housing and the clamp. If this piece is welded to the torsion tube, the steering box is still removable, but won't rotate on the tube.

Porsche Box Swap—The ZF worm-and-roller steering box from a Porsche 356 has been a favorite of the off-road racer for a long time. Compared to the stock VW box, the Porsche offers better durability. It features a strong, cast-iron housing and cover, four mounting studs compared to the VW's two, a tapered and splined pitman arm, and tapered roller bearings to take thrust loads.

But now they're getting scarce and very expensive. Although the box is available new from aftermarket sources, some parts, especially the worm drive, are no longer available from Porsche. The stock roller is strong—so strong that it hammers indentations into the worm drive. After a few races, the brinnelled worm causes steering action to be "lumpy" or notchy, which can get very exciting in the turns. Because the worm is unavailable, you're forced to install used parts or custom, hard-chromed ones as replacements.

On the plus side, the Porsche unit bolts right onto the VW torsion tubes without modification. If bolted into the stock location, bump steer is no problem. Tie-rod ends from a '68-or-later VW fit the pitman arm.

You will, however, have to modify the tie-rod end of the Porsche pitman arm. Cut off the end and weld on a VW end to maintain proper geometry.

GM Saginaw Box Swap—Some racers install a GM Saginaw worm-and-recirculating-ball steering box. This cast-iron box offers the same strength and smooth operation as the Porsche units. But it's readily available from most junkyards at low prices.

Most of these manual steering boxes

I know this Chevy steering-box setup works because the owner of this competition Baja Bug has beaten me more than once. When using a Chevy box, you'll have to make your own mounts. Get it right or geometry will go way off scale.

Wright rack-and-pinion steering is a popular item on many off-road race cars. Here it's used with a Sway-A-Way torsion-bar adjuster for a widened front end. Photo by Ron Sessions.

originally lived in front-end-heavy GM sedans. So the steering ratio—usually about 28:1 or six turns lock-to-lock—may be a bit on the slow side. Quick-ratio 16:1 gears, such as those found in the Corvette, are available from your GM dealer. Use of these gears will get the ratio down to about three turns lock to lock—close to the stock VW and Porsche ratio.

The drawback with the Chevrolet box is that you must fabricate a new mount and set up the steering geometry from scratch. The mount and necessary bracing adds weight to the front end.

Because of its size, the Saginaw box is mounted behind the torsion tubes instead of in front of them. This requires using a specially fabricated leading pitman arm, which will then swing in a different arc than the trailing steering arms at the knuckles. Unless located correctly, this setup will give horrendous bump steer. See page 66 for tips on how to control bump steer with proper geometry.

Rack-and-Pinion Steering—Out of the need to find a heavy-duty off-road steering box, aftermarket rack-and-pinion units have been developed. The goal was to make a steering unit without the durability problems of the stock VW box, the high parts and machining costs of the Porsche ZF box, and the mounting, bracing, and geometry problems of the Saginaw box. Rack-and-pinion steering offers the advantages of simple design, operation and repairability. It is easy to mount, and has no pitman arm, so it requires no modifications to retain proper geometry.

Many racers, especially those running single- or tandem-seat buggies, have switched to rack-and-pinion steering. Among those currently offering rack-and-pinion steering boxes are The Wright Place, AMS, FUNCO and Stilleto. The Wright Place offers units in two ratios—1-1/2 and 1-1/8 turns lock-to-lock. They cost well over $200, but reasonably priced parts are available to rebuild them.

One problem with rack-and-pinion steering is *feedback*. Felt as a shock to the steering wheel when a front wheel hits a bump, feedback is more pronounced with this type steering because of its more direct design. Low internal friction makes it as efficient in transmitting rotational movement of the front wheels to the steering wheel as it is in transmitting steering-wheel rotation to the front wheels. Got that?

To keep the steering wheel from "jumping" out of their hands, racers install hydraulic dampers connected to and parallel with the tie rods. While the dampers increase steering effort slightly, they do damp shock loads and resulting rapid movement of the steering linkage.

Most rack-and-pinion units mount to a bracket welded onto the top torsion tube. The unit is positioned to the rear of and slightly above the tube. On tube-frame buggies, there's plenty of room to get the unit positioned. But on sedans and Baja Bugs, the stock fuel tank must be removed and the frame head may interfere. Some builders beat a little depression in the frame head to make room for the steering gear.

Before welding on the mounting bracket, find the mounting location for the steering gear that gives the least *bump steer*. As with any type of steering, rack-and-pinion setups can suffer from a lot of bump steer if not located correctly. To check for bump steer, remove the torsion leaves and reinstall the torsion arms. With the mounting bracket clamped to the torsion tube, check front-wheel toe as you move the suspension through its full travel. If toe changes, you have bump steer.

Cecil Wright, maker of the Wright rack-and-pinion box, says that a properly located unit will give only 1/8-in. of bump steer. Zero bump steer is ideal. On wide front ends or on cars using extra-long torsion arms, the stock mount may have to be trimmed to get the box as close to the torsion tube as possible.

Center Steer—The stock steering box can be mounted anywhere on the torsion tube. Before the days of rack-and-pinion steering, it was common practice for single-seat-buggy builders to center the steering gear on the tube.

Because the stock pitman arm has offset holes for the ends of the unequal length tie rods, its geometry is wrong for center steer. The fix is to take a "pie-slice" section out of the arm. Reweld it so a line through the holes is perpendicular to and the holes are equidistant from the steering-shaft center line. See the sample drawing.

Steering Dampers—A steering damper cushions shock loads and vibration in the steering. It acts like a small shock absorber.

A steering damper was first used on the '60 Beetle. If you're using an older-model front end, there's no mounting provision on the torsion tube for a damper. Fortunately, clamp-on steering-damper mounts are relatively inexpensive. With a clamp-on damper mount, setting toe-in is easier than with a rigid welded mount. You merely loosen the clamp while rotating the tie-rod end to get the desired toe. Then retighten the clamp.

Use the stock damper mounts if possible. One is enough on the Volkswagen, Porsche and GM boxes. Due to

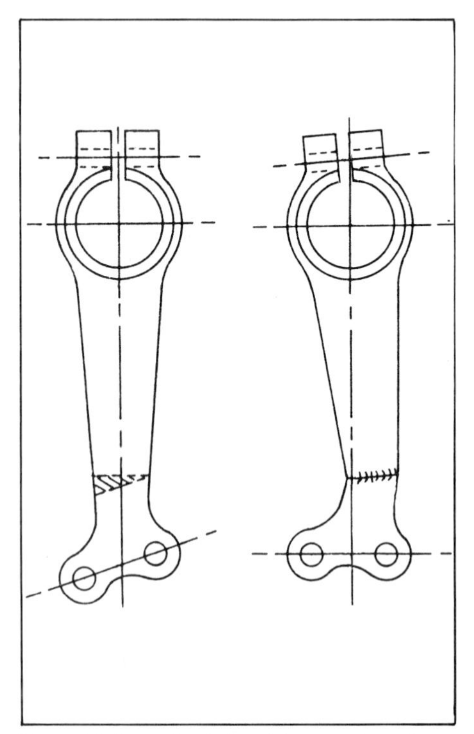

If modifying stock box for center steer, end of pitman arm must be changed to retain proper geometry. Pie-slice taken from arm gives desired alignment of tie-rod-end holes. To keep stock arm length, cut two pitman arms to make one modified arm.

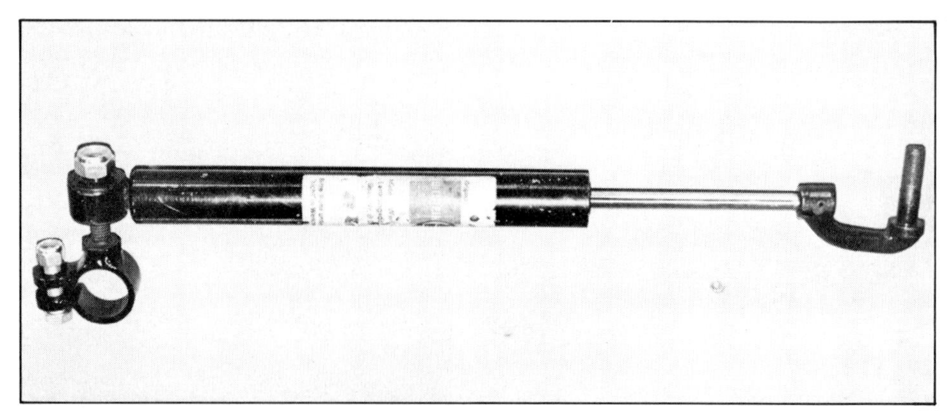

Steering damper reduces feedback through the steering box to the steering wheel. Stock single damper may not be enough. When adding an extra steering damper, this little clamp will come in handy. You can clamp it on a piece of tubing to mount the damper.

its higher feedback, use two dampers on the rack-and-pinion unit.

Power Steering—As horsepower, suspension travel and top speeds increased, off-road racers began experimenting with power steering to reduce steering effort and improve response. Today, most competitive cars use a power-steering system if class rules allow it. Typically, the V-belt-driven power-steering pump is operated off a small accessory-drive pulley bolted onto the larger crank pulley with a special long bolt. A variety of compact- and mid-size-car pumps have been adapted to the task—Toyota and Honda come immediately to mind. Also, fork-lift power-steering pumps have been successfully adapted.

How pump pressure is used varies with the system. Unique Metal Products' system uses a steering-column-mounted servo that directs a ram to assist a rack-and-pinion steering gear. Another setup uses a column-mounted fork-lift servo. Cars that have been converted to a GM Saginaw manual-steering box for strength need only replace the manual gear with a power gear and plumb a pump.

Aside from the cost, drawbacks to using power steering are few. If using a fully integrated rack-and-pinion unit in a car with the stock VW floorpan, the pan will have to be clearanced to clear the servo. And regardless of power-steering type, a broken drive belt or leaky pressure line will cause steering effort to increase dramatically. It's just one more thing to go wrong—the price you pay for increased complexity.

Power-steering pump is run off small crank pulley added onto conventional crank pulley on this Type-4 engine.

Unique Metal Products' power-steering setup uses column-mounted servo. Servo directs boosted pressure to ram attached to a Wright rack-and-pinion steering gear.

Column-mounted power-steering servo is adapted from a fork-lift truck.

ALL ABOUT BUMP STEER

Bump steer is toe change with wheel travel. Total bump steer is the difference between front-wheel toe measured at full jounce compared to that measured with the car sitting at ride height.

Any time you change the steering box or relocate the stock box, there is a chance you may increase bump steer. Consequently, mounting location is extremely important.

Here's how gear location affects bump steer: With toe set at zero at ride height, moving the steering gear up increases toe-in in bump and toe-out in rebound. Move the steering gear down, and the opposite happens: toe-out increases in bump and toe-in increases in rebound. Move the steering gear to the rear and toe-out increases both in bump and rebound. Finally, moving the steering gear forward increases toe-in both in bump and rebound. OK? Now you know.

chapter 4 REAR SUSPENSION

When an off-road car goes on its nose, it's not always because of a big bump. The usual cause is inadequate rear-suspension travel. Photo by Judy Smith.

As explained briefly in the Transaxle Chapter, VW rear suspensions fall into two major categories: *swing axle* and *semi-trailing arm.* The swing-axle suspension was used on Type-2 and Type-3 models through 1967. It was used on Type-1 manual-transmission models until '68. The semi-trailing-arm suspension is used thereafter.

The semi-trailing-arm suspension is popularly known as the *IRS*—independent rear suspension. Actually, both the swing-axle and semi-trailing arm suspensions are independent rear suspensions. The rear wheels of both suspension types operate independently. It's just that IRS is a *general* designation, and semi-trailing and swing axle are *specific.* I'll stick to the popular, though not totally correct, IRS designation.

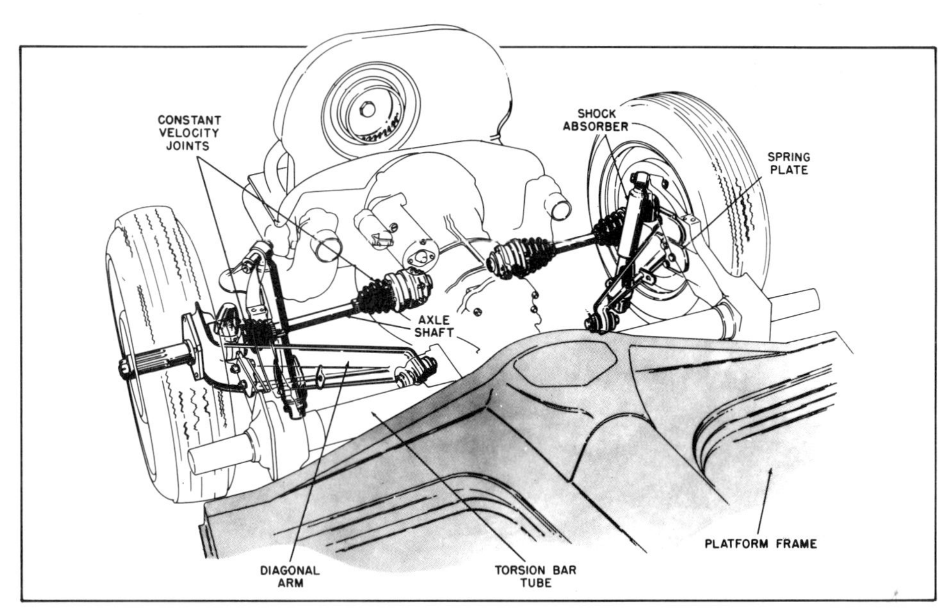

VW semi-trailing-arm rear suspension—*IRS*. It was introduced on the '68 Beetle with Automatic Stick Shift—*ASS*. A year later, it made it into manual-transmission Beetles as well. Drawing courtesy of Volkswagen of America.

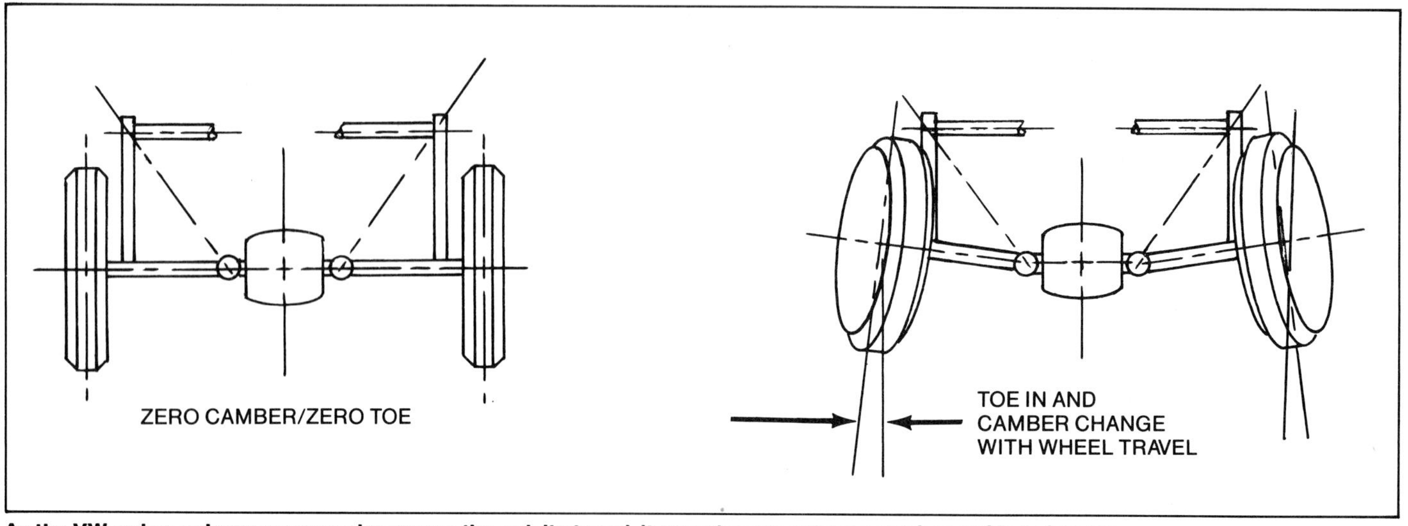

As the VW swing-axle rear suspension moves through its travel, it experiences extreme camber and toe changes.

Confused? Well, it gets easier. Both suspension types are sprung using similar torsion-bar setups. The swing-axle and IRS suspensions use a pair of transverse, side-to-side, torsion bars in a single large-diameter tube at the rear of the floorpan.

The torsion bars are splined at both ends. The inboard end of each bar fits into a splined socket at the center of the tube. The outboard ends of each bar fit into the splined end of a *spring-plate* assembly.

A spring plate is similar to a front-suspension torsion arm. It transmits the action of the wheel to the torsion bar and *helps* locate the wheel. Unlike front-suspension torsion arms that do the total wheel-locating job, spring plates act only as *radius rods.* They locate the rear wheels longitudinally, or front to rear, handling braking and power-on and -off thrust loads.

Swing-Axle vs. IRS—The major difference between the swing-axle rear suspension and the IRS is how the wheel is located laterally, or side to side. The resulting geometry is different for each type.

With the swing-axle suspension, each axle is housed in a rigid tube. The spring plates and axle tubes combine to locate the rear wheels. The axle splines straight into the wheel hub, but is driven at the transaxle through a spade-and-socket-type universal joint.

The tube is rigidly fixed at the wheel and pivots at the transaxle. Consequently, the wheel *swings* about the transaxle pivot, creating a high roll center. Because the axle tube moves through an arc, there is an extreme camber change with wheel travel—negative camber with *jounce* (suspension compressed) and positive camber with *rebound* (suspension extended). Because of the spring plates, the VW swing axle also *toes-in* considerably with both jounce and rebound.

Those are the bad features of the swing axle. The simplicity and durability of the swing-axle-type suspension are very desirable for off-roading.

Rather than a rigid axle tube, the IRS uses a *semi-trailing-arm* arrangement to complete the wheel-locating job. This semi-trailing arm consists of the spring plate and a *control arm* or *diagonal arm.* The term *semi-trailing* is used because the combination of the spring plate and the control arm isn't *fully trailing,* as it is on the front suspension. The imaginary axis the spring plate and control arm pivot around is *not* perpendicular to the car. This axis runs through the spring-plate and semi-trailing-arm pivots.

The front-suspension torsion arms are fully trailing. There is no camber change with wheel travel. The torsion-arm-pivot axes are horizontal and 90° to the vehicle center line. The semi-trailing arm of the rear suspension pivots at a bracket mounted behind of the torsion tube, but near the torsion-tube center.

Draw a line through the spring-plate pivot—the torsion-bar end of the spring plate—and the semi-trailing arm, and you have a pivot axis that is also horizontal. But it is not 90° to the vehicle center line, like the front-suspension torsion arms, or parallel to it like with the swing axle. The angle is 77-1/2°. Unlike the front suspen-

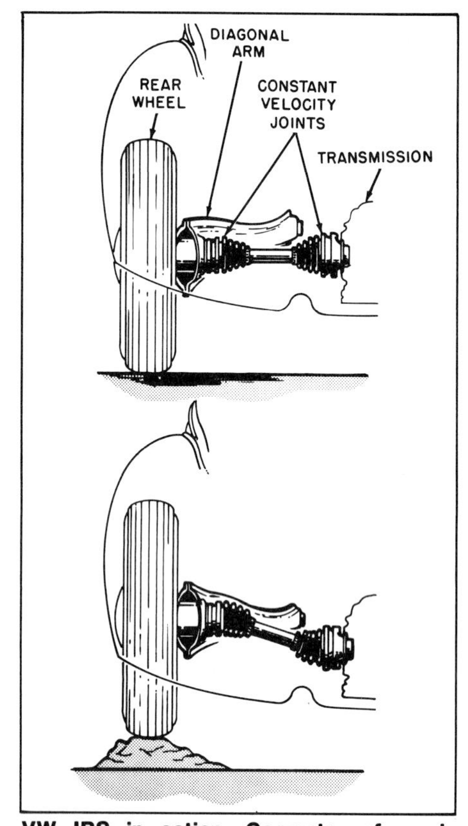

VW IRS in action. Geometry of semi-trailing arm gives moderate camber and toe changes with wheel travel. High-angle axle-shaft joints are key to long-travel IRS. Drawing courtesy of Volkswagen of America.

Tires of this swing-axle off-road car are in a positive-camber attitude. High-rate torsion bars, increased preload and big tires have contributed to the camber problem. Off-road race cars with this setup are notoriously twitchy and take extra driving skill. Photo by Tom Monroe

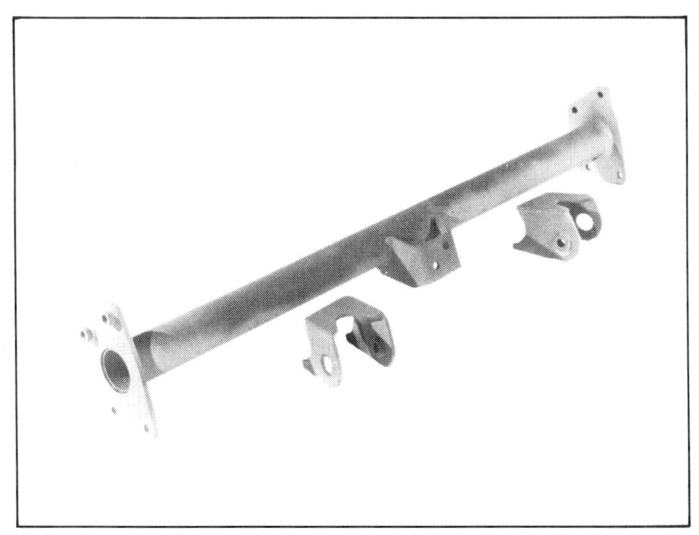

Chenowth's kit for converting a swing-axle rear suspension to an IRS. Besides the brackets on the torsion-bar housing, other parts you'll need for the conversion are diagonal arms, IRS spring plates, and, of course, IRS axles, CV joints and stub axles. Photo courtesy of Chenowth.

sion, this gives some camber change. But it doesn't give extreme camber change like the swing axle with a 0° axis, or parallel to the car center line. The IRS gives less camber change than the swing-axle suspension and a slight amount of *toe-out* as the wheel travels in either direction.

The axle on the IRS is fitted with two constant-velocity (CV) universal joints—one at the transaxle and one at the wheel. So the only function the axle serves is to transmit engine driving and braking forces to the wheel. It does not help locate the wheel.

Camber and Toe—The VW rear suspension is simpler than the front because there's no steering involved. Therefore, of the three measurements of wheel orientation, caster is not a factor. However, camber and toe-in are critical to handling, especially when you increase ride height and suspension travel for going off-road.

As with the front suspension, toe is the *difference* between the right- and left-side tire-tread centers, measured at the front of and at the rear of the tire. The measurement is taken at wheel-center height.

Toe-in is a condition in which the tires are closer together at the front than at the rear. If a line is drawn through the longitudinal center line of each toed-in wheel, the two lines will meet at a point in front of the vehicle. Toe-out occurs when the rear of the tires are closer together than at the front. A line drawn through the center line of two toed-out wheels will meet at a point behind the vehicle.

Most independent suspensions, whether they be front or rear, are designed with a small amount of initial toe-in. This is designed to compensate for the toe-out or "spreading apart" of wheels that occurs at road speeds as suspension components deflect or comply. Ideally, a car with initial toe-in will have zero toe at speed. I'll explain how to adjust toe-in later in this chapter.

Camber is the angle at which the tires sit relative to true vertical when viewed from the front or rear. *Positive camber* means the tires are farther apart at the top than at the bottom. *Negative camber* means the tires are closer together at the top than at the bottom. Zero camber is when the tires are straight up and down.

Is Swing Axle or IRS Better?—Most of you will be stuck with the rear suspension your car has. If you have that option open, such as when building a car from scratch, you may ask, "which suspension is better?"

Swing-Axle Pluses—The swing-axle suspension offers lower *unsprung weight,* which is the weight that "bounces" with the wheel. This lets the tires follow road surface more closely. This results in less *inertia* for the shock absorbers to overcome.

The swing-axle suspension also has fewer parts. As GM's Boss Kettering said, "Parts left out cause no problems and cost nothing." This is even more true in the dirt. Fewer parts mean fewer parts to break, a blessing when it's a long walk home. There's an economic consideration, too. The swing-axle suspension generally costs less because of the old rule of supply and demand. More are available.

Swing-Axle Minuses—In the negative column, the swing-axle suffers from excessive camber change. To restrict positive camber at full rebound, special spring-plate stops are needed to limit rebound travel. Also, longer axles and wider flanges should be used to reduce this camber change for given amounts of wheel travel.

Other swing-axle disadvantages stem from the use of the axle to locate the suspension. If you plan to go racing with a swing-axle setup, the axles should be checked for cracks, shotpeened, and polished at the transaxle end, and their *fulcrum plates* carefully matched. To keep the axle tubes from bending under high loads, axle-tube stiffeners, or overtubes, should be used.

IRS Pluses—If you could ask a suspension engineer, he'd tell you the IRS, or semi-trailing-arm rear suspension, is the better. The IRS can be modified for additional ground clearance and wheel travel without the extreme camber change of the swing-axle. With longer travel, lower-rate torsion bars can be used. The lower-rate bars provide a less-harsh ride. All the big-money racers use IRS.

IRS Minuses—The IRS suspension has its drawbacks. Because the system has more parts to wear out or fail, repair and maintenance are needed more frequently. Boss Kettering wouldn't like that. And because off-roaders are encouraged to use every inch of suspension travel possible, the stock axle-shaft joints are forced to operate at angles they weren't designed for.

The fix is to replace the stock units with Type-2 or expensive Porsche Turbo CV joints. Another way racers make the axle shafts work at high angles is by using Spicer or late-model Datsun Z-car U-joints and splined axle shafts. For details on this hardware, see page 38.

Other IRS components are at a disadvantage in severe off-road use. The semi-trailing arm and *stub axles*—short axles in the hub carriers—must be replaced or reinforced so they'll withstand the side loads experienced off-road. Otherwise they will fail.

REAR-SUSPENSION PREPARATION

Most off-road suspension preparation is common to both the swing axle and the IRS. Read the following general section first. For modifications unique to each type, read the appropriate section.

For serious off-road use, rear-suspension stiffness and ground clearance must be increased. One way to increase both is to preload the torsion bars. You can preload torsion bars by rotating the spring plate on the splines of the torsion bar or by rotating the torsion bar on the inner splines. Most racers do both, though either will load the spring plates against their lower, or *rebound,* stops.

Although both the IRS and swing-axle torsion bars are adjusted the same way, the degree of preload is more critical with the swing-axle. Preload affects ride height. Because of swing-axle geometry, a ride-height change affects camber and toe-in. This, of course, affects the car's handling.

When the swing-axle suspension absorbs a bump and the wheel travels from full rebound to its upper limit, *full jounce,* the rear wheels go through maximum camber change. The rear camber changes from positive, through zero, to negative and back again. This isn't much of a problem if the car is going straight, but it can get your attention if the car is sideways or in a turn.

Positive camber is the enemy here. Every time the car gets into the air and the weight is off the rear wheels, the suspension extends fully, which is full rebound. When the wheels contact the ground, they are at maximum positive camber or *tucked under.* Keeping this in mind, let's talk about increasing rear-suspension stiffness.

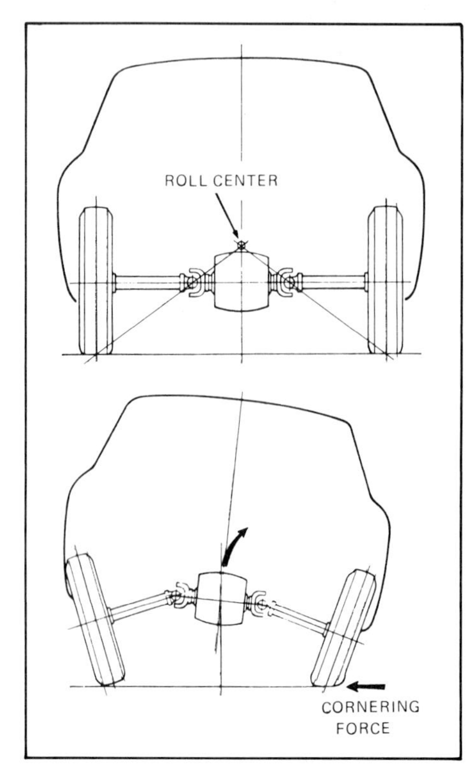

Positive-camber effects of a swing-axle are unsettling when going straight. They get downright scary in the turns. Jacking effect increases ride height and positive camber.

Increasing Stiffness—At stock settings, the VW rear wheels are near vertical with no passengers or driver. This puts the wheels close to zero camber most of the time. Handling is satisfactory on smooth surfaces.

The stock springing is too soft and ground clearance inadequate for off-road. After a while the suspension will beat itself to death. Consequently, you must stiffen the rear suspension some way. As with the front suspension, these methods include adjusting in more preload or installing larger-diameter torsion bars.

One thing common to all modifications is that the rear suspension must be partially disassembled. The simplest and most inexpensive way is to remove the each spring plate from its torsion bar, set in more preload, and reassemble. If it's too much or not enough, disassemble and reset the spring plates until you get it where you like it.

If your finances allow, install torsion-bar adjusters. Most race cars are equipped with both front- and rear-suspension adjusters so they can be set up quickly for different courses. Adjusters eliminate the need to remove the spring plates and/or torsion bars every time preload is adjusted. The suspension must be disassembled to install adjusters. For details, turn to page 77.

Larger diameter, higher-rate torsion bars for stiffer suspensions are also available in a variety of sizes. Setting ground clearance or preload with larger diameter bars is done the same as with the stock bars. Refer to page 76 for details.

Preloading Considerations—To save money, you can preload the rear suspension yourself. Here are some things to consider first.

Just as the stock rear-suspension setting is too soft for off-road use, it's possible to make it too stiff. Think about how you intend to use your car. If you're prepping a buggy, you won't have to anticipate the added weight of a back seat full of people or camping gear, food and an ice chest. But weight can be added to a Baja Bug or sedan with a rear seat and an added passenger or two. This can easily add several hundred pounds to what the rear suspension is already supporting.

With the ideal preload setting, the suspension uses almost all its travel. It *bottoms out,* hitting its jounce stops, only on the harder bumps during normal off-road driving. A suspension that occasionally bottoms out is good in that it's telling you that you're taking advantage of all its travel, but not overworking it. When it repeatedly bottoms, you need to adjust in more preload or install a larger-diameter torsion bar—or slow down!

Torsion bars are a major contributor to rear-suspension performance, but other factors come into play. Shocks and tires are very important. When the suspension is fully compressed and the torsion bars wound up, they want to unwind and rebound. The combination of the tires acting like basketballs and the rebound reaction of the torsion bars will kick the rear of a car up after a large bump is hit.

Shock absorbers damp and control suspension movement in both directions—jounce and rebound. But it's severe rebound that can put a car on its nose. Rebound is more difficult to control. For this reason, most off-

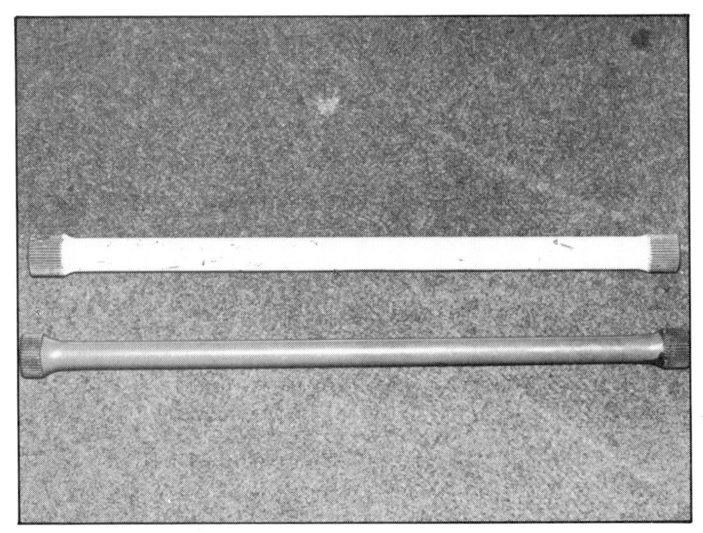

VW rear suspension has two torsion bars—one for each side. Bar at bottom is stock. Top larger diameter, high-rate bar will stiffen the suspension.

Removing a spring plate can be dangerous. Rope holds axle back out of the way while the screwdriver guides the plate over rebound stop. Chain, under jack to upper shock mount, keeps car from lifting off jack stands.

road shocks are valved for two or three times more damping on rebound than on jounce (compression).

Torsion-bar *spring-rate,* or stiffness, must be considered when choosing a shock or shocks. The stiffer the torsion bar, the more damping needed to control the suspension. That means using either stiffer shocks or multiple shocks.

Tires are a part of the suspension many people never consider. As you might recall from your high-school science class, every action causes an equal and opposite reaction. When a tire strikes a bump, the tire is compressed. How much depends on the severity of the bump. When the tires spring back like huge basketballs, the rear of the car kicks up. The more air pressure, the more kick.

The idea is to run tire pressure as low as possible, but not so low that the rims get bent on those unexpected holes or rocks. Different terrains require different tire pressures. Read about tires on pages 114 and 117.

PRELOADING TORSION BARS

As you may have found from the front suspension, it might take the better part of a day the first time you reset the torsion bars. You can reduce this to less than an hour after you've gained some experience.

Jack the rear of the car up and set it on jack stands. Make sure it's secure. Remove the rear wheels and shocks. If you have a sedan or Baja Bug, unbolt the emergency-brake cables. Do this at the brake-lever boot between the front seats.

Mark the position of the wheel-bearing housing or diagonal arm to the spring plate. The spring-plate holes are elongated. The position of the axle-flange or diagonal-arm bolts in these holes determines toe. If you don't make reference marks, you'll lose the original toe setting. The suspension will then require realignment.

Now remove these bolts. Loosen the smaller clamp on the rubber boot next to the transaxle. Pull out a little slack in the emergency-brake cables. Maneuver the axle and brake assembly back and out of the spring plate. Twist the unit so the bottom shock mount is to the rear and clears the spring plate. *Do one side at a time so you can see how the parts fit back together.*

Tie the axle assembly up and out of the way with some rope. On IRS cars, do the same with the diagonal arm. With these items tied back, you won't have to fight them while you're adjusting the spring plate.

Remove the four bolts that secure the spring-plate-hub cover and rubber bushing. **WARNING: The spring plate is still under extreme load and can slip off the stop with a dangerous force. The more preload you set in, the more dangerous the spring plate becomes. BE CAREFUL!**

At the stock setting, the force of the spring plate against its rebound stop is pretty easy to overcome. Use two large screwdrivers to pry the plate out and off the stop to unload it. When the spring plate has more preload on it, there's a safer and easier way to unload it.

Place a jack under the end of the spring plate. *Use a floor jack*—no jury-rigged bumper jacks or VW screw-jacks, please. Loop a stout towing chain under the jack body and bolt the ends to the upper shock mount as shown. The chain should be snug but not tight. This will tie the jack and the car together so the car doesn't lift off the jack stands when you jack the spring plate off its lower stop.

IRS spring plates are shorter than those used with swing axles. Adding preload to the IRS becomes a lot easier if you add leverage by "extending" the spring plate. Bolt an old spring plate end-to-end to the one you are adjusting. You can also drill a couple of matching holes in a 3/16-in.-thick steel plate and bolt it to the end of the spring plate to give the needed leverage.

Raise the spring plate off the stop far enough to see part of the hole in the rear-suspension submember. It's just in front of the rebound stop. When enough of the hole is exposed, insert a medium-size screwdriver under the plate and about an inch or two into the hole. The screwdriver will hang down at about a 45° angle.

Slowly release the jack. The screwdriver will guide the spring plate smoothly off the stop. Keep your fingers off the screwdriver shaft so they don't get pinched by the sliding spring plate. If the spring plate binds up part way off, just jack it up a little bit and release again.

The idea here is to move the spring

Once the plate is off the bottom stop and in the no-preload position, mark its position for a reference point when reinstalling. Note chalk line along top edge.

Tools for measuring angles in degrees. One at left is much more expensive than plastic model at right. Use either to set spring-plate angle/preload.

plate out far enough so it's off its stop and unloaded, but still connected to the torsion bar. Make sure the bar stays splined to its center socket. This way you can indicate the bar's original setting so you'll have a point of reference. Now, with the spring plate hanging down in the unloaded position, scribe a chalk line along the top edge of the plate onto the submember as illustrated.

Measuring Preload—The factory recommends setting preload with a special protractor/level. You should be able to pick up a protractor/level at almost any auto-parts store or welding-supply shop for a few bucks. Although it's not the special VW tool, it will do the same job.

Preload is set by adjusting the spring plate either up or down in relation to an imaginary horizontal line running through the car, front to rear. Use a protractor to measure the angles of these two lines. The top edge of the spring plate will give you one angle and the top of the tunnel, door sill or buggy frame rail will give you the other. The difference between these two angles is the amount of preload your car's torsion bars will have when the suspension is reassembled.

The more you rotate the spring plate down, the greater the difference between the two angles. By indexing, or rotating, both the inner- and outer-ends of the torsion bars in their splines, you can set preload accurately.

All marks or settings should be made with the inside rubber bushing in place. Also, any slack between the spring plate and torsion bar must be eliminated by lifting up the end of the spring plate with your hand. The bushings must not be worn. Otherwise, you'll get inaccurate settings. It is not necessary to remove the torsion bar to set preload. Just slide it out of its inner splines.

There are 40 splines on the inner end of the torsion bar, and 44 splines on the outer end. Turning the inner end of the bar one spline in either direction changes spring-plate angle 9° 0'. Similarly, turning the spring plate one spline on the bar alters the angle by 8° 10'. In case you don't remember, there are 60' in 1°. Therefore, it's possible to change spring-plate angle by multiples of 50' by turning the splines in opposite directions.

Setting preload this close is not important. It is important that preload is as close to equal as possible on both sides.

If you don't care for all this measuring, there's an alternative way to set preload. Use the bottom rear hole of the four bolt holes that hold on the spring-plate-hub cover as a reference point.

On a swing axle, position the spring plate on the torsion bar so the bottom edge of the spring plate is covering *half of the bolt hole. For an IRS,* set the spring plate *even with the bottom of the bolt hole.*

This amount of preload works well off-road. I've seen it done this way successfully on many cars. This won't give you a race-car ride, but it will increase rear-wheel bump travel and ride height so you can go off-road.

If you marked the position of the spring plate before slipping it off the bar, you now have two reference points: the stock setting and the setting covering half the bolt hole (swing-axle) or all of the bolt hole (IRS). If the suspension is too stiff when preloaded to one of these positions, you can reset it somewhere in between using the reference points.

A typical preload setting for off-road is about 4—8° more than the stock setting, depending on how much the rear suspension has sagged. If it has sagged a lot, a bigger angle change is necessary to get the same preload.

If you think your car needs more rear-suspension stiffness than 8° of additional preload can provide, I recommend using larger-diameter aftermarket torsion bars. Preloading the stock VW torsion bars much more than this will overstress them, possibly breaking the bars. With the bars highly preloaded, you will also have a hard time getting the spring plates back onto the lower stops.

How Much Is Right?—While making final preload adjustments, you may encounter a problem. When you come close to the desired setting and the next spline puts the setting way past what you want, try this:

Remove the torsion bar and rotate it exactly one-half turn—180°—and reinstall it in its inner spline and start

A great ballpark estimate for setting preload on a swing-axle car. Use the spring-plate-hub-cover bolt holes as your reference. Cover half the bottom rear hole (arrow) with the bottom edge of the spring plate. See the copy if you have an IRS car.

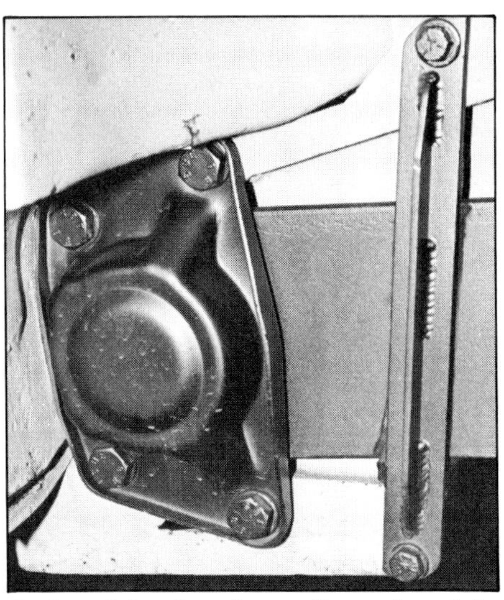

Spring-plate strap keeps the spring plate from slipping off the bottom stop.

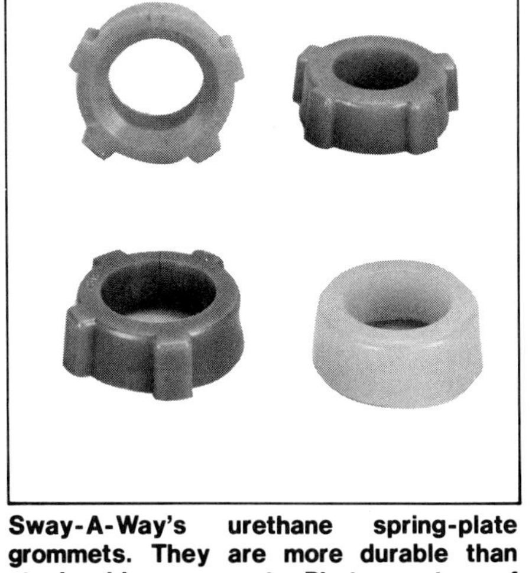

Sway-A-Way's urethane spring-plate grommets. They are more durable than stock rubber grommets. Photo courtesy of Sway-A-Way.

over. Because the bar has an unequal number of splines—40 at the inner end and 44 at the outer end—rotating the bar a half turn will "split the difference" of the setting.

The Positive-Camber Riddle—As I mentioned earlier, the IRS can tolerate suspension preload and ride-height changes without causing handling problems. But when you increase the swing-axle rear-suspension preload substantially, you bring out the swing axle's worst characteristics. The spring plates spend most of their time at or near their bottom stops. This means the tires are tucked under in the extreme positive-camber position most of the time.

The accepted compromise between extreme positive camber and lost suspension travel is to raise the bottom stop for the spring plate 1/4 in. Some suspension travel is lost, but this is more than made up for in improved, more-predictable handling.

Raising the bottom stop 1/4 in. is easily accomplished by welding a piece of 1/4-in. steel to the top of the bottom stop. One problem: This makes it much harder to get the spring plates back in place against the torsion-bar preload. For details on this procedure, see "Limiting Rebound Travel," page 80.

Spring-Plate Straps—A set of retaining straps will keep the preloaded spring plates from bending outward and sliding off the bottom stops. Fabricate the straps from 1-1/2-in.-wide, 3/16-in.-thick steel. Mount a strap from each upper stop to the stock lower stop as pictured. Weld a tab onto each stop and bolt the straps into position.

If you have more money than time, inexpensive kits are available for this purpose. The kits include tabs, mounting bolts and the straps.

Urethane Bushings—There are some other items to consider before you reassemble the torsion bars and spring plates. The greater preload or larger-diameter torsion bars puts more pressure on the rubber bushings, or grommets, on both sides of the spring plate. They won't hold up. Like other rubber suspension components, the bushings compress, wear and eventually crack. VW recommends using talcum powder on the bushings to reduce scuffing and noise. But off-road, the talcum powder will not last.

Sway-A-Way makes urethane bushings for the IRS and swing-axle suspension. The bushings compress slightly and resist scuffing and cracking. I recommend them highly. Sway-A-Way says their bushings will only fit their heavy-duty spring plates. If you are driving the car hard enough to damage the stock grommets, you should probably install heavy-duty spring plates anyway.

All Bars Are Not Alike—Make sure you have the *right* torsion bar in the *right* side and the *left* bar in the *left* side. New torsion bars can be installed in either side. But once used, the bar takes a permanent set, or twist. The bar is twisted in one direction only. Twisting a bar in the opposite direction will cause it to sag prematurely or break. To make it easy, the stock VW bars are stamped L and R at their outer ends. L is the left side (driver's side in U.S.) and R is the right side. You must mark aftermarket bars.

Install The Spring Plates—With the spring plate set for the torsion-bar preload you want, here's the easiest way to reassemble the suspension. Put the outer bushing in place. If you're using stock bushings, make sure the word oben appears at the top. Bolt on the spring-plate-hub cover. Snug all four bolts, but don't tighten them.

Jack up the spring plate so the lower edge is just above the bottom stop. You will need the chain between the jack and the shock mount now that you've adjusted in more preload. Evenly tighten the cover bolts to move the spring plate in over the stop.

The more preload or bigger the bar, the less the spring plate wants to cooperate. Help it along by jacking the spring plate up and down a couple of times while tightening the four bolts.

If the cover fits flush and the bolts are tight, but the spring plate is still not fully over the stop, pull it in with the spring-plate strap. I'm assuming you made provisions for this. On swing-axle cars, jack up the plate so the 1/4-in. stop, if used, will fit. Bolt the bottom end of the strap on loosely. Start the top bolt in. Then tighten the

Slotted holes in VW spring plate are for adjusting rear-suspension toe.

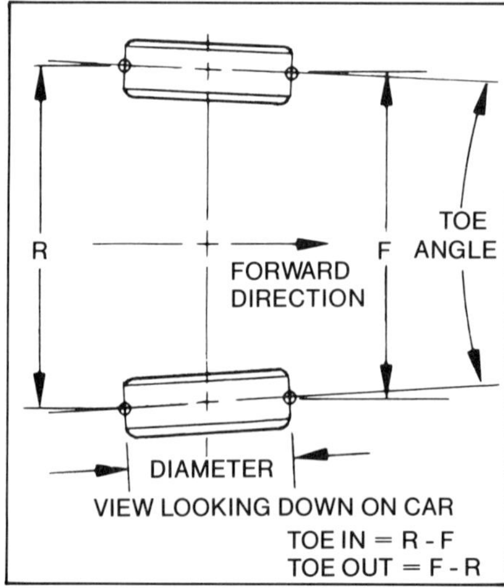

Toe is the difference in distance between the tires at wheel-center height, measured front and rear.

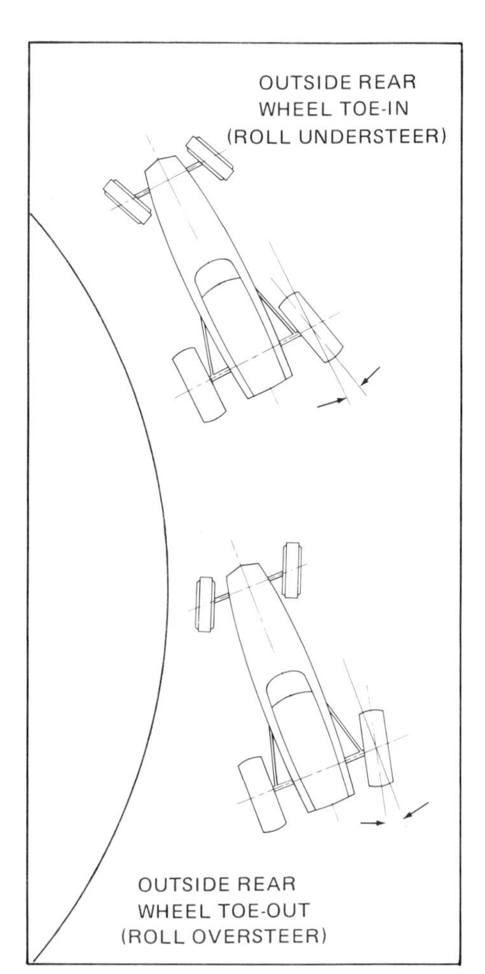

Effects of rear-wheel toe on handling. Toe-in promotes understeer or push. Toe-out is dangerous. It causes oversteer—the tendency of a car to spin.

bottom bolt. Lower the spring plate and tighten the top bolt. The spring plate will work into its proper position after the car is driven over a few bumps.

With the spring-plate cover bolted in place, untie the axles, and diagonal arms on IRS cars. Work the axle assembly forward into the end of the spring plate. Bolt the spring plate to the axle flange or diagonal arm.

On swing-axle cars, install these three bolts with their heads *toward* the transaxle or they'll hit the shocks. Snug, but don't tighten the bolts yet. Tighten them after setting toe. Now install the other axle assembly.

Set Toe—Holes in the stock spring plates are slotted for toe adjustment. Starting from zero toe, sliding the wheel-bearing housing or diagonal arm forward on the spring plate increases toe-in; sliding it back increases toe-out.

The ideal toe-in setting on a VW off-road race car, regardless of suspension type, is 1/4-in. On a swing-axle car, you can get close to 1/4-in. toe-in by using a hammer to knock the bearing housing all the way forward in the spring-plate slots.

Make a crude toe check with a tape measure before tightening the spring-plate bolts. Recheck toe-in after tightening the spring-plate bolts.

On a swing-axle car, toe-in is at a minimum with the wheels at zero camber. The wheels toe in as they travel up into negative camber.

On a swing-axle car, nothing will get you "on your head" quicker than the absence of an extra stop under the spring plates and a lot of rear-wheel toe-out. A car with excessive positive camber is bad, but toe-out makes handling unpredictable and unforgiving. Toe-out causes *oversteer*—the tendency of a car to lose traction at the rear and spin.

If you're using aftermarket spring plates such as Sway-A-Way's older plates for the IRS, the holes are *not slotted* to allow for toe adjustment. The reason was to keep the toe setting from changing during a race if the bolts loosened. Loose bolts would let the wheel move fore and aft, causing drastic toe changes—toe-in during power on; toe-out during power off and braking.

On Sway-A-Way's IRS plates with round holes, the holes are drilled so toe is fixed to a value they think is best. Of course, this assumed that two cars are exactly the same—they're not. Consequently, the *fixed* toe setting for one car will be different for another. That's why VW slotted the plates in the first place.

Sway-A-Way's spring plates now have slotted holes for toe adjustment. When you install one of these plates, set toe-in using the existing three holes, then drill the fourth one to preserve the toe setting for your car.

You may find there is not enough adjustment to reach the correct toe setting. In severe off-road use, suspension components may bend enough that the rear wheels toe-out, even with the axle-flange bolts all the way forward in the slots.

To remedy this, you must either find and fix the damaged component(s) or *shim the transaxle rearward.* Shims between the transaxle and its front mount will move the transaxle and engine rearward. You may have to make similar adjustments to the length of the gearshift-, clutch-, and throttle-control linkages.

Rubber Bumper/Suspension Stop—The rubber jounce bumper (bump stop) on the stock spring plate or IRS diagonal arm cushions the impact when the suspension bottoms. The bumper works fine on a street car, but is not desirable on an off-road car.

When the rubber bumper is compressed with the suspension at full jounce, the bumper wants to pop back into shape. In doing so, it tries to push the spring plate quickly back down to

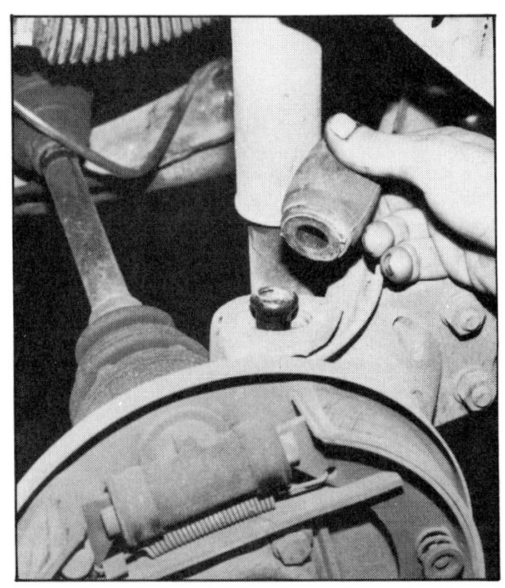

Stock rubber jounce bumper is too soft to be of any use on an off-road car, causing undesirable wheel hop on rebound. Remove the bumper.

With jounce bumper removed, upper lip of spring-plate cage (arrow) will clunk against spring plate at full jounce.

A jounce bumper can save components from being damaged. Sway-A-Way urethane bumper is mounted to a bracket on the tubular frame. Note notched spring plate for increased suspension travel.

the full-rebound position, thereby unloading the suspension. This full-jounce-to-full-rebound action increases the tendency of the rear of the car to get airborne after bottoming the rear suspension.

Most all race cars, especially those with aftermarket long-travel IRS, use a jounce stop or bumper. To save the spring plates from being repeatedly hammered against the spring-plate cages, urethane jounce stops are used. These urethane stops have a high-durometer rating to absorb shock without compressing much. One is shown above.

Rubber Boots—Take any twist out of the rubber transaxle boots used with the swing-axle suspension and tighten the clamps. Replace the boots if they're cracked or torn.

Checking The Shocks—Before you install the shocks, check their condition. Grip each one by its housing and compress it against the ground. What happens next depends on the type of shock—high-pressure gas or conventional double-tube.

A gas shock is easy to check. First, it should take a good amount of force to compress it. Afterwards a compressed-gas shock will return to full extension by itself. If a gas shock has little resistance to compression and doesn't return by itself, the seals are shot. The shock is trash.

An ordinary double-tube shock should have slight resistance to compression and will have to be pulled to extend it. If a double-tube shock offers negligible resistance in either direction, or if it has visible signs of fluid leakage, throw it away.

Regardless of shock-absorber type, discard the shock if it binds. Binding indicates a bent shaft.

If the shocks are OK, mount them. If the shocks have rubber bushings for mounts, be sure to "capture" the bushings by installing large washers between the mounting nut and bushing. The washers keep a shock from coming off its mount if the bushings fail. Washers are not necessary if the shock ends have "Heim joints"—spherical bearings—rather than rubber bushings.

Finish The Job—Push the emergency-brake cables back in the chassis and reconnect at the hand-brake lever. Now you can put the wheels on and take your car for a ride.

REAR-SUSPENSION BRACING

If you will be giving your off-road car a lot of hard use, you must reinforce some rear-suspension components and structure. For instance, if you have the transaxle out and have the urge to do some reinforcing, get out your welder. Like the front clip, the rear horns and torsion tubes are spot-welded assemblies. Strengthen the horns by simply running a continuous weld along all the spot-welded seams.

The small brace connecting the outboard ends of the torsion tube to the floorpan tends to break. The torsion-tube ends then flex at their welds at the tunnel. Using an 1/8-in.-thick steel plate 4—5-in. long, weld the ends of the torsion tubes to the pan, just inside the existing brace.

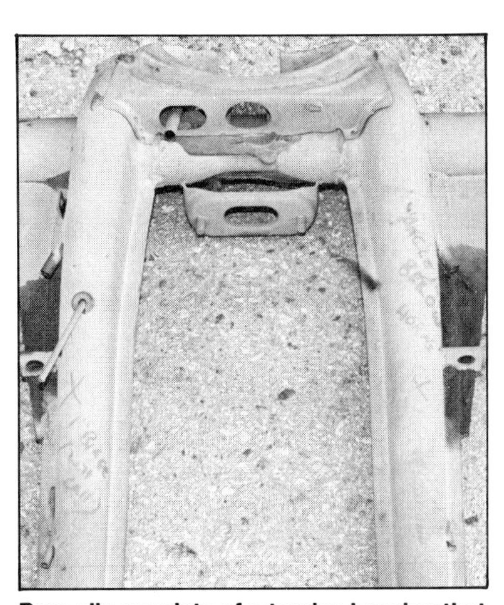

Rear clip consists of a torsion housing that runs across the back of the pan and two frame horns that the transaxle mounts between. Horns can be strengthened by rewelding spot-welded seams.

The part welded onto the end of the torsion tube that makes the spring-plate *support*—upper and lower spring-plate stops and rear-shock tower—is cast. You can strengthen the bottom stop by installing a strap between it and the top stop as previously shown. However, the bottom stop takes a beating with a stiffer suspension. It really needs further reinforcement.

Reinforce it by fabricating a gusset from 1/8-in.-thick steel plate to fit in

View from under the car. Small stock brace between end of torsion housing and floorpan is insufficient. Cardboard pattern simulates brace that should be added to augment the stock brace. This prevents cracking at the horns.

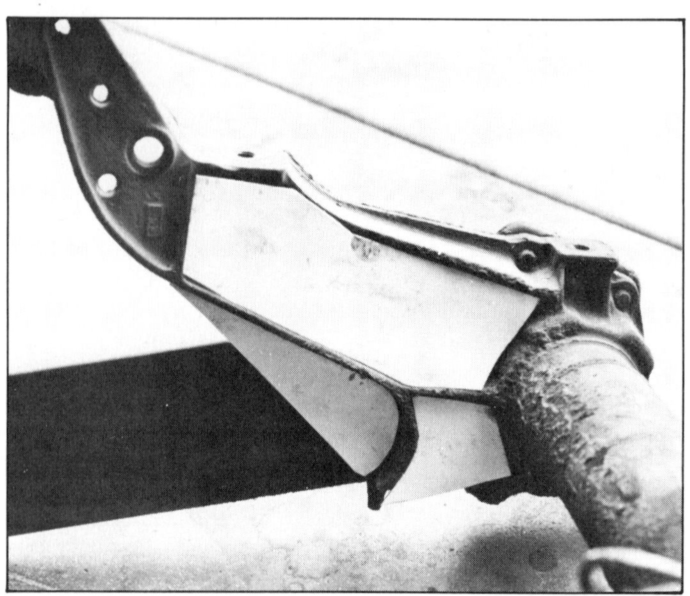

To strengthen the spring-plate supports, gusset and box across the reinforcing webs.

Cradle welded between the horns for the middle transaxle strap. Ends of horns are supported by removable braces. Skid plate adds rigidity to the assembly, too.

the curved area between the bottom stop and the shock tower, about halfway up. See the accompanying photo. Position the gusset in the center of the spring-plate support so it won't interfere with the spring plate. Weld the gusset in place.

From its back side, notice the reinforcing ribs that section the spring-plate support. Box in these sections with 1/8-in.-thick steel plate, An easy way to make these plates is to cut cardboard templates first. Use the patterns to transfer the shapes onto the steel plate. Cut out reinforcing plates and gussets with an oxyacetylene torch or a band saw. Smooth the edges with a grinder.

Skid Plate—There is a good way to reinforce the *rear horns*—sheet-metal extensions of the floorpan—and provide protection for the transaxle at the same time. Weld a transaxle skid plate to the bottom of the horns. This ties the two horns together. If you use this type of skid plate, you must also use a solid transaxle mount that includes the bottom half of the middle transaxle strap. If not, you'll have to weld in a strap, or cradle, for the middle transaxle strap to clamp the transaxle against. See page 32 for details on solid-transaxle-mount kits.

With the skid plate welded in place, there's no access to the front transaxle-mount nuts unless you cut slots in the skid plate. Prefabricated transaxle skid plates are available for about what the material will cost you. For more details on skid plates, see page 90.

SPECIAL TORSION BARS

First, a couple of basic rules: All things being equal, the larger diameter a torsion bar has, the *higher* its spring rate; the longer a torsion bar is, the *lower* its spring rate. To review the definition of spring rate and how it translates to wheel rate, turn to page 52.

Stock torsion bars used on '58—67 Bugs are approximately 21-3/4-in. long. All '57-and-earlier Bugs, plus mid-'68-and-later cars have longer 24-11/16-in. torsion bars. These longer bars project about 4-in. through the spring-plate hub cover.

Years ago, off-roaders replaced these with later type, shorter bars so larger diameter tires could be used without hitting the ends of the bars. But with the advent of longer spring plates, longer diagonal arms or bus transaxles with the reduction boxes laid down, the longer bars could be used with big tires without interference.

Longer torsion bars have the advantage of having less force buildup when twisted the same amount as a shorter bar of equal diameter. The longer bar will be less apt to sag and gives a softer ride over rough terrain.

Special larger-diameter aftermarket torsion bars are available. Bar OD is measured at the shaft or shank OD between the two splined ends. Splined ends have the same OD and spline count—40 on the inner end and 44 on the outer end—regardless of bar size. Torsion-bar shafts are ground to a specific size. A common stock diameter is 22mm; 30mm being a very large, high-rate torsion bar. After being machined and ground to size, the bars are heat-treated.

Because of the importance of these bars to the handling and durability of your off-road car, I suggest that you stick with bars of known quality. Sway-A-Way's are ground to a smooth finish, then coated with epoxy to protect them from corrosion.

Finish is important because paint will chip and allow the bars to rust. Rust can create a *stress riser*—a weak point that can cause the bar to break.

Don't think that if you live in the desert that steel won't rust. It will.

More Torsion-Bar Tips—As I've already mentioned, a torsion bar will take a *set,* or permanent twist, after it's installed. If you start with 12° of preload, you may end up with only 9—10° after doing some serious off-road driving. This amounts to a 2—3° set. If you want 12° preload, you'll have to reset the suspension. A real pain.

For this reason, a good set of used torsion bars is better than a new set. You don't have to go to the trouble of readjusting the bars after they've taken their set.

A ballpark figure for setting the large-diameter torsion bars is as follows: On a Baja Bug with 28mm bars and a multi-shock setup, set in about 12° of preload. With larger bars (29—30mm), you generally adjust in less preload. On a smaller-diameter bar (26—27mm), you'll have to adjust in more preload to maintain ride height. On the lighter buggies, start with a bar about 26mm or 27mm and adjust in 9—10° of preload.

remove or install the torsion bar. On Baja Bugs, there's enough clearance for the torsion bar without removing the fender bolts.

REAR-SUSPENSION-PRELOAD ADJUSTERS

Preload adjusters eliminate the need to disassemble the suspension every time the rear suspension needs fine tuning. They are available from the aftermarket. Sway-A-Way makes adjusters for both ends of the torsion bars—at the center of the torsion tube and at the spring plates.

Center adjusters have to be spliced and welded into the center of the torsion tube. Besides the welding, installation of this unit requires a lot of cutting and a good understanding of what to do. But once installed, adjustments are a breeze. There's an individual adjusting screw for each torsion bar. Each provides approximately 15° of adjustment.

Heavy-duty spring plates with integral adjusters merely bolt in place of the stock units. They provide about 20° of adjustment. Because of the additional complexity and weight, these outboard adjustable plates are seldom raced but are plenty strong for most other off-road uses.

Outboard adjusters are better than center adjusters in one respect: Quick adjustments can be made without raising the car and crawling under it. They can be used to change ride height and preload in an instant. Adjustable spring plates are available in varying lengths for cars with different wheelbases.

With either type of adjuster you can simply screw in or out the rear-suspension preload you need for either street or dirt. Both types of adjusters cost about the same, although more work is required to install the center adjuster. The center adjuster has several advantages. Its durability is good and it does not move with the suspension, so the center adjuster doesn't add to the shock absorber's work.

APPROXIMATE TORSION-BAR ADJUSTMENTS

Vehicle	Bar diameter (mm)	Preload
1400-lb Single-Seat Buggy	25, 26, 27	7—10°
1700-lb Two-Seat Buggy	26, 27, 28	8—11°
2000-lb Baja Bug or Sedan	28, 29, 30	9—12°

Rear suspensions using these bars and settings need more than one shock for control.

There is no set formula on the size of the torsion bar and exact preload setting. There are too many variables to consider, such as intended use, weight distribution, total weight and shock valving. Ask 10 racers and you'll probably get 20 different answers. The ride obtained with two identical torsion-bar setups will vary considerably.

Most competitive racers install a center torsion-bar adjuster so they can easily change torsion-bar preload for different terrain from race to race. Many do this by feel and have long since lost track of the exact amount of preload used at any given time or any given race.

On stock-bodied sedans, it may be necessary to remove several bolts at the forward edge of the rear fender and pull the fender out slightly to

Center torsion-bar adjuster. When welded into the center of the rear torsion housing, it allows for preload adjustments of both torsion bars. Photo courtesy of Sway-A-Way.

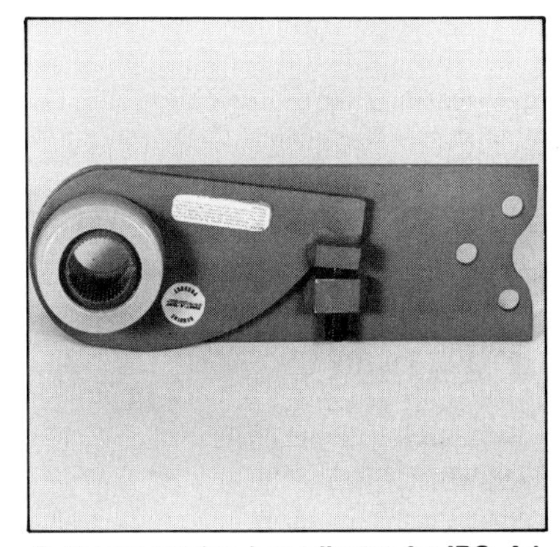

Outboard torsion-bar adjuster for IRS. Adjusts each torsion bar up to 20°. Mounted on 0.250-in.-thick HD spring plate. Although this spring plate has fixed holes at the diagonal-arm connection, new-design Sway-A-Way plates have slotted holes for adjusting rear-wheel toe. Photo courtesy of Sway-A-Way.

Adjustable spring plates with lower shock mounts for the swing-axle suspension are expensive, but convenient and reliable. Preload is changed 1/2° for every full turn of the adjustment screw. Photo courtesy of Sway-A-Way.

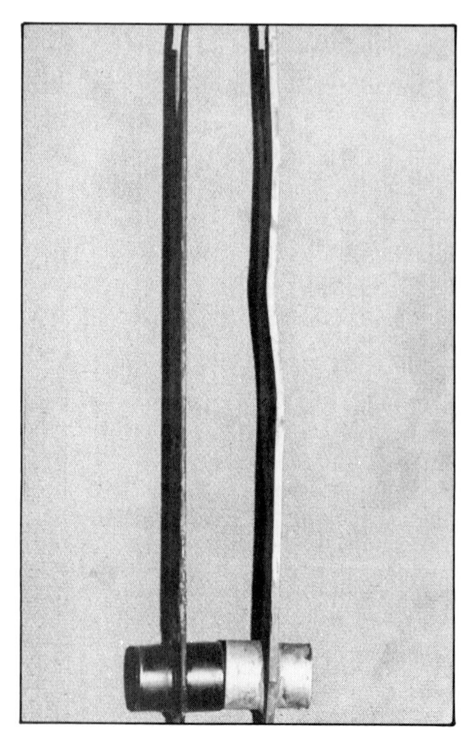

Spring plate at left is stock. One at right is a thicker replacement unit that was tweaked from a hard impact and still held its shape. Had the stock plate been on the car, it would have been Z-shaped, leaving the car with extreme toe-in.

Shocks used with long-travel racing suspensions should be angled to match wheel arc. Idea is to use most of shock-absorber stroke with minimum of shock-body movement.

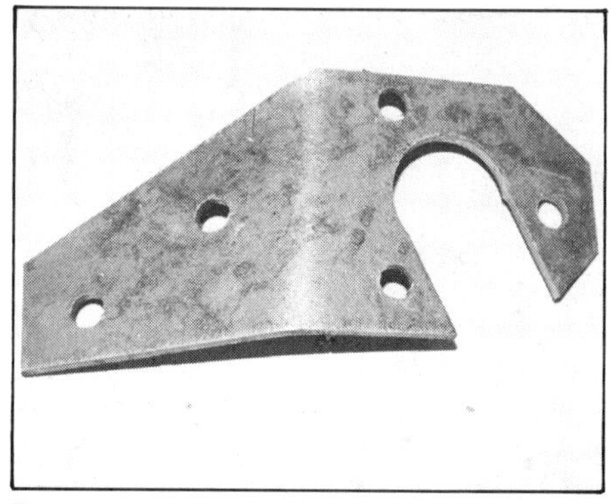

Pre-cut, bent and drilled swing-axle shock mount. Plate bolts up using the same bolts that hold the swing axle to the spring plate.

HEAVY-DUTY SPRING PLATES

A stiffer suspension with increased wheel travel gives the spring plates a beating. Other than the normal twisting, the stock spring plates flex too much in severe off-road use, especially with high-rate torsion bars. Heavy-duty spring plates are usually twice as thick as stock plates—0.250 in. versus 0.125 in.

The thicker spring plates are more rigid. This rigidity helps save the transaxle side covers from damage due to side loads. If you intend to drive your off-road car hard, but don't plan to install the outboard torsion-bar adjusters, I strongly recommend using these heavy-duty plates. They'll set you back the better part of $100. The same HD spring plates with three lower shock mounts cost twice that much.

SHOCK ABSORBERS

Before spending your money on special torsion bars and HD spring plates, remember you'll need additional shocks to damp the stiffer suspension. Shocks cannot be mounted to the body, so you must incorporate the upper shock-absorber mounts into a roll cage for strength. See page 88. Although it's common practice to weld in the lower mounts or drill mounting holes in the spring plates, bolt-on lower shock mounts are available.

If you decide to install additional shocks, the big question is where to install the mounts. If you find advice is slim to none and you can't get a close look at setups that work, here are some tips.

Mount the shocks as close to the wheel as possible. This provides maximum shock-absorber effectiveness because the *velocity* of the shock piston should be as close as possible to *vertical wheel velocity.* The ideal shock location is one that gives maximum shock travel for total wheel travel.

The top of the shocks should be tilted inboard for tire clearance when the suspension is at full jounce (upper wheel-travel limit). This is more important on a swing-axle car than one with IRS because of the swing-axle's extreme negative camber at full jounce.

The shocks should also be tilted slightly forward at the top because the wheel travels in an arc. Ideally, you want the shocks to be 90°, or perpendicular, to the spring plate with the suspension about 1 in. from full jounce. The shocks are more effective in this position. Multiple shocks should be mounted so they won't touch each other through full wheel travel.

High-pressure-gas, reservoir shocks dissipate heat better than conventional types. Here, racer uses air adjustable (right) along with two Fox Shox reservoir gas shocks.

Full-race Baja Bug uses Giese Racing air suspension. Single, large-diameter air shocks replace both torsion bars and hydraulic-type racing shocks.

On this Class-5 Baja Bug, a steel plate is run from rear roll-cage support to tubular frame to provide top shock mounts. Shock lugs or short sections of heavy tubing are welded in for shock bolts.

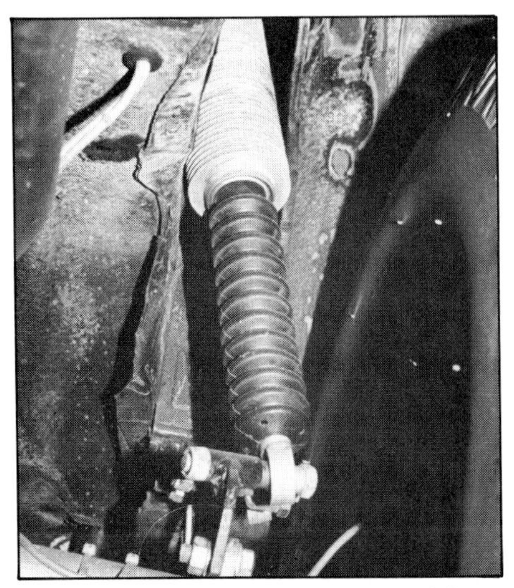

When long-travel shocks are used in a sedan, holes must be cut in the rear fender-wells to allow clearance. Shock upper mounts are then fabricated off the roll cage or roll-bar rear supports.

Reduction-box, swing-axle suspension with the lower shock mounts built into the laydown adapter where they should be.

I recommend using shocks that have slightly longer travel than theoretically needed. They afford more of a margin of error if the suspension has more travel than you planned. A shock that runs out of travel before the suspension does will be destroyed. Try to locate the shock mounts so you don't use the last inch or so of shock travel at full jounce or rebound.

When you find a shock-mount location you think will work, tack-weld the mount in place. If you want to change its location, simply break the mount loose with a chisel and hammer and relocate the mount.

Position the shock mounts so their center lines or the mounting bolts at both ends are parallel to the *suspension-pivot axis.* The swing-axle axis is described by a line through the spring-plate and axle-tube pivots. The IRS pivot axis goes through the spring-plate pivot and the semi-trailing-arm pivot. Installing shock mounts in this manner maximizes shock-mount durability by minimizing the twisting action on the mounts, particularly at the wheel. This is a must with rubber-bushed shocks.

The best way to prevent errors while installing shock absorbers is to use a "setup shock." It's best to do this with the torsion bars removed and the suspension loosely assembled. Test with a worn-out shock the same length you are going to use.

If it's a high-pressure-gas shock, carefully drill a small hole in the end of the shock tube opposite the piston to relieve the pressure. **Wear safety goggles while drilling a gas shock.** You can now use this dummy shock while you run the rear suspension through its full travel to check for shock-mount locations.

If you use the swing axle with bus reduction boxes laid back, you must fabricate the lower shock mounts. Usually, this is a plate that is bolted or welded onto the reduction-box housing. You can buy multiple-shock lower mounts from John Johnson Racing Products.

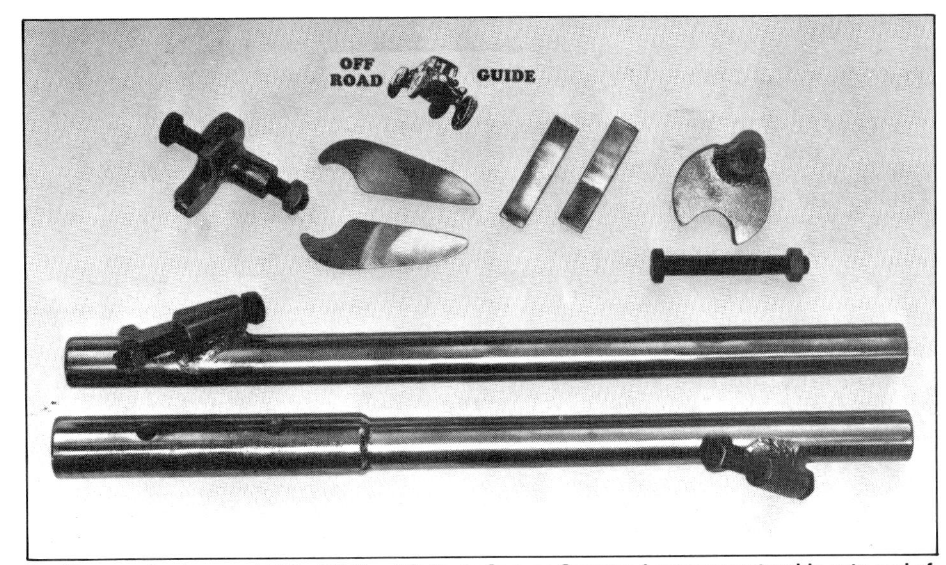

Double rear-shock-mount kit by Off-Road Guide in Canyon Country: Lower mount welds onto end of stock lower shock mount. Upper mounts are on bar that is routed across vehicle and welded to top of stock shock towers. It fits '68-and-earlier Type 1. Photo courtesy of Off-Road Guide.

Regardless of suspension type, I recommend using aftermarket shock *lugs* for the lower mounts. These can be welded into position on the spring plates, reduction-gear housings, diagonal arms or whatever. Also known as *spuds* or *bosses,* these inexpensive mounts are available through most off-road suppliers.

When setting up a multi-shock suspension, contact the different shock manufacturers or their distributors: KYB, Bilstein, Fox Shox/Phoenix Phactory, Rough Country, Gabriel

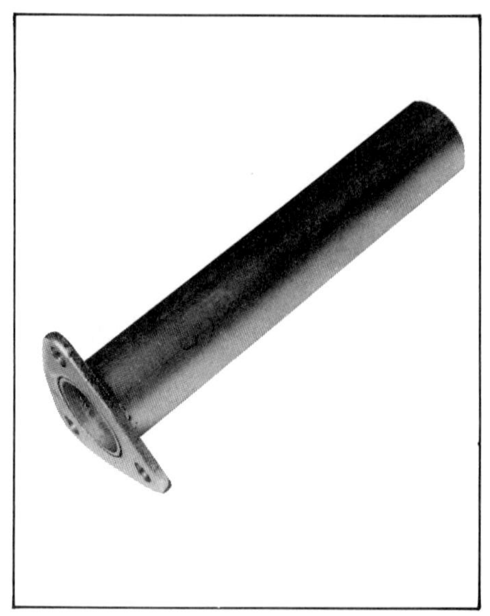

Axle overtube for reinforcing stock swing-axle tube. It slides over the stock tube and bolts to the wheel-bearing housing with the spring plate. Photo courtesy of Chenowth.

Spring-plate retaining strap with 1/4-in. stop (arrow) welded to it. Strap was turned around to show stop. When the stop is in this position, it allows the spring plate to drop down to the stock stop.

and so on. After giving them the weight of your car, suspension preload, torsion-bar diameter and length, wheel travel and desired shock travel, they should be able to advise you on shock-absorber valving, length and number of shocks to use. See the listing at the back of this book for addresses and phone numbers. For more information on shock-absorber basics, see page 48.

TAKING THE SWING-AXLE SUSPENSION OFF-ROAD

I've discussed the design features of the swing-axle suspension and its suitability for off-road use. I've also covered general off-road suspension-preparation procedures that apply to both the IRS and swing-axle rear suspension. What follows are specifics on modifying the swing-axle suspension for off-road.

Axle Overtubes—Swing-axle tubes bend with rough use. One way to reinforce them is to install axle overtubes. An overtube is a heavy-wall steel tube, about 0.187-in.. This tube fits snugly over the outboard end of the stock axle tube from the boot to the bearing-housing flange.

To install an overtube, a pin must be driven out of the rear of the wheel-bearing retainer, and the retainer pressed off. A buildup of paint or undercoating on the stock tube OD or a bent tube will make it difficult to slide the close-fitting overtube over the stock tube. When reassembled, the spring plate is sandwiched between the overtube flange and the bearing retainer.

Another way that most racers use to reinforce axle tubes is to replace the stock axle tube with a heavier-wall aftermarket-replacement tube. As with the overtube, the retainer pin must be driven out and the bearing housing pressed off to make the switch. The new axle tubes can be used to support lower shock mounts. Stock axle tubes are not strong enough to support shock absorbers.

Limiting Rebound Travel—As mentioned in the preload section, the best compromise between positive camber and wheel travel is to limit suspension rebound with extra 1/4-in. stops under the spring plates. Using a spring-plate strap allows you the luxury of welding the 1/4-in.-thick stop to the strap. You can then easily remove the stop when setting preload or when the stop isn't needed.

When installing the 1/4-in. stop, first install the spring-plate strap. Jack up the spring plate so there's room for the stop between the existing bottom stop and the bottom edge of the spring plate. Using 1/4-in. stock, make the stop about 3/8-in. wide and as long as the strap is wide.

To determine the exact location for the stop, lower the spring plate onto the new stop. Tack-weld the ends of the stop to the strap. Now, jack up the plate again and remove the strap with its tack-welded stop. Finish welding the stop and reinstall the strap.

If you find that you don't want the extra stop for certain situations, just flip the strap over and the stop is out of the way. The strap will continue to do its job of keeping the spring plate in place. It will also keep the bottom stop from being beaten downward by the spring plate after miles of hard off-road driving.

TAKING THE IRS OFF-ROAD

If you're preparing the VW IRS for off-road, read the preceeding general suspension-preparation section first. Then read this section for information unique to the IRS.

The IRS has weak points that need attention. But they're worth putting up with to get the predictable handling it offers.

Spot-Welded Diagonal Arm—The diagonal arm is the heart of the IRS. Like most VW chassis components, it's a spot-welded assembly. For off-road use, run a weld bead around all the seams.

Vulnerable Shock Mount—The IRS diagonal-arm shock mount hangs below the arm, exposing it to rock damage. Run a 1/8-in.-thick, 1-1/2-in.-wide steel strap from the bottom of the mount to the front edge of the diagonal arm. Brace the center of the strap to the arm. This will not only strengthen the shock mount, but will also act as a mini-skid plate to deflect rocks.

Diagonal-Arm-Brace Kits—To prevent bending, the diagonal arm should be braced by adding gussets and webs. You can make these pieces using 1/8-in.-thick mild steel. Make patterns from cardboard first. Or, if you have more money than time, you can buy a reinforcing kit that includes precut pieces.

Diagonal-Arm-to-Spring-Plate Connection—The connections between the diagonal arm and the spring plate should be reinforced for serious off-road use. The stock setup has *three* bolts that go all the way through the arm and spring plate with nuts on the other side. Some late-model VWs come with a *fourth* bolt that threads into the diagonal arm.

If you have trouble keeping the bolts between the diagonal arm and

Popular 1600cc racing classes allow use of 1-in. longer IRS diagonal arms. V-Enterprises offers these strong and reasonably priced stock arms that have been lengthened and straightened.

Extra spring-plate-to-diagonal-arm bolts can be installed here to strengthen the connection. Just drill two holes through the assembly where indicated (arrows) and install Grade-6 or better bolts.

Secondary rear torsion-bar setup allows primary bars to be set with less preload, in effect, making a progressive-rate suspension. Small bumps are handled by primary bars, but as wheel travel increases, arms contact stop (arrow) and brings secondary bars into play.

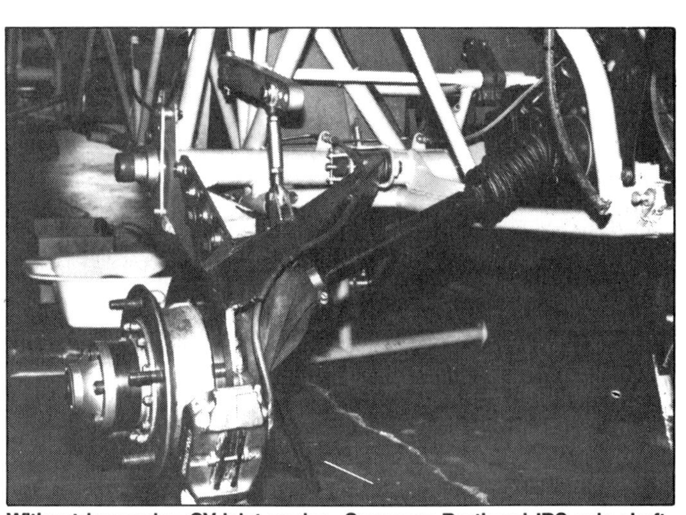
Without increasing CV-joint angles, Summers Brothers' IRS axle-shaft setup gets 20 in. of travel by moving CV joints farther apart with extra-long shaft. Outboard CV joint rides on full-floating hub. Note secondary torsion-bar outboard connection.

spring plate tight, try using larger bolts. You can also add as many as two extra bolts, as shown. Drill out the holes for 1/2-in. bolts. Use Grade-6 bolts or better with locknuts. Regardless of the type of diagonal arm or spring plate, be sure to install the bolts so they will clear the spring-plate supports. Otherwise wheel travel may be restricted.

Diagonal-Arm Pivot—The bushings for the diagonal-arm pivot take a beating in off-road use. Stock rubber bushings can easily be replaced with Sway-A-Way's metal-sleeved urethane bushings. The urethane bushings are more durable and will reduce deflections at the pivots.

You may have trouble with the Allen, or socket-head, bolt securing the diagonal arm to the torsion housing loosening. A simple fix is to weld a 1-in.-long piece of tubing to the pivot bracket and around the bolt head. Make sure it's big enough to fit over the head of the bolt. Drill a 1/16-in. hole through the tube. Then install a cotter key through the hole so the bolt can't back out.

Special Diagonal Arms—Although IRS diagonal arms need reinforcing for severe off-road use, there's a point of diminishing return. The more metal you add for reinforcing them, the more weight you add. This is *unsprung weight*—weight that "bounces" with the wheel and has to be controlled by the shock absorbers.

The more unsprung weight your car has, the slower the suspension can react to surface irregularities. In simple terms, the tires spend more time in the air—truly "off-road." This is farther off-road than the tires should be because they can't corner, accelerate or brake while they are not in contact with the ground.

The high-buck solution is to discard the VW arms in favor of custom-fabricated tubular-steel arms. These arms are usually constructed of 4130 (chrome-moly) steel. Even with extra lower shock mounts built in, a custom arm is significantly lighter than a rein-

Keep the stock-VW pivot bolt from backing out by welding a section of 1-in. tubing over the Allen-bolt head. With the bolt installed, drill a hole through the tubing. Install a cotter pin so it bears against the bolt head.

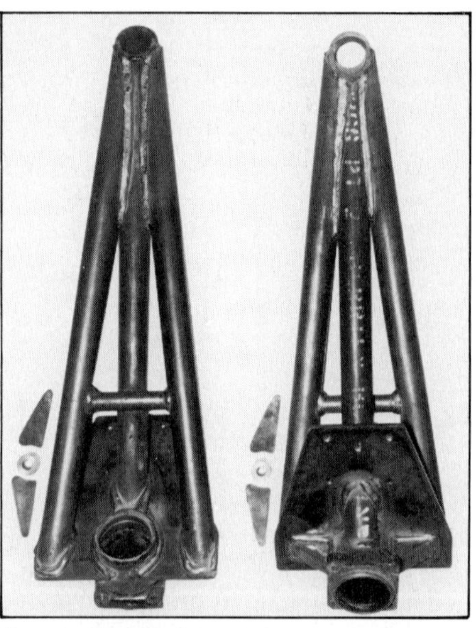

Custom-fabricated, tubular-steel, diagonal arms for long-travel rear suspension. These Probst Racing Team-designed arms are available in lengths 1-, 3-, and 5-inches longer than stock. Longest fully reinforced arm actually weighs 1-lb less than stock arm. Photo courtesy of Berrien Buggy.

forced stock arm of equivalent strength.

Using custom arms opens up another possibility of increased wheel travel. By using 2-, 3-, 4- or 5-in.-longer arms, wheel travel can be increased from under 8 in. to almost 14 in. With the longer arms, wheelbase is automatically increased.

With custom arms, the stock spring plates will no longer reach the wheel-bearing housing. Most racers substitute modified swing-axle spring plates to make up the difference. Swing-axle plates are longer than IRS spring plates and actually must be shortened to fit. You'll have to fabricate new bump stops for the arms as well. On a tube-frame buggy, these are usually urethane-covered and mounted off the roll cage. Remember that long-travel shock absorbers will also be needed.

Keep in mind that these arms are for race cars. At a couple of hundred dollars apiece, custom arms are "overkill" for a recreational buggy or street-driven Baja Bug.

TORSION-BAR SPRING RATE (in-lb)

Diameter	Bar Length (in) 21-3/4	24-11/16	26-9/16
22mm (stock)	588	497	464
25mm	979	828	773
26mm	1145	968	904
27mm	1333	1126	1052
28mm	1541	1303	1217
29mm	1774	1499	1400
30mm	2031	1716	1603

EFFECT OF PRELOAD ON TORSION BAR

Because of leverage, rear-suspension stiffness depends on the center-to-center distance from the torsion bar to the axle.

In a stock suspension, the distance between the axle and torsion bar is approximately 16-1/2 in. In a swing-axle reduction-box setup with the boxes laid back, or in a long-travel IRS, the distance is more. I use 20 in. in the following example.

To find the amount of force 1° of preload produces in a torsion bar, divide the torsion-bar-to-axle distance into the torsion-bar spring rate.

Bar (mm)	Rate (in-lb/deg)	Distance (in.)	Change (1°) (lb)	Distance (in.)	Change (1°) (lb)
27	1333	16-1/2	81	20	66
28	1571	16-1/2	93	20	77
29	1774	16-1/2	108	20	89
30	2031	16-1/2	123	20	102

EXAMPLE: 10° of preload on a short (21-3/4 in.) 28mm bar will exert about 930-lb of force on the axle of a stock-length suspension setup.

1541 in-lb ÷ 16-1/2 in. = 93 lb/degree
93 lb/degree X 10° = 930 lb

Whereas that same 10° on a long (26-9/16 in.) 28mm bar produces 740 lb

1217 in-lb ÷ 16-1/2 in. = 74 lb/degree
74 lb/degree X 10° = 740 lb

With 20 in. and 10° on a long (26-9/16 in.) 28mm bar, preload per 1° of twist is 610 lb

1217 in-lb ÷ 20 in. = 61 lb/degree
61 lb/degree X 10° = 610 lb

Chapter 5 FRAMES, ROLL CAGES & SKID PLATES

Crunched roof illustrates the need for a well-designed and built roll cage. John Vanatta and George Jirka finished the AMSA Mindamar 6-Hour Gold Rush because of it. Their Class-5 Baja Bug "took a lickin', but kept on tickin'." Photo by Jean Calvin.

The day of the home-built buggy is almost gone—not because it can't be done, but because in today's competitive market, buggy frames are relatively inexpensive. Some *tack-welded* kits can be had for a couple of hundred dollars. A fully welded chassis with a VW rear torsion-tube assembly will cost about twice as much as the tack-welded kit.

If you insist on building your own buggy frame, get a picture of the chassis you like. If you have the equipment and the talent, simply copy it. If you don't have dimensions, make forms out of heavy-gage wire. Lay the wire on a flat surface and trace the shape with chalk. Bend the tubing to suit.

You Get What You Pay For—Lest my previous remarks led you to believe that you can buy a quality off-road *racing chassis* for less than $500, let me assure you that this is not so! Inexpensive buggy frames are for light-duty recreational use only. For example, FUNCO mild-steel racing chassis go for well over $1500. If your plan is to build a top-notch off-road car, don't try to save money by using a "bargain-basement" frame.

Baja Bug/Sedan—Almost any buggy-manufacturer's brochure gives more than enough visual information on how a tubular chassis is put together. Therefore, I'll use most of this section to cover roll bars and cages, rear-shock supports and front and rear bumpers for the sedan and Baja Bug.

Roll-cage kits for the sedan and Baja Bug are relatively inexpensive. Still, the typical off-roader fabricates his own roll cage so he can have it the way he wants it. After reading this section, you'll be able to make your own cage, too.

Rules and Regulations—If you're thinking about racing, contact the sanctioning organization for the races you plan to run. Examples are SCORE International and the High Desert Racing Association. These organizations specify what you need in the way of roll bars, gussets, braces and supports to race in any given class. Send for a rule book. One can be had for a few dollars. Addresses and phone numbers are at the back of this book. Depending on where you live, there may be a local off-road racing organization. Most follow the SCORE rules more or less.

Basic Roll Cage—Most people build roll cages or roll bars, bumpers, extra shock mounts and the like, individually when they have the time, money and inclination. But, when it comes to building a roll cage, you'd better know what you're doing. If you lack welding experience, find a capable metal fabricator to build the basic cage. A car's roll cage is like the rough framing of a house. If it starts out

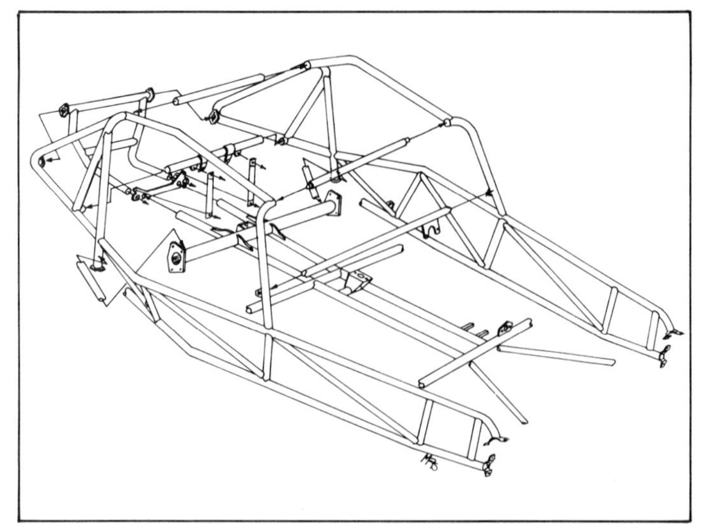

If you opt for a weld-it-yourself kit, here is what's inside the box. The parts must be assembled carefully to maintain proper alignment. Chenowth's "self-jigging" buggy kit makes it much easier. Drawing courtesy of Chenowth.

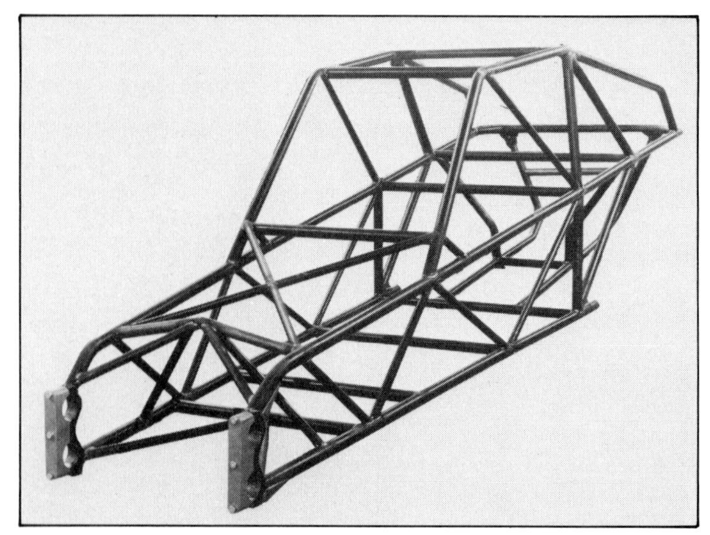

Full-race chassis: Berrien Buggy LCW 1001. Note the extensive bracing and triangulation. Photo courtesy of Berrien Buggy.

Fiberglass body panels make a tube-frame buggy feel more like a car. Quick-release fasteners allow for easy access to components. This is a five-piece body kit for Berrien Buggy 1001 chassis. Photo courtesy of Berrien Buggy.

Class-1 car was built by a friend who also happens to be a metal fabricator and welder. $15,000 later, the car was ready to race. Don't build a racing buggy from scratch unless you know what you're doing. And don't expect to get off cheap. Photo by George Jirka.

of alignment, it just gets worse the further along you get.

The Volkswagen chassis is different from most cars in that the body can be unbolted and lifted off in one piece. This can be done without disturbing the basic running gear or front and rear suspensions. But just because it is possible doesn't mean it is easy.

Removing the body of a stock sedan can be time consuming. Unthreading rusty bolts and disconnecting the steering column, master-cylinder-reservoir hose, speedometer cable, fuel tank and wiring harness can easily gobble up a day. Many race cars are modified with quick disconnects so the body is easier to remove for service or repair.

If you convert your sedan to a Baja Bug, you'll have to provide protection for the engine and body when the stock bumpers are removed. The cheapest way is to install prefabricated tubular front and rear bumpers. You can use them to mount skid plates, an oil filter, lights, license plates, reflectors and so on.

Sedan Roll Bars—If the plan for your car includes casual, low-speed off-roading only, with no serious racing, one roll-bar hoop behind the front seats is fine. To keep the bar neat looking and out of the way as much as possible, attach the tubing to the raised area of the body at the B-pillar, or door pillar. The closer you get the tubing to follow the inside contour of the car, the more out of the way it will be.

Roll-Cage Template/Form—Here's a neat trick. Anytime you can't get your car to the place that's bending the tubing for you, make a pattern out of heavy-gage wire. For that matter, if you see a bar or cage you like and the owner doesn't mind, copy it out of wire.

Chenowth's basic roll cage for the VW sedan. Note how the rear supports tie into the torsion-tube/spring-plate-support area for strength. Drawing courtesy of Chenowth.

To reduce weight, this racer built the rear bumper out of aluminum. Oh well, that's the breaks. Photo by Judy Smith.

Rolling Your Own—If you live out in the "sticks" where no tube benders can be found, you can bend a couple of roll-bar hoops with the torch-and-sand method. Before starting, round up a friend and some heavy gloves.

The idea is to pack the tube tightly with sand. This prevents collapsing of the section being bent. Start by plugging one end of the tube. There are two simple ways to plug. Tack weld an engine core plug, sometimes referred to as a *freeze plug*, over one end of the tubing. Or whittle out a tapered wooden plug using a short section of 2x2 or 2x4. Drive the plug in one end of the tube. Pack the tube tightly with *dry sand* from the other end.

If you use wet sand, steam will form as you heat the tube. This may cause an explosion, a potentially dangerous situation. To ensure it is dry, spread the sand out on a sheet of steel and heat it with a torch. Then pack it in the tube.

Pack the sand tightly using a hammer and a wooden dowel or anything that's a little smaller than the tube ID. Once the tube is filled and packed with sand, cap its open end in a manner similar to the other. I like to use a wooden plug here because it further packs the sand as the plug is driven in.

After taking measurements inside the car, lay out the shape of each piece to be bent. If you used heavy-gage wire to make patterns, trace around the wire with chalk, preferably on a concrete garage floor. Evenly heat the area of the tube to be bent with an acetylene torch. Slowly pull the tubing around to the correct angle.

Single roll-bar hoop in a sedan. Supports tie into the rear fenderwells.

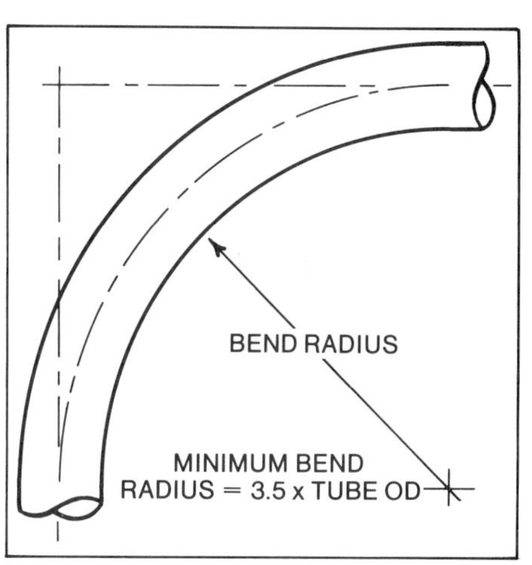

Steel tubing is strong, but don't weaken it by exceeding its minimum-bend radius.

Lay the tubing over the chalk pattern to check it. If you're bending a long tube, you may need to clamp one end in a vise. You heat while a helper bends or vice versa. After you have the bends you want, unplug the ends and poke out the sand.

Do's and Don'ts—For the main roll cage and diagonal supports, use at least 1-1/2-in.-OD mild-steel round tubing with a minimum wall thickness of 0.095 in. That's 13 gage. If you plan to race a heavy car of more than 2000 pounds such as a full-bodied sedan, Karmann Ghia or a VW Thing, some sanctioning organizations require 1-3/4-in.-OD, 0.120-in.-wall tubing. That's 11-gage wall thickness. Check the rules carefully.

Don't use square tubing. Round tubing has greater torsional strength than square tubing of equivalent size and weight. Totally out of the question is *pipe* such as that used for gas or water. Don't use threaded fittings or crimped bends. *No exceptions.*

Minimum-inside-bend radius of the tubing must be at least 3-1/2

Fish-mouthed end looks like the mouth of a fish. The tube end should butt against the side of another without gaps. Never cut to an approximate shape and fill the gaps with weld. Photo by Tom Monroe.

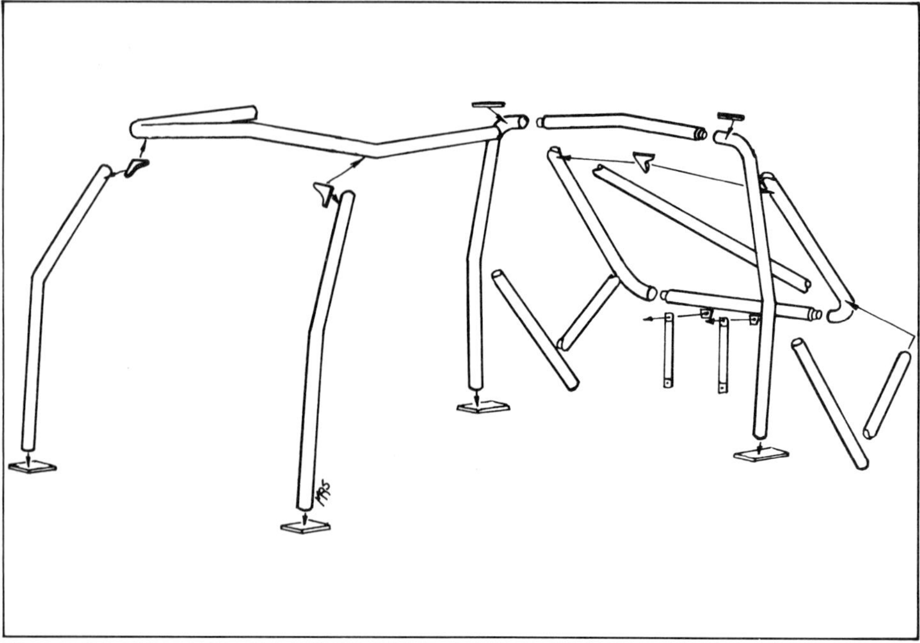

Chenowth's Class-11 sedan roll-cage kit includes mounting tabs, gussets and baseplates. Drawing courtesy of Chenowth.

times the tubing OD. Example: If you're using 1-1/2-in. tubing, no bend should be sharper than 3-1/2 x 1-1/2=5-1/4-in. radius. If you're using 1-3/4-in.-OD tubing, the minimum-bend radius should be 6 in. If you don't know what a minimum-bend radius is, look at the accompanying illustration.

The end of one tube that intersects and joins another tube must be *fish-mouthed.* The fish-mouthed end of the tubing is trimmed to fit snugly against the tube it joins. This will give it a fish-mouth shape. The gaps should be held to a minimum for welding.

Fish-mouthing is a cut-and-fit operation. You can use hand shears to cut thin-wall tube. A band saw, grinder or hole saw will be needed for 0.120-in.-wall tubing. When fitting the tubes in a sedan, make sure you leave enough room to weld all the way around the joints. Remember, in most cases the roll bar or cage is going to be assembled *inside* the car.

Never *butt-weld* roll-cage tubes together. By this, I mean never weld tubes end to end. Butt welds are weaker than an unwelded tube of the same length. If you must butt-weld a damaged section, first slip in a smaller-diameter sleeve under the joint.

You should never gas-weld or braze roll bars or cages. For best strength, use an electric welder. A TIG welder will give the prettiest weld, but a properly used MIG or an arc welder will be as strong. Make sure you get a 100% weld—all the way around—with good *penetration.* Penetration means that the *parent metal,* the tube, and the *filler*—the rod or wire—*flow* or melt together.

Mild Steel vs. Chrome-Moly Steel— In recent years 4130 —chrome-moly—steel has become popular in some race-car circles. The term *chrome moly* is derived from its two major alloys: chromium and molybdenum. Chrome-moly steel is about 25% stronger than mild steel of the same thickness.

Because it is stronger, you can use thinner-wall tubing to save weight. Many off-road race-car frames are made of it.

But it's not without drawbacks. Chrome-moly steel costs considerably more than mild steel. Because it's so strong and hard, it takes longer to bend and cut. It must be preheated immediately *before* welding. And chrome-moly steel requires *stress relieving,* after welding or bending. Otherwise, it may crack or break due to brittleness. To stress relieve the bent or welded area, *evenly* heat it to a cherry red in still air and allow to cool. *Do not quench with water.*

My advice is to leave chrome moly for the race-car-chassis builders who know what they're doing. Yes, chrome-moly steel is stronger than mild steel, but it's also more brittle. On impact, it's more likely to break rather than bend as with a mild-steel part.

For more information on metals and their properties, see HPBooks' *Metal Fabricator's Handbook* or *Welder's Handbook.*

Sedan Roll-Bar Mounting Plates— Attach the ends of the rear hoop or roll bar to the floorpan or to the *sills*—the raised part of the body below and just inboard of the doors, right next to where the body bolts on. Because the sills are boxed-in members, they're stronger than the flat floorpan.

But because the sills are curved, it's more difficult to fit the bases of the hoops to them. The sills are also part of the body, so the roll bar becomes a permanent fixture. If the body is wrecked beyond repair and you want to replace it, you'll have to cut out the roll bar.

Whether you mount the bar to the floorpan or sills, it should have 3/16-in.-thick, 4-in.-square steel mounting plates welded between the ends of the bar and the floorpan. The plates will distribute loads to and from the roll-bar tube and the relatively thin body sheet metal. Make sure the sharp edges of the baseplates are rounded

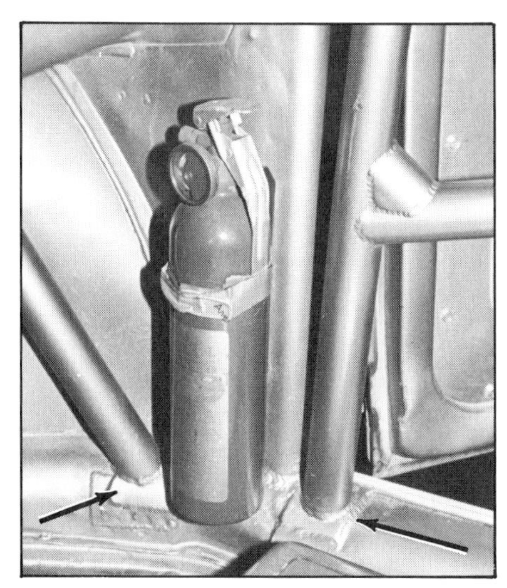

It's important to spread the load between the roll cage and body sheet metal. Here the bars tie into the body sill.

Roll-bar-hoop baseplate is bolted through the floorpan. Always use doublers on the opposite side of the sheet metal.

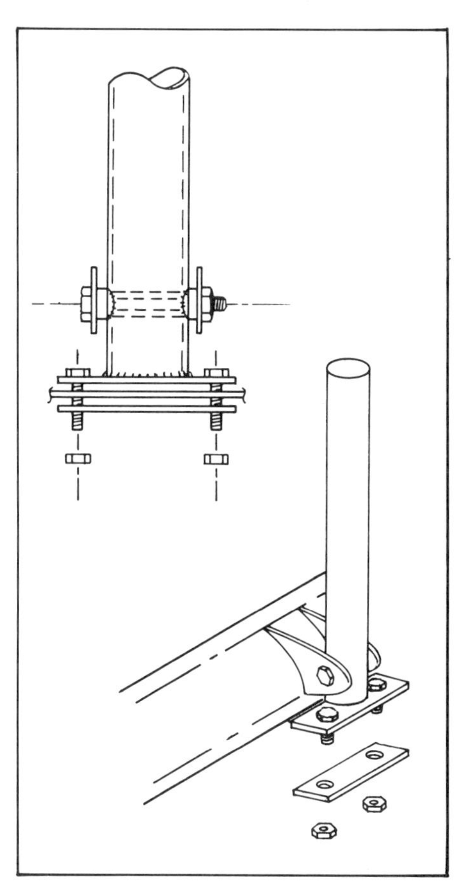

Ideal way to support hoop. Base plate is bolted to the floorpan. Hoop is captured by through bolt and ears welded to the body sill.

Rear support welded to the floorpan and tied into the fenderwell and fire wall. Support also doubles as a top shock-mounting point. It may look bulletproof, but the body will probably crack at the seams after rough off-road use.

off. Otherwise, they may *shear* through the floor or sill on a sudden impact.

If you opt to mount the bar to the floorpan, you can either bolt or weld it at the baseplates. Bolting is OK as long as *doubler plates,* the matching plates on the bottom side of the floorpan, are used.

Bolt through the plates, floorpan and doublers using four 3/8-in. (minimum) bolts—two for each mounting plate. Attaching bolts should be Grade-5 or better. But remember, once you add roll-bar supports, the assembly becomes too bulky to remove from a sedan without removing the body. So you may as well weld it in place.

If you weld or bolt the bar to the floorpan, weld a couple of 2-in.-wide, 1/8- or 3/16-in.-thick steel reinforcing tabs from the hoop ends or plate to the sill. If you have to remove the body from the floorpan, cut these tabs loose first.

Rear Supports—To support the roll-bar hoop, run two tubes of the same size to the rear of the car. The heavy-gage inner-fenderwells are the strongest places to mount them. If they are rusted, repair them by welding in new sheet metal first. If you've cut holes in the inner fenders to accommodate long-travel shocks, don't mount the rear supports near these holes. As with the roll bar, fit 3/16-in.-thick, 4-in.-square baseplates between the supports and the inner-fenderwells.

Another good place to anchor the rear supports is to the rear-deck area just inboard of the rear inner fenderwells. Like the roll bar, the roll-bar supports should be strengthened at their bases by welding mild-steel tabs between the inner fenderwells and the rear supports.

Still another way to brace the roll bar is to run each support tube back to the rear fire wall, and butt it against a top rear bumper-cage mount. This brace gets support from the fire wall and rear bumper cage, so it's another good way to support the roll-bar hoop. For more details, read the next section.

Rear Bumper Cage—Once you've removed the rear bumper and installed a Baja-Bug kit, the rear of the car is unprotected. To protect the rear bodywork and engine, it's a good idea to install a rear bumper cage. Unless you need a unique design to accommodate some added engine component, I suggest you buy a prefabricated cage.

Some prefabricated rear bumper cages use about 3-in.-square steel baseplates at the top mounts to bolt the cage to the fire wall. The area of these plates is too small to spread the load to the sheet-metal fire wall. When a skid plate is hit from the bottom or if the bumper is hit hard from the rear, the impact may bend or crack the fire wall or push the upper cage and baseplates through the fire wall. Therefore, when mounting the roll-bar supports, I suggest using 6-in.-square 16-gage (0.0598 in.) or

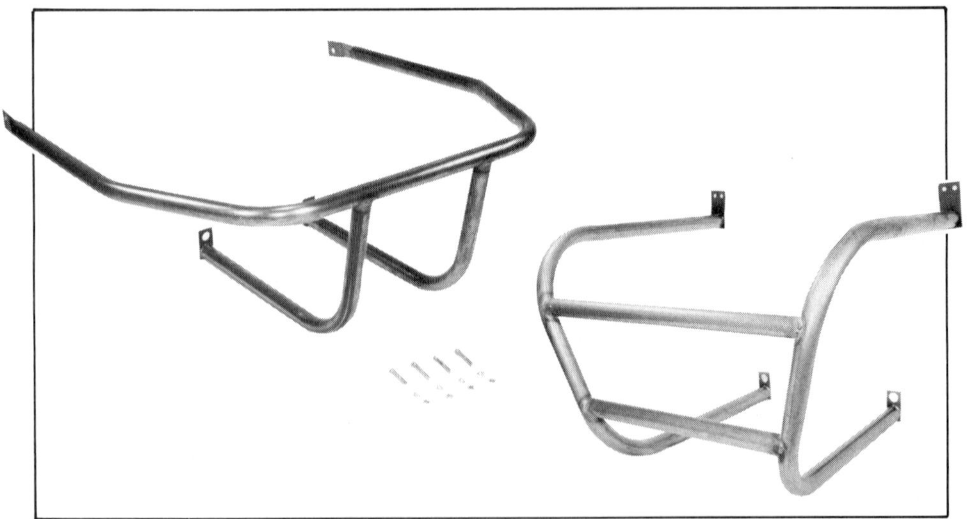

Recreational-type rear bumper cage (left) uses the upper shock-mount bolts at the spring-plate supports and the frame horns for mounting. Competition-type rear-bumper cage (right) mounts to the fire wall and roll-cage supports—the best. Photos courtesy of Bugpack.

Baseplates at the rear fire wall support the rear bumper's top tubes.

Rear part of this cage is actually a tube frame. It does not rely on the body for support. Cage extends through the fire wall to support the rear bumper. Upper shock mounts are incorporated into the cage/frame.

thicker steel doubler plates on the opposite side of the fire wall.

Drill holes in this doubler plate to match the holes for the rear bumper bolts. Bolt each doubler plate opposite the bumper baseplates, sandwiching the fire wall in between.

Drill a 1/4-in. hole through each corner of the double plate and through the fire wall. Using large OD washers on the engine side, secure the plate to the fire wall.

Now that you've spread the load over a larger area of the fire wall, run each roll-cage support tube from the corners of the roll-bar hoop to the center of the plates. Fish-mouth the ends of the support to match the contour of the hoop. Cut the opposite end of the support flush with the rear-bumper doubler plate. The object is to get 100% welds at both ends.

Upper Shock Mounts—A whole book could be written about shock setups because each one is a little different. Most race-car multi-shock setups use long-travel shocks. The shocks project through a hole cut in the inner-fenderwell and mount to well-braced rear roll-bar supports.

To use additional shocks at the rear suspension of a sedan, the rear roll-bar supports and shock mounts must tie into something stronger than body sheet metal. Two of the strongest support points are the rear shock towers or the ends of the rear torsion tube.

You can use the body for some lateral support, but solid vertical support from below is needed to keep the shocks from breaking off the upper mounts. Also run a brace, made from good-quality steel tubing or plate, between the two rear roll-bar supports, as close as possible to the rear fire wall.

Follow the basic rules for setting up shocks as described on page 78. You should have little trouble.

Diagonal Brace—A diagonal tube or X-brace in the middle of the roll-bar hoop will make the roll bar much stronger. If you plan to install a shoulder harness, the intersection of the X-brace is a good place to mount a horizontal tube to anchor the harness. A diagonal brace will restrict access to the fuel tank, battery, and anything else in the rear-seat area of the car.

Again, the decision depends on how hard you intend to use the car. If you're building a recreational car and plan to take more than one passenger along for the ride, rear-seat access is important. Racing demands safety first and access second.

Roll Cage—If you're going to run a stiffer front suspension, more support is needed for the front torsion tubes. What you really need is a full roll cage.

To transform an existing roll bar into a full cage, bend a roll-bar hoop similar to the rear hoop. Follow the front-door hinge-post line up from the floor past the dash. Continue up the outer edge of the windshield pillar, across its top, and down the other side. Mount the front hoop as you did the rear hoop.

Connect the front and rear hoops at their top corners. Run a tube front to rear, above each door opening. Run an additional tube between each roll-bar hoop on each side at or just below seat level. To strengthen the front hoop, run a tube between the front-hoop legs just above or below the steering column. This will give you a sturdy roll cage that you can use to brace the front suspension.

Front-End Supports—With a full cage to tie into, run tubular supports forward off the front hoop—two at the top and two at the bottom. Route the top tubes just under the dash, as shown. Route the bottom tubes through the fire wall to the lower torsion tube. Bend the top support tubes so there is ample tie-rod clearance with wheel travel.

Fish-mouth the forward ends of the supports to match saddle-type mount-

Keep the roll cage neat and tucked tightly against the body. But don't get so close that you can't make 100% welds. Not the easiest, but the best approach is to remove the body to weld the cage after it's fit.

Roll-cage front hoop should follow the windshield pillar. Note the gussets to strengthen the joints.

On the way back to the rear bumper and shock mounts, this cage picks up some strength from the rear torsion housing.

Route front supports forward from front hoop and through fire wall. Note radiused gussets which are less crack-prone than straight-cut gussets.

Front supports with clamps for the front torsion tubes. Opposite ends weld to the body or floorpan.

View without the stock fuel tank. Front supports are visible. Baja-Bug support struts are tied into the tunnel.

Don't pad the entire roll bar or cage—just the areas your body will contact.

ing clamps. Weld the saddle clamps to the tubes and weld the tubes to the front hoop. Bolt the clamps to the torsion tubes.

Bosses, or *spuds,* with plates welded onto well-braced upper support tubes, make excellent upper mounts for the front shocks.

Roll-Cage Padding—When you're out bouncing around in the dirt, elbows, knees, your noggin and other parts of your anatomy tend to bang into a roll cage. So pad the tubes in the vicinity of your body. Foam-rubber roll-bar padding is available from off-road suppliers.

If you have problems finding padding, it is the same stuff air-conditioning suppliers use to insulate the pipes between the compressor and evaporator. Plumbers also use this material to insulate water pipes.

Cut it and tape, glue or cable-tie the padding to the tubing. Only cover the tubes that need it. The padding may look great at first, but after some cuts and tears and lots of dirt, the inside of your car will begin to resemble an old boiler room.

SKID PLATES

Rear Skid Plate—The primary purpose of a skid plate is to protect the engine and transaxle. Both the engine and the transaxle cases are light castings—no match for sharp rocks. These are expensive items, so protect them. A skid plate must also provide a smooth surface so a car won't get hung up on brush, rocks and dirt.

There is one other requirement to keep in mind: *Never fasten the skid plate to the engine!* A skid plate must protect the vulnerable and fragile pushrod tubes. These light-gage steel tubes that house the pushrods run under the engine between the engine case and heads. When a pushrod tube is damaged, it leaks oil. If hit hard enough, the pushrod may also be damaged, which can lead to valve and piston damage.

Skid plates need not be alike. All a skid plate must do is meet your vehicle's needs. A lightweight sand rail for the dunes can get by with a light sheet of aluminum bolted to the underside of the floor pan or torsion-bar housing and rear of the engine cage.

Aluminum or Steel—When building a skid plate, lightweight aluminum is great if you have access to a shear capable of cutting it. The aluminum stock should be 3/16-in.-thick 6061-T6 or equivalent.

Steel that is 1/8-in.-thick will do if you don't mind the additional weight. Even though it's 30% thinner, the 1/8-in.-thick steel plate is nearly double the weight of the 3/16-in.-thick aluminum plate. The advantage of steel is obvious. You can build the plate any way you like with a common torch and welder.

If you don't care to build your own skid plate, ready-made skid plates are offered by buggy manufacturers and some mail-order outfits. Most of these are fabricated from 12- or 14-gage steel—0.1046 and 0.0747 in., respectively. It might be easier to adapt one of these to your car than to start from scratch.

A good plate will protect the following areas, in order of importance: engine case, pushrod tubes, valve covers and transaxle case—especially the transaxle nosepiece. One properly cut and mounted steel or aluminum skid plate will protect all of these components. See the accompanying illustrations for details.

A one-piece skid plate bolted to tabs welded in place is sufficient for most cars. In the rear-suspension section, I recommended a two-piece skid plate with the transaxle plate *welded* to and supporting the rear frame horns. But this is a racing-only modification. If the engine skid plate is welded in place, you have to remove the engine to clean the oil screen.

Where To Mount—On a buggy or Baja Bug, mount the rear skid plate to the frame horns and the *engine cage*—tubular rear bumper. For rear support on a sedan, use the original rear bumper rather than the engine cage.

Ideally, you should mount the plate with a minimum 1-in. clearance to the engine or transaxle. You can make sure this clearance is maintained when installing the plate by laying a 1-in.-square tube or wooden block on top of the plate. Remove the tube or block after you have the plate fitted.

A piece of rubber cut from an old tire and glued to the top of the skid plate will cushion the engine and transaxle from a blow that deflects the

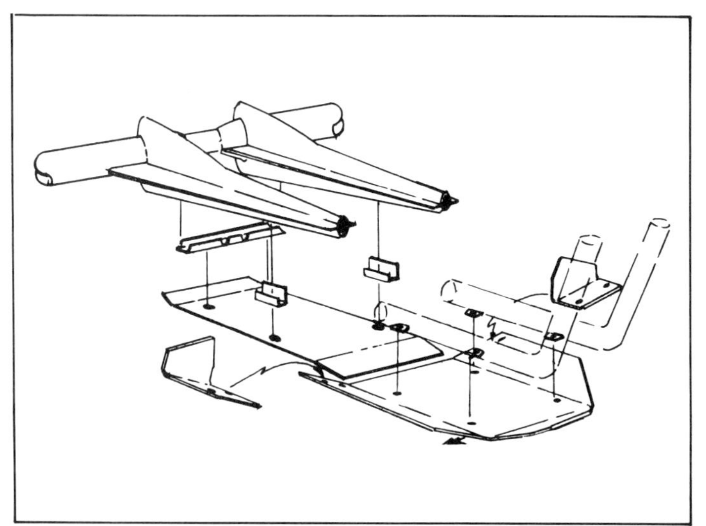

Chenowth's two-piece skid plate is made from 3/16-in. 6061-T6 aluminum. Skid plate cannot be used with the stock heater boxes. Drawing courtesy of Chenowth.

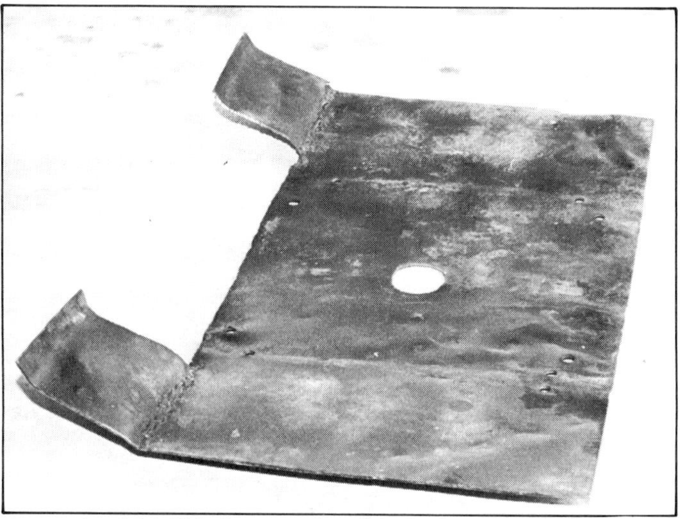

Pretty, it ain't! This homemade mild-steel engine skid plate is functional. Note hole for engine-oil drain.

This 3/16-in.-thick aluminum skid plate protects the engine case, pushrod tubes and valve covers.

Skid plate supports are attached to the stock bumper on this sedan. Photo by Ron Sessions.

Use locking nuts and bolts or put some silicone or weatherstrip adhesive on the skid-plate bolts. It'll keep the attaching bolts from becoming unattached. A loose skid plate can be a real drag. Photo by Judy Smith.

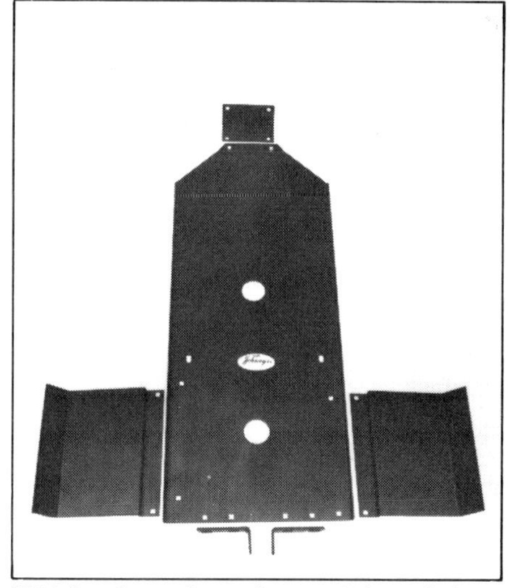

Skid plate for engine and transaxle is made from 12-gage steel. Note drain holes for fluid. Photo courtesy of Johnny's Speed & Chrome.

Class-5 Baja Bug torsion tubes and steering are protected by a front skid plate. Skid plate attaches to front bumper. Photo by Ron Sessions.

Deck plate should have smooth-side down so it can slide over objects. Plate should be longer so it will protect the bottom torsion tubes. Photo by Ron Sessions.

plate. Use weatherstrip adhesive to keep the rubber in place.

Drain-Plug Access—Many big-buck racers pull the engine after every race, so engine and transaxle drain-plug access is not important. However, you might want to be able to change engine and transaxle oil without pulling the engine or dropping the skid plate. If so, you'll need access holes for the transaxle and engine drain plugs. The risk of something catching on or getting through the holes is slight.

Take the time to mark the exact locations of the transaxle and engine drain plugs on the skid plate. One method that works well is to shove a 2-in.-long wooden dowel into the transaxle drain-plug recess and fit a socket over the engine drain plug. Coat the bottom side of the dowel and socket with paint.

Push the skid plate into position against the plug and socket. The paint will mark the locations for the access holes. Use a hole-saw or jigsaw to cut the holes. The transaxle drain-plug hole should be a shade larger than the drain plug—about 1-in. OD. The engine drain-plug hole should be large enough to get a socket through it and on the plug—about 1-1/4—1/1-2-in. OD.

Don't bother making the hole large enough to get the engine oil screen out. Rocks and debris by the pound would find their way through that 5-in. hole. You'll have to remove the skid plate or pull the engine to clean the screen.

Front Skid Plate—Some racing classes require a front skid plate. A front skid plate protects the front suspension and steering. Most recreational off-road cars don't need one.

If you are squeamish about the possibility of a close encounter with a large rock and what it might do to your car's torsion tubes, consider a front plate. It can help, especially if you're planning to drive at racing speeds. The idea is for the car to slide over objects rather than snag on them.

Front skid plates can be made out of the same stuff rear skid plates are made from—3-1/6-in.-thick 6061-T6 aluminum or 1/8-in. steel plate. I've seen them made from boiler plate too. Although prefabricated plates are available, most people cut and bend their own. Typically, the plate is bolted, riveted or welded to a tubular front bumper. See the examples.

Chapter 6 BRAKES

A car's wheels must be on the ground for the brakes to work. Like this airborne buggy, off-road cars spend a good deal of time in the air, or on surfaces with poor traction. Photo by George Jirka.

The amount of preparation needed for effective off-road braking should be determined by how you use the car. If you plan just to "play around" in the dirt, stock brakes are fine. If you intend to drive at high speed, you should make some changes to get better braking in the dirt. The more you modify the car and change the stock front/rear weight distribution, the more work you'll have to do to achieve the correct front-to-rear braking balance off the pavement.

The Volkswagen Beetle has been equipped with hydraulically operated drum brakes at all four corners since 1950. Though later versions of the Karmann Ghia, Bus and Type 3 featured disc brakes at the front, the Beetle and Super Beetle retained drums. Because most off-road Baja Bugs and buggies start out using original Beetle parts, I'll assume your brake system uses Beetle parts, too.

Braking Bias—Like all passenger-car brakes, the VW front brakes do most of the work under hard braking. Therefore, the front brakes have more lining area, larger wheel-cylinder bores, or both. The reason for this is simple. When the brakes are applied, weight transfers to the front tires, thereby unloading the rears. Generally, the amount of traction force a tire develops is proportional to the load it exerts against the ground. So more braking is *proportioned* to the front wheels under hard braking.

This stock brake setup acts a little differently off-road. This is compounded by the increased traction of bigger tires on the rear. With the good traction of the pavement lost, the front tires will lock up and slide easily. When the fronts lock up, stopping ability is impaired, and steering control is lost.

In a modified sedan or lightweight tube-frame buggy, even more rear-wheel braking is required. On these cars, weight is removed from the front end, thus further unloading the front tires.

To adapt Volkswagen brakes to off-road use, you must change the braking bias from the front to the rear. One way to do this is by using larger wheel cylinders at the rear or smaller ones at the front. Rear-brake wear can be reduced by installing wider linings. The low-cost approach, which is the one most sedan and Baja Bug owners take, involves visiting a VW-parts outlet or perusing the boneyards for these pieces.

If you have several hundred dollars remaining in your racing budget, you can install a special dual master-cylinder/pedal assembly. This setup can be adjusted to provide different front-to-rear brake bias for different race tracks or terrain.

FRONT BRAKES

Changing Brake Cylinders—If your funds are limited and you plan to stick with the stock drum brakes, I recommend reversing the braking bias from front to rear: Just use 17mm wheel cylinders at the front.

With the exception of very early models, VWs were equipped with 22mm front wheel cylinders. To get 17mm cylinders, you'll have to adapt *rear* wheel cylinders. The 17mm rear wheel cylinders from '57-and-earlier Beetles or '68-and-later Beetles will fit. If you have a choice, use the '68-and-later 113 611 053B cylinders. These will fit any '58-or-later Beetle front-brake assembly without modification. They're also cheaper than the earlier units.

To adapt the '57-and-earlier rear cylinders to most Beetle front-brake assemblies, you must swap the backing plate, adjuster and springs, plus make other modifications. These items are hard to find in the junkyards, so use the '68-or-later parts if possible.

To correct any mismatch of wheel cylinder to shoes, open up the slots on the ends of the cylinders or move the cylinder inboard or outboard on the backing plate. This may require machining or shimming the wheel cylinder to fit.

Reducing Front Brake-Lining Area—You may have heard that another way to bias braking to the rear wheels is to reduce front brake-lining width. Like installing smaller-diameter front-wheel cylinders, this modification is supposed to reduce the braking action of the front brakes. There's only one problem with doing this. It doesn't work.

Reducing lining width does have an effect on the front brakes—lining wear increases. Nothing else. If it were practical to use narrower brake drums with narrow linings, brakes would be more prone to overheat and fade. Don't change the lining width.

There are three practical ways to change brake bias: Install wheel cylinders of different bore sizes, install separate front/rear master cylinders with adjustable pedal linkage, or install different-diameter brakes. For the low-budget off-roader, changing wheel cylinders is the only way to go. Changing the brake diameter gets more involved and is much more expensive. The adjustable master-cylinder-and-brake-pedal assembly is somewhere in between.

Eliminating Front Brakes—You may have seen sand buggies with no front brakes. Under no circumstances should you do this to your off-road car. A sand buggy is a different animal with different priorities than your off-road car.

In the dunes, the front wheels wear farm-implement-type tires that are used only for steering. Paddle tires at the rear have the job of propelling and stopping the car. Even with paddle tires at the front, the brakes couldn't generate stopping force because three-quarters of the car's weight is over the rear wheels. Therefore, front brakes are jettisoned as excess baggage. Most sand buggies weigh little more than 1000 pounds, so the rear brakes easily handle the job.

Stock '68-and-later 17mm rear wheel cylinder (top), 19mm wheel cylinder (middle), and Type-3 22mm rear wheel cylinder (bottom). On ATE cylinders, bore size is cast in (arrow). On aftermarket cylinders, you must rely on the part number. Photo by Ron Sessions.

A LITTLE BRAKE THEORY

Without getting too far into hydraulic theory, here's a quick explanation of how wheel-cylinder-bore and master-cylinder-bore diameter affect braking. The VW hydraulic-brake system consists of a brake pedal, master cylinder, steel lines flexible neoprene lines and four wheel cylinders.

When the brake pedal is depressed, the master cylinder pressurizes the brake fluid to a uniform pressure per square inch throughout the brake system. This pressure applies a force to the wheel cylinders at each wheel.

The wheel-cylinder pistons force the brake shoes against the inside of their rotating drums. You can change the relative amount of force exerted on the brake shoes by installing wheel cylinders or master cylinders with different diameter bores.

If you increase wheel-cylinder bore, a given pressure will exert more force on brake shoes because the pressure has more piston area to work against. Doubling wheel cylinder piston area would double the force. It follows that decreasing wheel-cylinder bore decreases piston area and the force applied to the shoes.

Changing master-cylinder-bore size produces the reverse results. Fluid in the master cylinder must resist the force of the piston. Increasing piston diameter by increasing bore size distributes this force over a larger area, thus reducing hydraulic pressure. Reducing master-cylinder-piston size distributes the same force over a smaller area, thus increasing hydraulic pressure. Again: *Increasing* master-cylinder bore *decreases* hydraulic pressure and increases pedal effort; *decreasing* master-cylinder bore *increases* pressure and reduces pedal effort.

Changing master-cylinder bore isn't necessarily the change to make for off-road because the intent is not to change total hydraulic pressure. The helpful change is to proportion more braking to the rear wheels, where the weight and traction are.

REAR BRAKES

To complete the process of chang-

Sand buggy can get away without front brakes. Don't think for a minute that your off-road car can too. Photo by Ron Sessions.

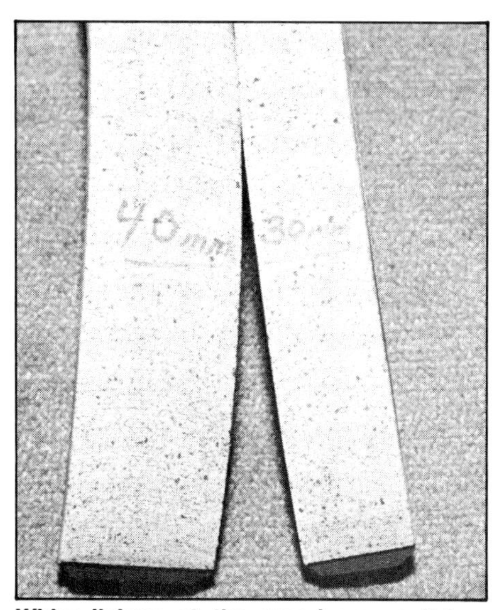

Wider linings at the rear increase lining durability and fade resistance, but will not make the brakes more effective. Here 40mm-wide rear linings from a '68-or-later Beetle are compared with 30mm-wide rear linings from earlier Beetle.

This face has to be machined 0.55 in. when installing Type-3 rear brakes on a '67-or-earlier sedan.

ing brake bias to the rear, use larger-diameter rear wheel cylinders. I recommend using 22mm wheel cylinders. Wider 40mm-wide linings can be used at the rear to increase lining durability and fade resistance.

The largest rear wheel cylinders available for a Beetle are the 19mm units used in '58—'67 models, so many racers substitute 22mm front wheel cylinders from a '65-or-later Beetle (part 131 611 057). The 40mm-wide rear shoes (part 113 609 537C) can be adapted from any '68-or-later Beetle. To make the swap complete, use '68-or-later rear backing plates.

The best rear brake system, without resorting to costly disc-brake setups, is an adaptation of the rear-brake assembly from a '65-or-earlier VW Type-3 squareback, notchback or fastback. Actually, linings and wheel cylinders can be used off any year Type-3. But to get the more desirable 5-bolt wheel-lug pattern, you must use a '65-or-earlier rear brake drum.

Type-3 45mm-wide linings are half again as wide as the early Beetle 30mm linings. And the Type-3 has a 22mm wheel-cylinder bore, the largest of all VWs. To install these in a '67-or-earlier Beetle, machine about 0.55-in. off the surface of the drum that the axle nut tightens down against. Type-3 brake assemblies bolt right up without machining on '68 swing-axle Beetles and all IRS units.

MASTER CYLINDER

When using larger-diameter rear wheel cylinders, you may have to use a bigger master cylinder if the front wheel cylinders are left unchanged. This is because the bigger wheel cylinders require more brake fluid as they move the shoes. Consequently, the brake pedal must be pushed farther to actuate the brakes. Excessive pedal travel is disconcerting. Check the brakes before changing master cylinders. It may not be necessary.

Bore size, in millimeters, is cast onto the side of the master cylinder. If you have a 17mm master cylinder, use a 19mm. If it's 19mm, use a 22mm. The 22mm master cylinder is from a '66-and-earlier bus, VW part 211 611 011J. **Note: All else being equal, a larger master cylinder requires more pedal effort. If you get carried away and install too large a master cylinder, pedal effort may become too high. This will become very apparent to your leg after a few hours of driving.**

Tandem Master Cylinders—All '67-and-later VWs use a tandem master cylinder, with two pistons in one bore operated by a common pushrod. Required by U.S. law to prevent a single leak from causing a total hydraulic brake failure, the tandem master cylinder divides the hydraulic system. The front and rear brakes are operated by separate systems. In the VW, the front piston operates the rear brakes and the rear piston operates the front brakes.

Although this sounds like a good idea for off-roading, where rocks can puncture a brake line easily, the VW tandem master cylinder presents another problem. It uses a divided remote reservoir. As a car bounces around, the connections between the master cylinder and its reservoir can loosen and leak.

Most off-road racers replace the tandem master cylinder with the pre-'67 single-piston master cylinder, which has the single reservoir attached directly to it. However, if you have a '67-or-later Beetle and the brakes are working fine, use the existing tandem master cylinder until it poses a problem. In the meantime, use worm-drive hose clamps at the reservoir connections to minimize leakage.

If you use a tandem master cylinder and switch the wheel cylinders to different sizes, you must also switch the brake lines at the master cylinder. Why? The rear piston of the master cylinder, which normally pressurizes the larger-ID front wheel cylinders, has more stroke than the front piston. When you move these large-ID wheel-cylinders to the rear, and move the small-ID cylinders to the front, master-cylinder piston stroke will be too much for the front brakes and not enough for the rear brakes. The result

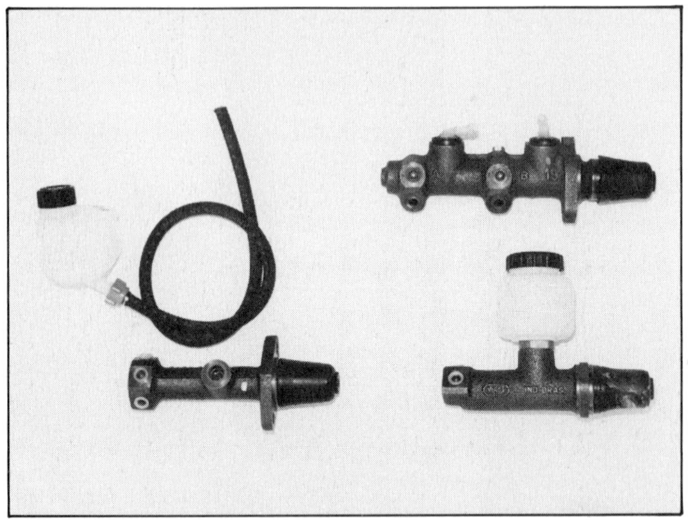

1966-and-earlier Type-1 17mm master cylinder with remote reservoir (left), '67-and-later Type-1 19mm tandem master cylinder (top) and '61-'66 Type-2 22mm master cylinder with integral reservoir (bottom). Of the VW-type master cylinders, the 22mm Type-2 unit is preferred because of its large bore size, integral reservoir and simplicity. Photo by Ron Sessions.

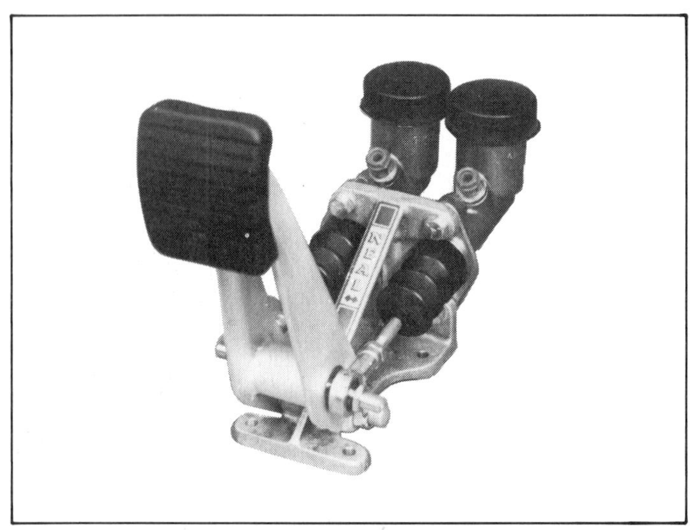

Neal dual-master-cylinder assembly. One cylinder controls the front brakes and the other cylinder controls the rears. Assembly combines the master cylinders, brake pedal and bracket into a single unit that's easy to install. Photo courtesy of Neal Products Inc.

will be uneven braking action—unless you switch the lines.

Pedal Assembly—The stock VW brake-pedal assembly is OK for off-road. But sometimes the pedal gets much too slippery, even with a new rubber pad. Remedy this by removing the rubber pads and spotting beads of weld on the metal brake- and clutch-pedal pads. The same thing can be accomplished by tack-welding *expanded metal* to the pads. Expanded metal is the grid-like flooring used in many industrial applications.

Aftermarket Pedal/Master Cylinders—Many off-roaders have replaced the stock brakes and pedal assemblies with aftermarket hydraulic clutch-, throttle- and brake-pedal units. With these brake-pedal assemblies, the pedal and master cylinder(s) are mounted on a common bracket. They can accommodate both single and dual-master cylinders of different types and bore sizes.

For example, Neal offers 5/8-, 3/4-, 7/8- and 1-in. Girling, Howe and Airheart master cylinders. Ja-Mar manufactures their own master cylinders in similar sizes. The pedal assemblies can be bolted into ready-made brackets that weld or bolt into place.

As mentioned on page 95, tandem master cylinders offer independent front and rear braking. However, the aftermarket dual-master-cylinder setups differ from the stock tandem master cylinder.

Aftermarket units have two mechanically separate cylinders. Consequently, they can be adjusted to change brake bias, or distribution. This virtually eliminates the need to change wheel cylinders to change brake bias. And precise adjustments can be made.

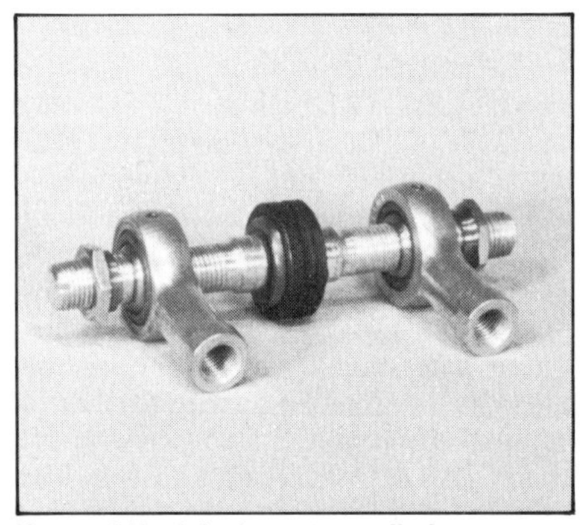

Heart of Neal dual-master-cylinder assembly is this equalizer bar. Brake bias is adjusted by moving the center bearing from one side to the other on the threaded rod. Photo courtesy of Neal Products Inc.

For instance, the Neal dual-master-cylinder-and-pedal assembly uses an adjustable *equalizer bar* that connects the brake pedal to two master-cylinder pushrods. The pushrods attach to each end of the equalizer bar.

The threaded bar joins the pedal through a spherical bearing between the pushrods. The bar can be adjusted by threading the bearing from side to side in its housing in the pedal. When the bearing is centered on the bar, the pedal force is equal at both master-cylinder pushrods. Equal force produces equal hydraulic pressure in the front and rear brake circuits.

AMS rear disc brake installed. Photo by Ron Sessions.

Adjusting the bearing off-center with the equalizer bar changes pressures, or brake bias, in the two brake circuits. This is similar to a teeter-totter or see-saw with a big kid on one end and a small one at the other end.

Brake bias is *inversely* proportional to the relative distance the bearing is from each pushrod. For example, adjusting the center bearing *away* from the front-brake master-cylinder pushrod on the balance bar *increases* the pressure in the rear-brake circuit. This reduces pressure in the front-brake circuit by the same percentage.

If you're into math, here is how to find brake bias stated as a percentage:

$P_R = 100 \times d_F/(d_F + d_R)$

or: $P_F = 100 \times d_R/(d_F + d_R)$

where P_R = % of total brake pressure to the rear

P_F = % of total brake pressure to the front

d_R = distance from bearing to rear pushrod

d_F = distance from bearing to front pushrod

Pulling some numbers out of the air let's say d_F = 1-1/2 in. and d_R = 1 in. The percentage of pressure going to the rear brakes is 100 x 1-1/2 (2-1/2) = 60%. Rather than agonizing through the division and multiplication to get front-brake bias, it's simply 100% − 60% = 40%.

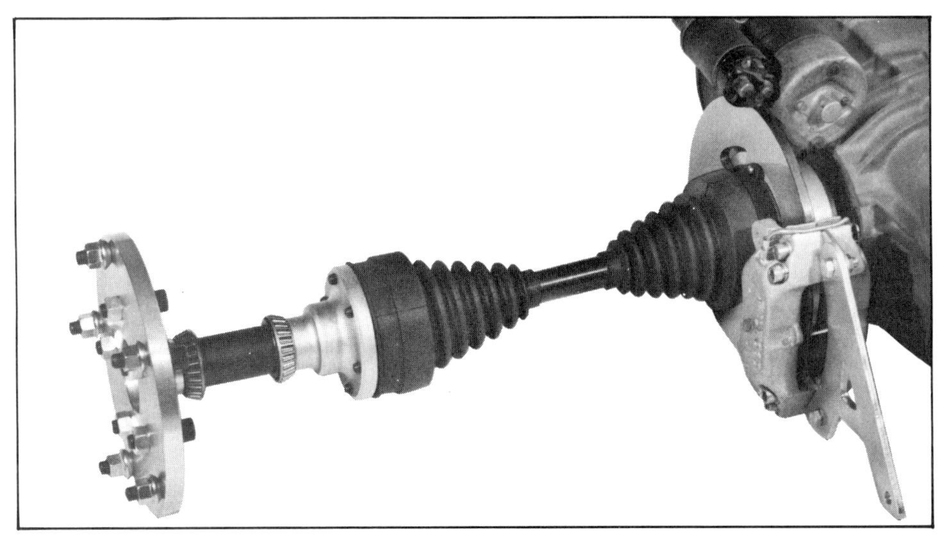

Summers Brothers' inboard disc brake/heavy-duty drive-shaft kit assembled. With the brake caliper and rotor at the transaxle, unsprung weight is reduced about 30 lb per wheel. Photo courtesy of Summers Brothers.

The '67-and-earlier 5-bolt drum provides a much stronger method of mounting the wheel than the '68-and-later 4-bolt unit.

DISC-BRAKE INSTALLATIONS

Unless you have racing aspirations, plenty of money, and the time and experience to install a disc-brake system correctly, you are better off sticking with the stock drum brakes. For street-driven, dual-purpose Baja Bugs and buggies, the stock drum brakes work fine. Complete disc-brake setups can cost anywhere from a few hundred dollars to well over a thousand.

Disc brakes do have their advantages. Drum brakes trap dirt inside, which accelerates lining wear during a race. Disc brakes throw off these abrasive particles. If your plans include running through water and mud, discs will dry out much faster. And disc brakes are self-adjusting, reducing the amount of pedal change in service or even during a long race.

The other advantage is primarily a racing consideration. Aftermarket aluminum-caliper disc-brake setups are lighter than stock brakes. This reduces unsprung weight. Inboard rear disc brakes, such as those offered by the Summers Brothers, mount alongside the transaxle side covers. This reduces rear-suspension unsprung weight by totally eliminating the brakes as unsprung weight.

Don't put a set of stock Type-3, bus, or Karmann Ghià discs up front. These units are *heavier* than the stock drum brakes.

If you use disc brakes, remember that you will probably have to change to a larger-bore master cylinder as well. Although the stock 5/8- or 3/4-in.-bore master cylinder works fine with drum brakes, disc brakes may require a 7/8- or 1-in.-bore master cylinder. Disc-brake calipers use large-diameter pistons requiring more fluid to operate them.

If you're contemplating using disc brakes, contact the manufacturer for recommendations on master-cylinder-bore size, caliper-piston OD, disc OD, and lining size and material.

OTHER BRAKE CONSIDERATIONS

Lug-Bolt Patterns—Through '67, type-I brake drums have a five-bolt wheel-lug pattern. The '68-or-later drums are four bolt and should be swapped with five-bolt types for regular off-road use. All '66 and '67 five-bolt front drums fit '68-or-later front brakes and spindles if you use '68-or-later inner wheel bearings.

Studs or Bolts?—Though not necessary with stock wheels, it's preferable to use studs and nuts to mount the wheels rather than bolts. It may seem like a lot of trouble to drill out the threaded holes in your drums and press in studs, but it's well worth it.

Brake-drum studs are pressed in after the threads are drilled out. Setup is stronger than the stock bolts and changing tires is a lot easier. Tack-weld the heads to the drum to prevent the studs from rotating.

Flexible brake hoses are vulnerable to damage from brush and rocks. This racer uses a tension spring and loop arrangement to keep the front brake hose up out of harm's way. Photo by Ron Sessions.

Woven rear-brake linings have high coefficient of friction and improve operation of rear drums in dirt.

Steering brake lets the driver brake one rear wheel at a time. When used with a little throttle, a car can turn much sharper than normal. This Ja-Mar single-handle steering brake is mounted on tabs welded to the frame. Shift lever is behind it. Photo by George Jirka.

With the thicker bolt flanges used on many aftermarket wheels, stock bolts aren't long enough to engage fully into the drum. Pressed-in studs are stronger than the stock bolts because of their bigger diameter: 14mm versus the stock VW 12mm. And, wheels are much easier to change with studs, especially when using heavy off-road rear tires and wheels. If you've ever tried to balance a VW tire and wheel on your foot while trying to align the wheel and drum to thread in the first lug bolt, you need no convincing.

Studs are available through various aftermarket sources. Make sure they are long enough for the wheels you're using. Measure the thickness of the wheel flange and add that to the thickness of the lug nuts to get an approximate figure. When the wheel is bolted on, you should have *at least* two threads showing past the lug nut.

Flexible Brake Hoses—Stock flexible brake hoses are fine but are no match for sharp rocks and tree branches. After about five years of sun and fun, the hoses crack. Excessive heat from exhaust systems, welding or oven-baking in a paint booth can hasten the demise of a flexible brake hose.

Provide additional protection for the brake hoses by slipping pieces of rubber hose over them while you have them disconnected. Leave the hose loose on the brake line so it will roll with brush and debris rather than snag it.

If your pocketbook allows, install braided-steel lines. Not only are braided-steel lines less susceptible to damage, but they provide a firmer brake pedal due to less flex upon brake application.

STEERING BRAKE

What Is It and How Does It Work?—If the front wheels are to steer, the tires must develop traction. Anyone who has ever driven in the dirt, or snow for that matter, knows that a car doesn't necessarily turn when you turn the steering wheel. This problem can be accentuated by skinny front tires and an extreme rear weight bias. If your car has this problem, one way to get around it is to steer with the rear wheels.

Many of us have marveled at how tanks and bulldozers negotiate turns. They merely lock up one track and pivot. The other track supplies the go.

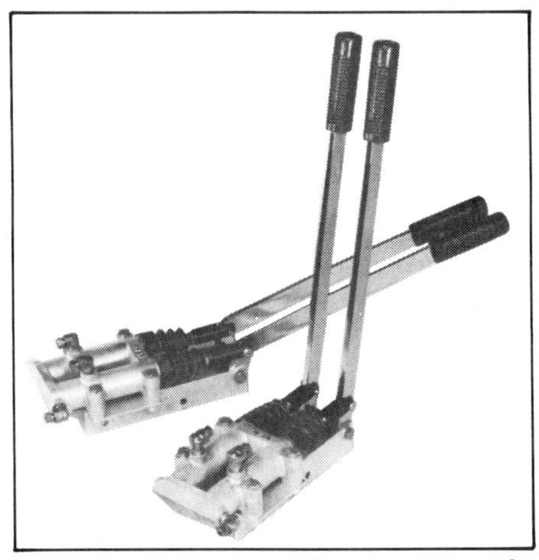

In the heat of competition, some people push when they should pull or vice-versa. Dual-handle steering brake removes the guesswork. Photo courtesy of Neal Products, Inc.

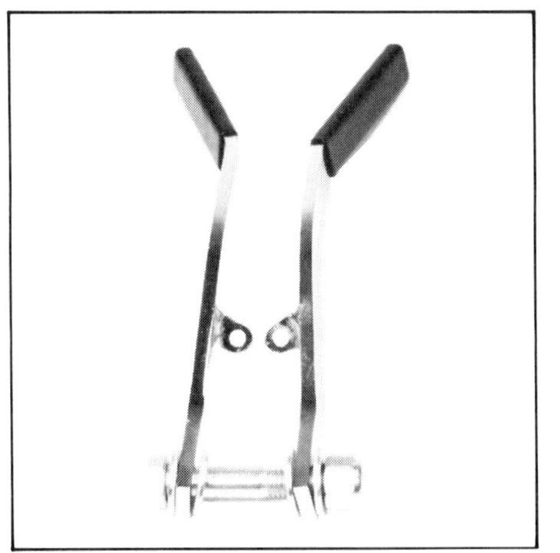

Simple mechanical steering brake. One emergency-brake cable mounts to each handle. Photo courtesy of Bugpack.

Ja-Mar Park-Lok locks all four wheels when applied. An ideal situation for using one of these is when you park a car on a steep incline to survey the situation ahead. Photo courtesy of Ja-Mar Inc.

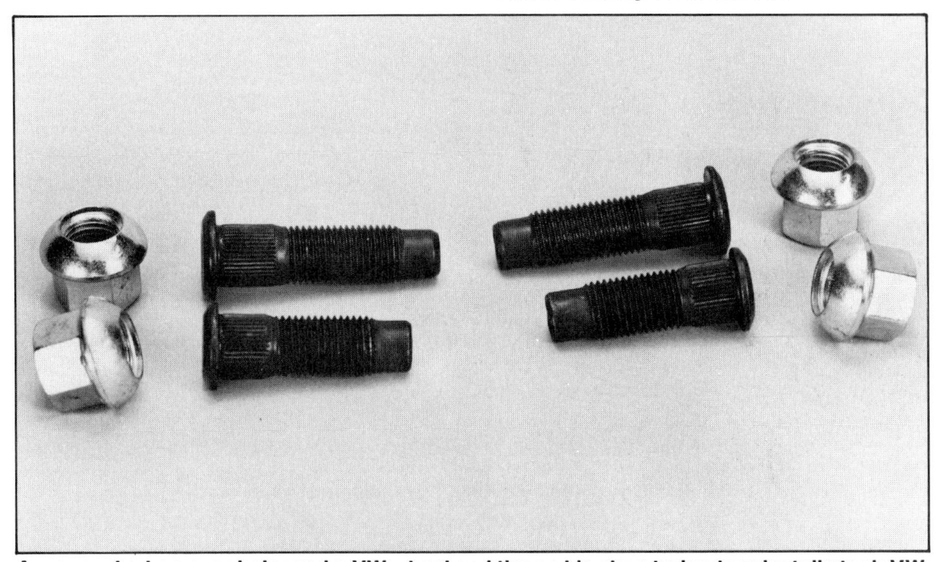

Anyone who has ever balanced a VW wheel and tire on his shoe trying to reinstall stock VW lug bolts will appreciate these. Photo courtesy of Sway-A-Way.

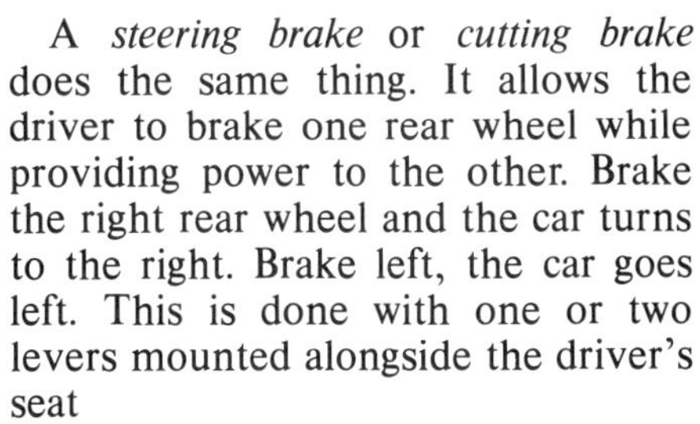

A *steering brake* or *cutting brake* does the same thing. It allows the driver to brake one rear wheel while providing power to the other. Brake the right rear wheel and the car turns to the right. Brake left, the car goes left. This is done with one or two levers mounted alongside the driver's seat

Tube-frame buggies have very light front ends, so a steering brake is a valuable piece of equipment. But the driver must know how to use it. On the heavier Baja Bugs or sedans a steering brake is less effective, but still has some advantages.

One big advantage will show up if you ever get *high-centered* with one rear wheel off the ground, or ever get stuck where one wheel has no traction. In those situations, simply lock up the freewheeling tire with the steering brake. This forces the transaxle to drive the other wheel. In effect, a steering brake is sort of a poor man's limited-slip differential.

Mechanical vs. Hydraulic—Each rear brake can be operated mechanically or hydraulically. Mechanical steering brakes use two levers, one attached to each emergency-brake cable. Pulling up on the left lever applies the left rear brake. Pulling the right lever applies the right rear.

Mechanical steering brakes are simple and inexpensive. But, if you've ever experienced total hydraulic-brake failure and tried to limp home using the emergency brake, you know that a cable-operated brake does not operate smoothly. There's a fine line between slowing down gracefully and locking the brakes completely. This is why mechanical steering brakes have been largely replaced by hydraulic steering brakes.

A hydraulic steering brake does the same thing, only with more finesse. The mechanical unit operates only one brake shoe at each wheel, but a hydraulic steering brake operates both shoes. The hydraulic unit is also smoother and requires less effort.

Hydraulic steering brakes plumb into the brake line at the T-fitting where the line splits to go to each rear wheel. When not in use, the steering brake has no effect on the rear brakes. But by operating the handle(s) you can brake one rear wheel at a time.

As with master cylinders and wheel cylinders, larger-diameter steering-brake pistons are needed for disc-brake applications.

Single vs. Dual Handle—Hydraulic steering brakes come in two different designs: a single handle with a push/-pull operation, and a double-handle unit where a pull on either lever brakes the corresponding wheel.

The advantage of the double handle is that there's no question which handle works which wheel. A disadvantage is its width—more than twice as wide as the single handle.

The single-handle steering brake fits more easily between the seats. But unless you're familiar with using it, you may find yourself pushing it when you should pull it. Turning in a direction opposite to that intended could be disastrous.

chapter 7 DRIVING LIGHTS

Driving at night can be a lot of fun when your lights are working right.

Ask a dozen people what kind of off-road driving lights to use and you'll probably get a dozen different answers. When you get wide differences of opinion, it generally means that the difference between the products is not that great, or not well understood. If you question these same people more closely, you'll probably find that they've had experience with just one brand of lights. An off-roader's preference for a brand of lights often comes from what his friends use, a good deal he got, or how much contingency money a manufacturer puts up.

If you look closely at some race cars, you may notice a surprising number of lights with one make of housing and another make of bulb. Most housings are compatible with other brands of bulbs. Many of these bulbs are used for aircraft landing lights.

Manufacturers who put up contingency money are generally more concerned that their decals and light covers appear on a car than what bulb is used. Also consider that a manufacturer can get his lights used on almost any car through direct sponsorship. A racer usually "sees the light" when sponsorship money is at hand.

With this in mind, my advice to the weekend off-roader is to be more concerned about getting a good buy on good name driving lights. Be wary of paying top retail price for a particular brand that's supposedly "the right way to go."

One thing is for sure: The faster you go, the more range you need from the lights. Make this your top priority. *Overdriving* your lights is a good way to get into serious trouble.

Types—Off-road driving lights fall into three major categories: fog lights, pencil-beam lights and good old driving lights. These names refer to the intended task of the light. All use tungsten filaments. The major difference between them is the type of lens used.

Of these light types, all can be further described as sealed-beam or replaceable-bulb, and halogen gas or non-halogen.

Sealed-Beam vs. Replaceable Bulb—A sealed-beam light combines the filament, reflector and lens into a single *sealed* unit. All air is withdrawn from the sealed-beam unit so the tungsten filament operates in a vacuum. If one component fails, the complete unit must be replaced. Conventional automotive headlights sold in the U.S. have been this type since about 1939.

A replaceable-bulb light features a filament in a separate bulb that plugs into the back of the light. Because the filament operates in its own

Sealed beam (left) and halogen replaceable bulb (right). Sealed beam combines filament, reflector and lens into a single sealed housing. Halogen bulb can be replaced separately. Don't get fingerprints on the bulb.

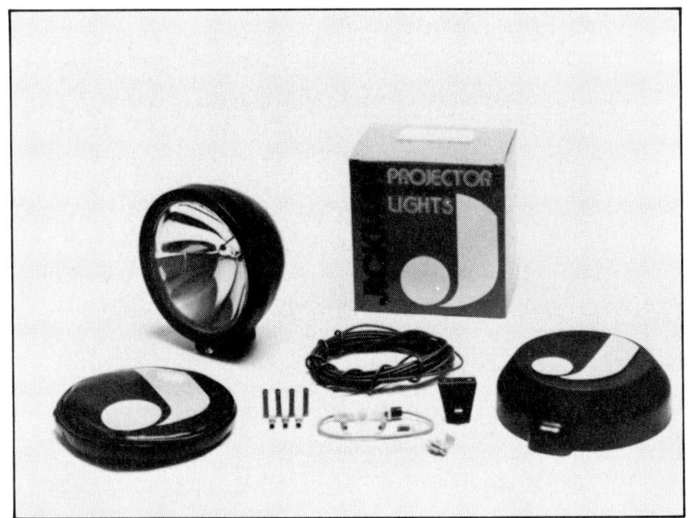

Jackman Protector Lights are sealed-beam units boasting 300,000 candlepower and 2.5-mile reflective range. Photo courtesy of Jackman Wheels, Inc.

This racing buggy sports four driving lights and two fog lights. If the driver turned on all six lights at once, the car's charging system would be seriously taxed. Make sure your car's generator or alternator has enough output to run driving lights. Photo by Mike Rehler.

atmosphere, the area between the reflector and the lens is not evacuated. Traditionally, replaceable-bulb headlights have been offered in European cars, but have been illegal for passenger-car, street use in the U.S. due to federal lighting laws.

Atmosphere—The major difference between a conventional or non-halogen bulb and a halogen or quartz-iodine bulb is the atmosphere the tungsten filament operates in. In a conventional non-halogen bulb, the filament is in an *evacuated* glass tube—a vacuum. In the halogen bulb, the filament lives in the same glass tube, but is surrounded and shielded by a halogen gas—usually iodine or bromine.

As electrical current passes through the tungsten filament, resistance builds and the metal is heated to incandescence—white light. Without either the vacuum or halogen-gas shielding, oxygen would reach the filament, causing it to oxidize completely at its high operating temperature.

In a conventional non-halogen bulb, a small percentage of the tungsten filament vaporizes every time it is used. The vaporized tungsten is deposited on the inside of the bulb, so it darkens with age. Eventually, the filament wears so thin that it breaks. When the filament breaks, electrical current no longer passes through and things go dark.

Halogen gas has the unique ability to redeposit the vaporized tungsten back onto the filament as the bulb cools. This prolongs filament life. And it prevents the bulb from darkening, thereby giving a more constant intensity.

The amount of current used generally determines how bright a light is. Don't be confused by candlepower ratings. The brightness of a light as measured in candlepower can vary considerably depending on lens type.

A more effective yardstick of a light's performance is its *reflective range* rating. Reflective range is the distance at which light from a given lamp can be seen. If you need more information on the reflective range of the lights you want to buy, contact one of the manufacturers listed in the appendix of this book.

Generally speaking, halogen-shielded bulbs operate at considerably higher temperatures—about 1100F (593C). Consequently, halogen lights put out considerably brighter light than their conventional non-halogen counterparts. The light from halogen lamps is a bluish-white, whereas that

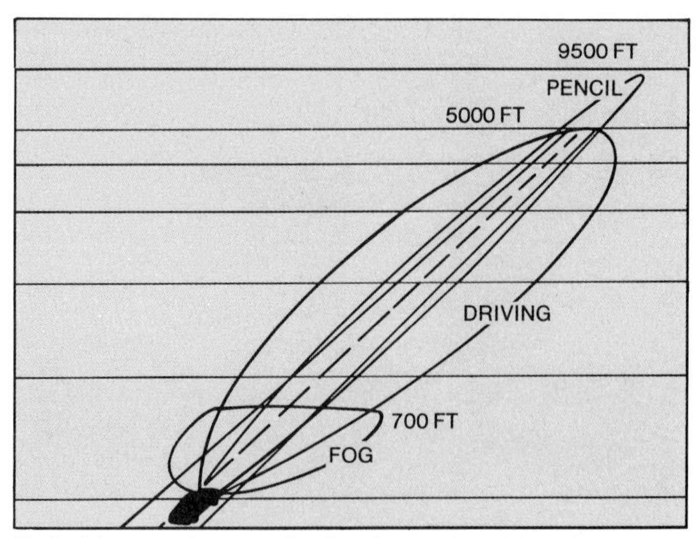

Typical beam patterns and reflective ranges of fog, driving and pencil-beam lamps. Drawing courtesy of SEV Corporation.

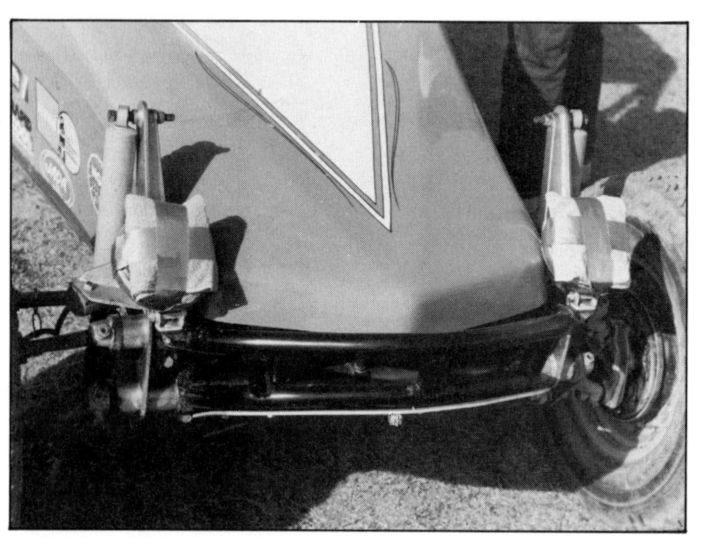

Someone stole your light covers? Well how about a couple of towels from the motel and some racer's tape. Return the towels by checkout time, please. Photo by Ron Sessions.

from conventional non-halogen incandescent lamps is more yellow-white. All else being equal, halogen lights draw no more current than non-halogen lights.

The only drawback with halogen lamps, besides cost, is that they typically don't last as long as non-halogen types. But in severe off-road use, vibration and rocks make this a moot point anyway. Regardless of lamp construction, most off-road driving lights don't last as long as lights used on smooth pavement because the bouncing around breaks the filament. Even on smooth pavement, most driving lights have a maximum service life of about 150 hours.

Before you go out and buy a set of the nastiest flame-throwers you can find, make sure your car's generator or alternator has enough output to handle them. See the electrical-system section starting on page 110 for amperage-per-light recommendations and generating-system capabilities.

Tungsten-Halogen Quirks—One thing to keep in mind when using halogen lights: Never touch the glass part of the replaceable bulb with your fingers! Your skin will leave a slight trace of body oil, causing the element to fail prematurely.

Fingerprint smudges inhibit heat from radiating from the bulb, causing the element to overheat. These smudges also form a greasy film on the reflector, reducing its effectiveness. Always hold a new bulb in a cloth or paper towel when installing it. If you do get your greasy paws on a bulb, clean it with alcohol before turning on the light.

Also, the halogen lamp won't tolerate an overcharge of current—*voltage surge*—from the generating system if the voltage regulator malfunctions. A tungsten element in the conventional sealed beam often will. Why? Because the tungsten element in a halogen lamp is thinner than that used in a conventional non-halogen bulb. With less mass to begin with, the halogen element is more easily vaporized by a power surge that momentarily overheats it. All the halogen gas in the world can't redeposit vaporized tungsten back onto a filament that's melted down.

Lenses—The lens "bends" or *refracts* the light beam into the shape that determines the directional characteristics of the light. Basically, it's the lens that determines whether the light is a pencil-beam spot, driving beam—a true driving light—or a fog light.

Stay away from pencil-beam spots because they are only worth their money on long, fast, straight roads. They can do more harm than good on rough and curving roads because they offer no *peripheral*—side—lighting. The pencil beam is long and narrow.

You're better off sticking to a driving light that gives you good distance, but also lights up the area in front of and to the sides of your car. Carefully check the beam-pattern illustrations in the manufacturer's catalog before choosing the light that's right for you.

How Many Lights Do You Need?—Good question. Depending on your needs, you may want to use two or more lights. Start with two and see how they work. Add more as necessary. Racers often run four driving lights and two fog lights, although rarely all at once. Two of the driving lights are used as spares. Overkill is waste, and it overworks the electrical system. Remember, it takes engine power and fuel to run the generator or alternator. Wiring these lights to switch on individually or in pairs keeps your options open.

Other Considerations—Depending on the manufacturer, it's usually less expensive to replace a driving-light bulb than it is to replace a sealed-beam driving light. From this standpoint, halogen lamps with replaceable elements look very good, indeed. However, consider the total cost. Bulbs aren't the only things that go bump in the night. In off-road use, lenses are the first thing a rock sees. Consider their replacement cost as well, and make sure they're available.

You may want to look into using clear plastic covers over the lenses. These are made from plexiglass, Lucite or Lexan, and are available from some off-road accessory suppliers.

If you're racing, consider this, too. If a rock hits a sealed beam, the light will go out because the filament will

A popular location to mount driving lights is atop a sedan's roof or buggy's roll bar. Always cover lights not being used. If you don't have light covers, be resourceful. This Class-11 racer taped Frisbees over his lights, padded by kitchen sponges. Photo by Ron Sessions.

One way to control glare is with shields. Make sure positioning of shields does not create blind spot. Photo by Ron Sessions.

vaporize the instant oxygen reaches it. A tungsten filament heated to incandescence can only operate in a vacuum or inert-gas atmosphere. But if the lens breaks on a replaceable-bulb light, the bulb will continue to shine unless it, too, is hit by the rock.

Fog Lights—In some parts of the country, fog and mist can seriously obstruct vision. These conditions require the use of fog lights. Likewise, fog lights are also effective in dust.

Unlike driving lights, fog lights should be mounted low—bumper height or lower. Because they will be vulnerable in this location, always keep them covered until needed.

Fog lights differ from other driving lights in how the lens is made. A quality fog-light lens has an intricate configuration of prisms, with scores of different cuts to control glare. On some less-expensive fog lights, the outside of the lens is shuttered. The shutter shades direct light aimed upward, reducing glare above the road caused by light bouncing off water or dust particles.

Regardless of whether a shutter or special lens is used to control glare, the idea of a fog light is to light the particles *from underneath.*

Most fog lights are available with clear or amber lenses. Although there is no major difference between the performance of these two lenses, the amber light creates less glare and is more restful to the eye. The clear lens puts out slightly more light.

MOUNTING

Roof Mount—There seems to be no consensus on driving-light mounting locations. One of the most-popular spots to mount lights is on the roof of a Baja Bug or atop the roll cage on a buggy. The main advantage of this location is that the light won't be blocked by the terrain. The lights are up out of the way from brush and rocks and probably will be damaged only by a rollover.

Compared to the body damage from a rollover, the loss of a couple of lights is of small concern to the Baja Bug or sedan owner who does casual off-roading. However, a buggy, which *should* suffer little or no damage in a slow rollover, will smash some expensive driving lights if they are mounted on the roof. If you're going to race, you could be out of business should this happen at night.

Glare—Before you mount the lights on the roof, make sure you take provisions to reduce glare. I always get a kick out of the guy with a plastic sun roof in his pickup truck and all his off-road lights mounted atop the roll bar. You just know if he ever turns them on at night, he won't be able to see out the windows.

A similar problem occurs when you mount bright lights right above the windshield of a bug or sedan. The glare from the lights reflects off the glass, off the dirt on the glass, and off dust in the air. This will definitely hamper night vision.

Sedan cowl/door-post mount puts lights up and more out of the way from brush and other race cars.

Sheet-metal shields that fit on the underside of the lights will correct this to a certain degree. Although the lights are at the side, the buggy pictured on this page shows an effective use of shields. If you insist on mounting the lights on the roof or top of the roll cage, use light shields mounted horizontally under the lights, combined with a flat-black hood. The black hood will further reduce glare.

Cowl/Door Post Mount—A lot of off-roaders mount their lights on the sides, on or near the top door hinge on a Baja Bug, or low on the front roll-bar hoop on a buggy. Lights in this location are less prone to rollover

This area gives some protection and little glare but "flicker" can be a problem. Don't mount lights on the front fenders unless Baja kit is sturdy enough to support them.

The easiest and best way to add off-road lights is with readymade side mounts. Compare mount used with mount on front fender. Light is not on fender because it vibrated, causing flicker.

This bumper mount puts the light where it should be, behind the fiberglass where it has some protection. Mount the light to the frame or front bumper. Photo by Ron Sessions.

damage and cause less glare. Use shields to reduce glare.

Fender Mount—On a sedan or Baja Bug, another place for driving lights is atop the front fenders. The problem is that many fiberglass fenders aren't rigid enough to support the lights. The flicker from lights mounted on wobbly fenders can be very annoying. This is probably the reason this setup is seldom used. If you do mount lights on fiberglass fenders, use stabilizing struts to control flicker.

Bumper Mount—The best spot, and the one I recommend, is to mount driving lights in place of the stock headlights on a bug, or right above or in behind the front bumper. They can also go on a buggy's shock towers. Lights in these locations produce much less glare. Low-mounted lights also give more illumination around the immediate area in front of the car. The problem with lights mounted at the front is giving them enough protection. The best way to protect them is to recess them inside the body.

Protective Covering—Almost all lights come with either plastic or cloth covers. A good way to increase protection for the lights is to cut several pieces of cardboard the same size as the lens and put them between the lens and cover. Heck, I've even seen upholstery foam, shop rags, and Frisbees taped over driving lights. Use whatever's handy. A good rule is to cover the lights when not in use. Violate this rule, and a stone will find your light before its time. Just make sure the lights are uncovered *before* they're turned on.

Rubber Mounts—One of the more popular off-road lights comes with a durable rubber mount. The idea behind this is to lessen the shock on the tungsten filament and increase its life. Some racers don't like to use these lights because they claim the rubber mount allows the light to vibrate or flutter just enough to bother their vision. But, if your needs involve getting more life out of your lights, consider them.

Wire Them Individually—While racing off-road, if a light gets knocked out of line, you'll sometimes find it easier to drive if you turn that offending light off so it won't distract you. Consequently, it's a good idea to wire each light or pair of lights through a separate switch. Label the switches.

With high-draw lights, use a relay between each switch and the light(s) that switch operates. A relay reduces the electrical load at the switch, thereby reducing switch arcing. Because the relay can be mounted close to the lights, resistance is reduced, giving full voltage to the lights.

Baja Bug/Sedan Considerations—If you have a Baja-Bug kit, it's a good idea to install stock-VW headlights in the holes provided for them in the fenders or nosepiece. Even if your bug is not street legal because it has high-intensity lights, no license or muffler, you'll probably end up at one time or another having to drive down the highway to get home, or to go for some help in case of a breakdown. At night, the chances of visiting with the Highway Patrol ranges from almost nil with stock headlights to good with roof- or cowl-mounted, high-powered driving lights.

Driving Lights and The Law—Many of you drive your off-road car to work everyday, maybe getting out in the dirt on weekends only. If your car is primarily used on the street, I highly recommend checking with the local law-enforcement authorities before mounting driving lights. These laws vary widely from state to state. In some states, it's OK to have roof, cowl or fender-mounted lights on a street-driven car as long as the lights are covered or disconnected.

The idea here is to prevent you from using your flamethrowers in a high-beam duel with an oncoming car. The theory is if you have to stop and get out to remove the covers or connect the wires to the lights, the approaching car will have long since passed. You may find the Light Mounting Chart on the next page to be helpful.

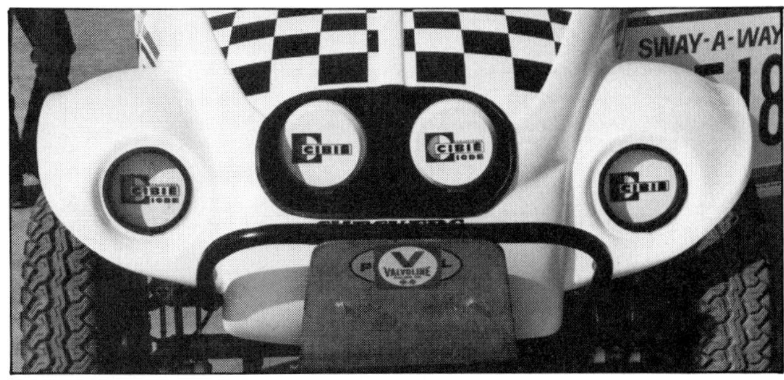

In my opinion, the ideal off-road setup. Every car manufacturer locates headlights here. The same reasoning applies off-road. Lights are up front with no glare, and low for maximum lighting at the front of the car. Photo by Ron Sessions.

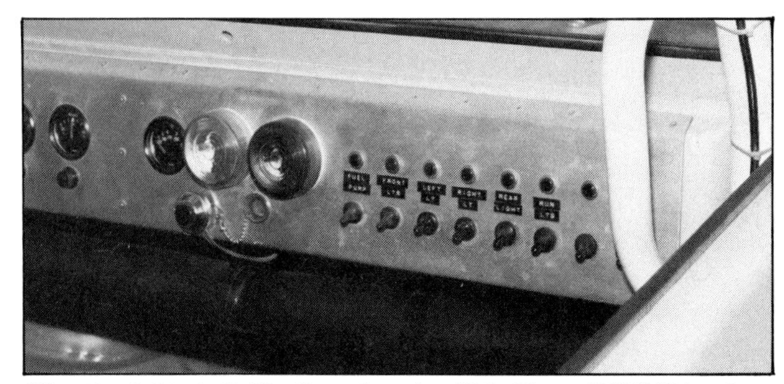

Wire the lights individually or in pairs. This Class-5 VW Thing has all switches neatly labled. Photo by Ron Sessions.

Light-Mounting Recommendations

This chart indicates the mounting specifications preferred by the AAMVA (American Association of Motor Vehicle Administrators) and should be consulted prior to mounting auxiliary lights. Chart courtesy of KC HiLites, Inc.

	Spot Lamps	Fog Lamps		Aux. Passing Lamps		Aux. Driving Lamps	
State	Max. No.	Max. No.	Mtg. Hgt. (in.)	Max. No.	Mtg. Hgt. (in.)	Max. No.	Mtg. Hgt. (in.)
Alabama	1	2	12-30	1	24-42	1	16-42
Alaska	2	2	12-30	2	24-42	2	16-42
Arizona	1	2	12-30	2	24-42	2	16-42
Arkansas	2	2	12-30	2	24-42	2	16-42
California	2(1)	2	12-30	2	24-42	2	16-42
Colorado	2	2	12-30	2	20-42	2	16-42
Connecticut	2	2	12-30	2	24-42	2	16-42
Delaware	2	2	12-30	1	24-42	1	16-42
D.C.	2	2	12-30	2	24-42	2	16-42
Florida	1	2	12-30	2	24-42	3	12-42
Georgia	1	2	12-30	1	24-42	1	16-42
Hawaii	—	—	—	—	—	3	12-42
Idaho	2	2	12-30	2	24-42	2	16-42
Illnois	1	—	—	—	—	3	12-42
Indiana	2	2	12-30	1	24-42	1	16-42
Iowa	1	—	—	—	—	3	12-42
Kansas	2	2	12-30	2	24-42	2	16-42
Kentucky	—	—	—	—	—	—	—
Louisiana	Prohibited	2	12-30	2	24-42	2	16-42
Maine	1	2(2)	—	2(2)	—	2(2)	—
Maryland	1	2	12-30	—	—	2	16-42
Massachusetts	1	—	—	—	—	—	—
Michigan	2	—	—	—	—	2	24 Min.
Minnesota	2	2	12-30	2	24-42	Prohibited	—
Mississippi	1	—	—	—	—	2	24 Min.
Missouri	1	—	—	—	—	3	12-42
Montana	2	2	12-30	2	24-42	2	16-42
Nebraska	1	—	—	—	—	2	24 Min.
Nevada	2	—	—	—	—	2	16-42
New Hampshire	2	—	—	—	—	3	12-42
New Jersey	1	—	—	—	—	Prohibited	
New Mexico	2	2	12-30	1	24-42	1	16-42
New York	—	—	—	—	—	—	—
North Carolina	2	—	—	—	—	2	—
North Dakota	2	2	12-30	2	24-42	2	16-42
Ohio	1	2	12-30	2	24-42	2	16-42
Oklahoma	2(3)	2	(4)	—	—	—	—
Oregon	1	—	—	—	—	3	12-42
Pennsylvania	1	—	—	—	—	Prohibited	
Rhode Island	2	—	—	—	—	2	(4)
South Carolina	1	2	12-30	1	24-42	1	16-42
South Dakota	1	—	—	—	—	3	12-42
Tennessee	—	—	—	—	—	—	—
Texas	2	2	12-30	2	24-42	2	16-42
Utah	2	2	12-30	—	—	—	—
Vermont	—	—	—	—	—	—	—
Virginia	2	2	—	1	—	1	—
Washington	2	2	12-30	2	24-42	1	16-42
West Virginia	1	2	12-30	1	24-42	1	16-42
Wisconsin	2(3)	2	(4)	—	—	—	—
Wyoming	2	2	12-30	1	24-42	1	16-42

(1) Cannot exceed 32cp or 30 watts
(2) Total of two fog or aux. lamps permitted
(3) Must be mounted at height between 30- and 72-in.
(4) Below headlamp centers

Even on a simple off-road buggy, electrical wiring can become a jumbled mess. This dash wiring is a clean installation. Photo by Ron Sessions.

Probably the least-understood area, and the cause of most headaches for the off-roader, is the electrical system. No matter how hard one tries, it's impossible to see those little electrons dancing through the wires.

If you're building a tube-frame buggy from the ground up, you have the problem, or opportunity, of wiring it from scratch. It's a problem if you don't know what you're doing when it comes to things electrical.

Fortunately, basic wiring kits with instructions are available to guide you through the procedure. Many of these kits include suggested routings, handy cable ties, a variety of connectors, a fusebox, 10-, 12- and 16-gage color-coded wire, and even a roll of electrical tape to make the job as easy as possible.

Wiring a car from scratch is really an opportunity to select components according to your car's needs. And you can use the more desirable 12-volt components from the start.

If you're modifying a sedan, the big question is how much of the original wiring and components are salvageable or needed. As a VW owner, you're fortunate that your car is simple and easy to work on. When you do have trouble out in the dirt, you're not too concerned about such things as a cigarette lighter, cruise control or the dome light. Therefore, you can concentrate on the parts and wires that concern basic operation.

I'll cover the early 6-volt systems first so I can devote more space to converting to and improving the 12-volt system.

A Little History—In case you're unfamiliar with VW history, here's a bit of background. VW switched from a 6-volt to a 12-volt electrical system on the '67 sedan. Because many dune buggies and Baja Bugs are based on '66-and-earlier components, many owners must deal with an obsolete 6-volt electrical system.

In all fairness, the 6-volt systems did the job when they were new, and would work fine today if they weren't at least 15 years old. Time takes its toll on wires, insulation and electrical connectors. High electrical loads from driving lights require the wiring to be up to the task.

Breathing Life Into the 6-Volt System—If your car is an early 6-volt VW that's used for basic dirt-road transportation only and you don't want to spend the time and money to convert it to 12 volts, here are some parts that improve its performance.

A big help is a starter relay. Use a simple Ford-type relay that bypasses part of the electrical system—except

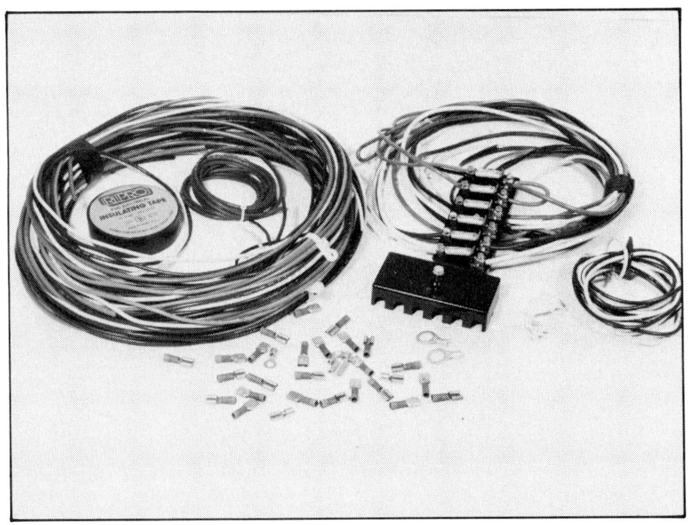

If you're working from scratch, basic wiring kits can get you started. This Bugpack kit includes enough color-coded wire, connectors, a fusebox, instructions and even electrical tape to wire any buggy to street-legal specifications. Photo by Ron Sessions.

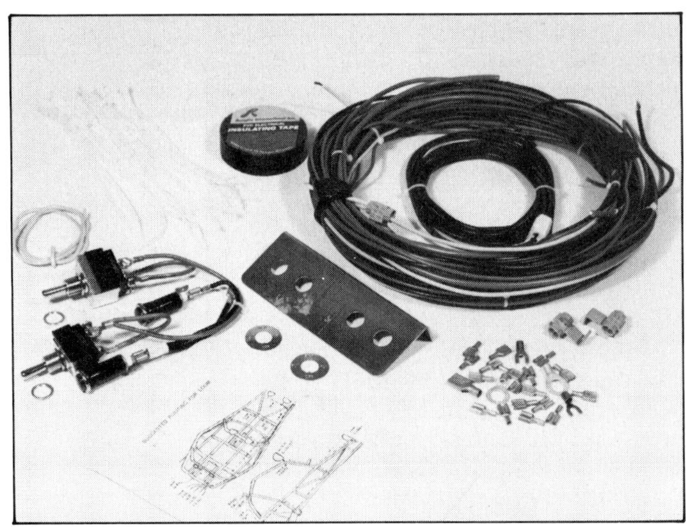

Another wiring kit from Bugpack. This one has a loom for a "pre-wired" dash panel and driving lights. Photo by Ron Sessions.

Ford-type starter relay sends as much juice as possible directly to starter by bypassing rest of electrical system. Handy item when you only have 6 volts to work with.

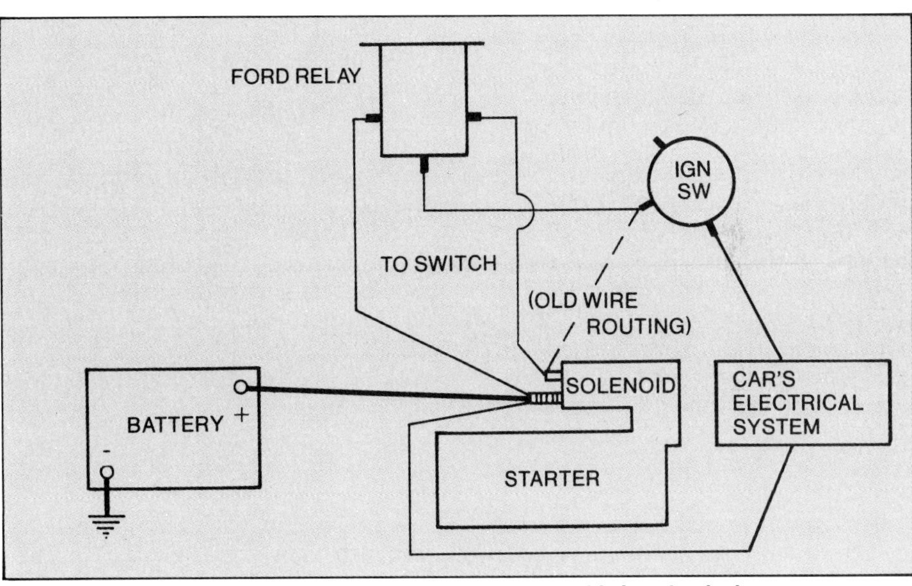

Wiring diagram for installing Ford-type relay. Mount solenoid close to starter.

for the lights—when the ignition switch is turned to the start position. Using the accompanying drawing as a guide, wire the relay between the ignition switch and starter solenoid. These are available from all Ford dealerships and most auto-parts stores.

Next, replace the 10- or 12-gage wire between the ignition switch and the starter solenoid, and the 16-gage wire between the ignition switch and coil. If in doubt, match the new wires to the old ones for length and gage.

Clean and tighten all connections in those circuits. Don't forget the battery connections and the positive terminal at the starter solenoid. Put a blob of silicone on these connections as insurance against them shaking loose. It also provides good corrosion protection at the terminal.

I recommend replacing wires for two reasons. First, as wires age, they weaken at their terminals and may corrode. This increases resistance. Second, insulation dries out, splits and cracks, exposing bare wires just begging to be short-circuited. These problems apply to 12-volt systems too. But, they are more critical for 6-volt systems, which must draw about twice as much current to do the same amount of work.

Even after upgrading a 6-volt system, you should not try to operate extra driving lights or accessories. These items will seriously overtax the 6-volt generator and regulator.

Also, with a 6-volt system, if you have an engine problem that requires much cranking, you won't have as much time before the battery dies. Once again, the decision to keep the 6-volt system depends on how you intend to use the vehicle. If you don't intend to venture too far off the beaten path, and don't plan to race the car, a good 6-volt system is fine.

6-TO-12-VOLT CONVERSION

Converting a 6-volt system to 12 volts is not an easy job, but it's well worth the time and money. This is another area where there is some difference of opinion. I'll offer some alternatives.

The things you'll need are a 12-volt battery—less expensive than a 6-volt anyway—a 12-volt generator or alternator and fan cover, 12-volt coil and 12-volt voltage regulator. It's possible to rewind a 6-volt generator for 12 volts, but it's not worth the time and effort.

If the following items are retained, you must convert them to 12 volts: headlights, a turn-signal flasher, bulbs for tail and stop lights, turn signals, dome light, license-plate light and instrument and indicator lights.

Depending on how much money you want to save initially, 6-volt parts that may be retained include the starter, radio and wiper motor. A 6-volt horn and fuel gage will work just fine on 12 volts. The horn will sound louder than normal. The 6-volt fuel gage will work on 12 volts because it uses its own voltage regulator to maintain voltage to the bimetal spring.

Starter—On a play buggy or street-driven Bug, the cheapest and easiest way to deal with a 6-volt starter is to leave it in and use it. With 12 volts coursing through its veins, it will turn the engine over much faster. If you only crank it for short periods the starter won't get very hot, so it will last a long time. I used a 6-volt starter on my pre-run car for more than two years. When I sold the car, the starter was working fine.

If you want to make sure the 6-volt starter-switch circuit will survive, replace its solenoid with a 12-volt type. If you don't replace the solenoid, you'll need to add the Ford-type relay previously described. The relay will supply power to the 6-volt starter solenoid. Otherwise, expect a behind-the-dash electrical fire in the starter-switch circuit.

There are problems that might occur from using a 6-volt starter with a 12-volt battery: starter failure and flywheel damage. You're feeding a 6-volt starter 12 volts, so there's always the chance that it will fail due to excess electrical and mechanical loads.

The first thing that is likely to fail is the 6-volt solenoid. If you have to crank the engine for a long time, the 6-volt solenoid will overheat and seize. If the solenoid seizes so the starter drive is partially engaged, the

Four bronze starter bushings. At bottom is stock 6-volt bushing. Right one is stock 12-volt bushing. At top is stock 6-volt bushing and special sleeve inside it to neck down bushing ID when switching from 6-volt starter to 12-volt.

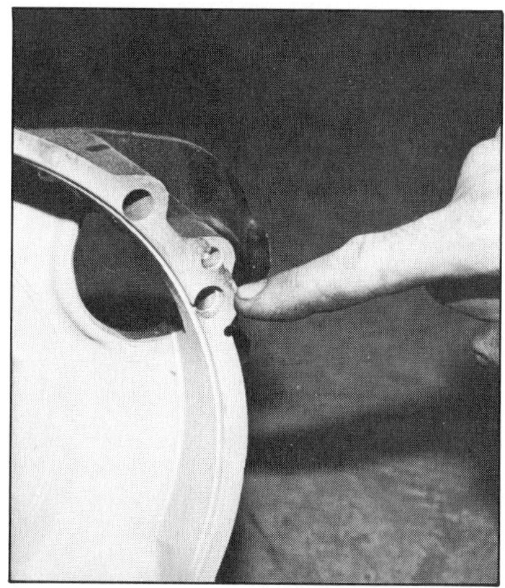

Starter bushing goes here. It is installed from opposite side. With engine installed, bushing can be removed only with threaded bushing puller. With engine removed, bushing can be driven out.

teeth on the flywheel may also be damaged. If you hear a grinding noise while cranking the engine, stop. That noise may be the starter-drive pinion grinding flat spots in the flywheel ring gear.

So, what do you do? If you're on a hill or incline, try starting the car by popping the clutch on the roll. If you suspect that some of the ring-gear teeth are milled off, but the starter is working fine, push or roll the car forward or backward with 4th gear engaged. This will rotate the flywheel, perhaps to a section of undamaged ring-gear teeth.

If you're out with another car, try starting your car by having the other car push- or pull-start you. If you're all alone and don't want to leave the car until you can repair or replace the starter, grit your teeth and try using the starter again. Sometimes it'll work properly and you can get home.

When a 6-volt starter is energized by 12 volts, it engages the flywheel with more "snap." Over a period of time, the flywheel ring gear will wear on the leading edge of its teeth. Eventually, this will require a new flywheel.

VW doesn't use a separate ring gear on the flywheel, so any ring-gear damage requires replacing the flywheel. But the alternative is buying a new 12-volt starter to begin with, so why not use the one you have until it fails? If it fails, pull the engine and install a 12-volt starter and flywheel.

Again, I wouldn't run this setup on my race car, but I wouldn't hesitate on a pre-runner or a play car. So you be the judge. If you decide to install a 12-volt starter, there are a few other pieces you'll need—a starter bushing and flywheel. You'll also need a clutch to go with the 12-volt flywheel.

Starter Bushing—The 12-volt starter drive has a smaller-diameter shaft than its 6-volt counterpart—0.43 in. vs 0.49 in. Therefore a special sleeve must be used inside the old bushing, or a smaller-ID bushing substituted for the 6-volt bushing. This bronze bushing installs with a press-fit in the transaxle bellhousing.

The sleeve saves you the trouble of removing the original bushing. But if the bushing is worn out, you'll have to replace it. Bushings and sleeves are available through most VW parts outlets for a few dollars. There's a special tool for extracting the old bushing in-car. If you can't find a shop that has this tool, you'll have to pull the engine to change the bushing.

Obviously, the time to make the decision on what electrical system and starter to use is when you're building the car.

Flywheel—The 6-volt starter is used with a 180mm, 109-tooth flywheel. The 12-volt starter is used with a 200mm, 130-tooth flywheel. When you switch to a 12-volt starter, the "6-volt" flywheel has to go. Also, the clutch used with the 180mm flywheel won't fit a 200mm flywheel. So add the cost of a 200mm flywheel and clutch to your conversion.

Yes Martha, there's mud in the desert. The owner of this Baja Bug will need all the action his wiper motor can generate. Photo by Ron Sessions.

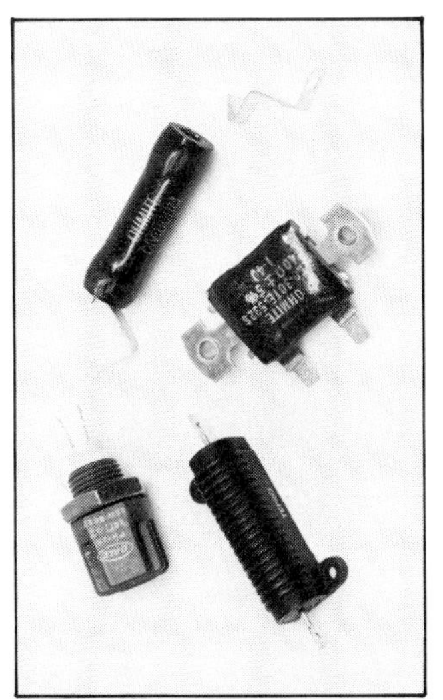

Dropping resistors. Most of us have never seen one before. They can be used to drop voltage from 12 to 6 volts. Make sure you get one with the right specs. Photo by Ron Sessions.

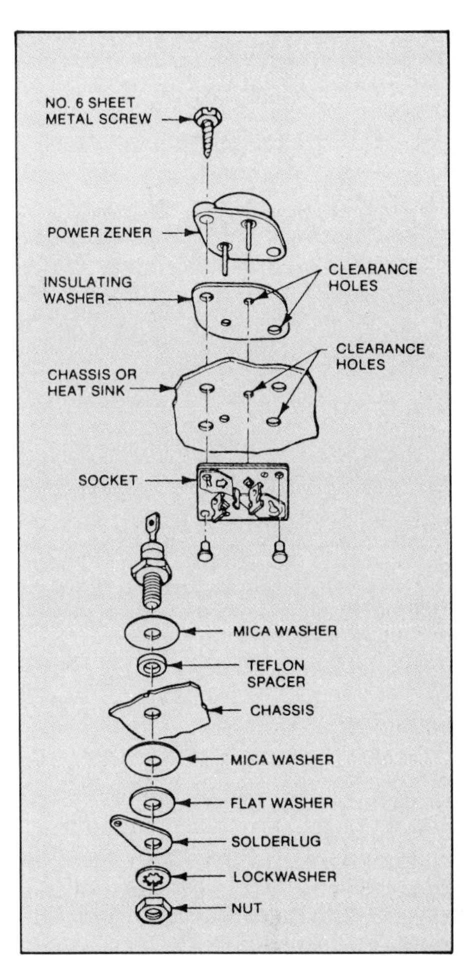

Zeners must be very carefully mounted if they are to work and last. Mica washers insulate against shorting yet allow mounting on metal for good heat transfer.

Ignition Coil—Using a 6-volt ignition coil with a 12-volt battery will cause prematurely burned ignition points. When converting a 6-volt car to 12 volts, always replace the 6-volt coil with 12-volt type.

Radio and Wiper Motor—Retaining the 6-volt radio and wiper motor in your '66-or-earlier sedan converted to 12-volt operation is another story. A purist would say, "Replace them with 12-volt units." This is obviously the right way to do it, if you have the money.

Surprisingly enough, a windshield wiper has a number of uses, even in the desert. An early morning mist or a splash from a stream crossing or mud puddle can make a sedan's windshield hard to see through. A few moments of wiper operation can remedy the situation. Under these conditions, a 6-volt wiper motor, used sparingly, may last a long time.

Also, when 12 volts are applied to a 6-volt wiper motor, it will run at twice its design speed. In fact, it will run so fast that it may sweep past its park switch and *not shut off.* Even if it does shut off, running it at near double its design speed will dramatically shorten its life.

What to do about a radio depends on your plans for the car. If you're building a racer, a radio is excess baggage. The dust and vibration encountered in serious off-roading is tough on a radio, especially if yours is an old 6-volt tube-type. For a street-legal Baja Bug, a radio is very desirable. If you use the car every day, it might be worth your while to invest in a new 12-volt radio with the features you want—FM, tape deck, CB and so on.

The Resistor Solution—If the plans for your car fall somewhere between serious off-road and occasional off-road, here's a compromise. Many auto-parts, electronic-surplus and radio-repair shops, carry *voltage-drop* resistors.

A 4.5- or 5-ohm, 20- or 25-watt resistor will permit most 6-volt wiper motors to operate on 12 volts. And a 7.5-ohm, 10-watt resistor will enable you to keep the original 6-volt radio—if it's a transistor type. If the original radio is a tube type, only a big, expensive resistor will make it work. You may find it cheaper and easier to get a 12-volt radio.

When installing a resistor, wire it in *series*—directly in-line between the battery or control switch and the component. It will get hot, so don't crowd it. Resistors range from a few dollars to well over $10, depending on quality. The best resistors maintain a constant voltage within a few percentage points.

For most of us, off-roading is a fair-weather sport. But there is one situation where dropping resistors exhibit sub-par performance. In cold weather, a wiper motor is under an increased load, causing it to draw more current. As it draws more current, electrical resistance goes up. Combined with the resistance of the dropping resistor, voltage at the wiper motor drops *below* 6 volts. The result is a sluggish wiper motor.

This can really be a nuisance if the wiper has to sweep away snow. If you drive your VW off-road car in sub-zero weather, this nuisance may be enough to cause you to look into another way to drop voltage—with a zener diode.

The Zener-Diode Solution—A zener is a special-function silicon diode used to drop voltage. Compared to a dropping resistor, a solid-state zener is less susceptible to a voltage change with temperature change. It's just the thing for dropping voltage to the wiper motor of an Eskimo's Baja Bug.

Zener diodes are expensive and difficult to install. A dropping resistor doesn't care what direction electricity flows through it; a zener works in only *one* direction. Hook up a zener backwards and it'll burn up the instant the switch is turned on.

An alternator will produce over 45 amps, even at idle. Could be just the ticket for running those flame-throwers your generator can't handle.

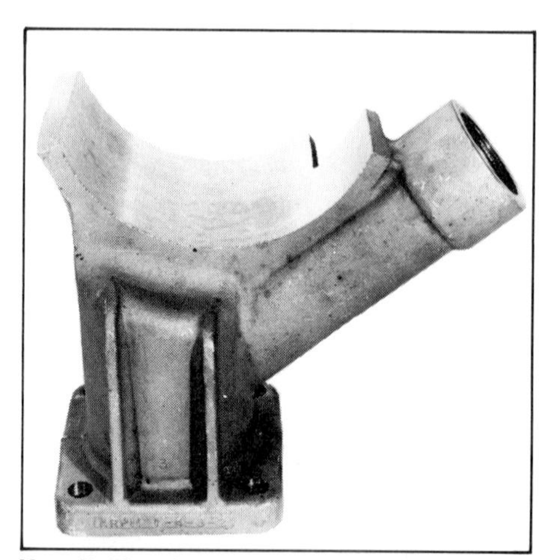

If switching from a 6-volt generator to a 12-volt type, you'll probably have to change the generator stand. This is a 12-volt stand. Photo courtesy of Bugpack.

Sealed switch. Dirt and dust are sealed out, keeping the switch contacts clean. Photo courtesy of Bugpack.

Also, a zener must be mounted to a solid surface to provide an adequate heat sink. It must not make electrical contact with its mounting surface so it must be insulated with special mica washers. For these reasons, dropping voltage with a zener diode is best left to an expert.

GENERATOR/ALTERNATOR

VW has used both generators and alternators for the charging system. I'll cover generators first.

Generator—A VW-sedan generator puts out about 30 amps—180 or 200 watts for 6-volt versions and 360 watts for 12-volt versions.

The ignition will draw about two amps. Off-road lights, in the normal 150,000- to 200,000-candlepower range, will draw about 7—12 amps each. To be safe, the stock generator should not be asked to handle more than two driving lights along with the engine, taillights, dash lights, and so on. For more output, use the '69—'71 VW-bus generator 211 903 031 D. It is rated at 38 amps.

Remember that generators come in different ODs. If you're swapping generators—6-volt to 12-volt—expect to swap generator stands as well in most cases. One exception is the 12-volt generator for the early Porsche 912, Bosch GR-26. It will fit the VW 6-volt generator stand. But it will also carry a Porsche price tag.

Alternator—To get more than 38 amps, use an alternator. Alternators became standard equipment in all Type-1 sedans beginning in mid-'73. VW alternators come in 45-, 50- and 55-amp ratings. Some have *integral*—built-in—voltage regulators and some have remote regulators.

When swapping the car's original generator for an alternator, don't use the old regulator. It won't work. Get a new regulator to go with the alternator. You'll also need a new alternator stand or universal stand, which bolts in place of the generator stand on '61-or-later models.

In addition to higher output, alternators also have the benefit of producing full output at idle. And because the brushes don't carry full field-winding load, brush life is nearly double that of a generator. Complete alternator kits cost well over $100.

Voltage Regulator—Both 6- and 12-volt generators and '73—'74 alternators use *electro-mechanical* voltage regulators. These have a set of contacts that open and close to regulate voltage. Unfortunately, when a car bounces around off-road, the critical regulator settings can get knocked out of adjustment. Dust contamination between the point contacts is another problem. For this reason, carry a spare regulator. All racers do.

Beginning in late '74, VW used alternators with an integral, solid-state regulator. Its advantage is no moving parts, so it's more reliable than the point-type regulator.

CONTROLS

Switches—Control switches can be toggle, rocker or push-button types; I prefer toggle switches. Regardless of the type selected, use good-quality, *sealed* switches. If the switches aren't sealed, dust will get to the contacts and prevent them from working. You can get sealed switches from most off-road parts houses for under $10. A second, more expensive choice is brass marine switches.

If you select a non-sealed switch, mount the switch so the knob or toggle is slanted slightly downward. Seal *all* but the toggle with silicone to keep dirt and water away from the contacts. Mount the switches within easy reach of the driver.

Slipping a 3—4-in. length of rubber hose over the end of the toggle is an easy way to make a frequently used toggle switch easier to find in a hurry. Many racers use such a toggle switch for the horn. Flick your wrist one way and the horn goes on. It stays on until you swat the switch the other way.

INSTRUMENTS

An odometer is a handy tool for figuring cruising range. If you haven't filled the hole in the left spindle with a bolt to strengthen it, keep the stock VW speedometer. The VW speedo will work fine off-road. Large OD off-road front tires will make the speedometer and odometer slightly pessimistic, but once you've factored in the error, it should present no problem.

This on- and off-road Baja Bug features complete instrumentation plus two small idiot lights behind the steering wheel. VDO tachometer is made specially for off-road race cars. Aluminum instrument panel is by Unique Metal Products.

Instrument panel of this competition Baja Bug makes trouble hard to ignore. Two big clearance lights, shaded from the sun, are first indicators of trouble. Gages are backup units. Horn switch at right can be activated by a slap of the hand. Quick push of the knob below floods engine and driver's compartments with Halon-13 fire-extinguishing agent.

If you filled the speedometer hole in the spindle or used an aftermarket spindle, forget using a speedometer. There is no other area on the car—such as the transaxle—that can be easily modified to mount a speedometer drive gear.

A tachometer may be impressive but is virtually useless off-road. In most units, the tach needle bounces too much to give an accurate reading. Off-road driving doesn't require running the engine consistently at or near its rev limit, so a tach isn't really needed. A good driver can tell when to shift by the pitch of the engine.

If you are really worried about overrevving the engine, use a simple rpm *limiting rotor*—centrifugal ignition cutout—in the distributor. Limiting rotors are available with cutout speeds ranging from 5400 rpm up to 7300 rpm.

The two most important signals you need come from the voltage regulator and the oil-pressure sending unit. Surprisingly, when you're bouncing around off-road, an ammeter and an oil-pressure gage are the poorest ways to detect trouble. You're more likely to feel any problem through the seat of your pants before you happen to glance down at a gage.

The importance of detecting an oil-pressure loss is obvious. Low or nonexistent oil pressure means serious engine damage. The importance of knowing if the charging system is working is not so obvious. You can drive a long way on the battery. But if a generator/alternator drive belt has been lost, you won't get far without the cooling fan operating. Again, serious damage will occur to the air-cooled engine if the problem is not detected quickly.

Idiot Lights, Bells and Whistles—The term *idiot light* refers to a warning light that goes on when the system it is monitoring becomes ineffective. Many automotive publications spend a lot of time bad-mouthing idiot lights and promoting "proper gages." But a bright yellow warning light indicating low oil pressure is a much better warning device than the needle of a gage pointing to zero. Chances are you'll feel the engine start to "go" before you notice the low gage reading.

Off-road, you need something to "grab your eye." Dust and vibration can often do in even the best-quality gages. And who has time to ponder whether the problem is oil pressure or the oil-pressure gage?

Enter idiot lights. Either there is enough generator/alternator output or oil pressure or there isn't. When something goes sour, one of the warning lights goes on. If small lights are good, big lights are better. Round clearance lights for trailers and RVs work well. They are large and bright. Get a yellow light for oil pressure and a red light for the charging system. Make sure the lights are shaded so you can see them in direct sunlight.

Wiring the Generator/Alternator Light—If you have a sedan with an existing generator/alternator warning light, merely disconnect the two wires from the back of the old light and connect them to the clearance light. If you're wiring a car from scratch, run one warning-light lead to the 15/54 terminal on the ignition switch and the other lead to the 61 terminal on the voltage regulator. Use 16-gage wire.

Electro-mechanical voltage regulators have contact points just like a distributor. Keep the contacts clean.

This connection must be made whether you use a "super" warning light or not. The generator/alternator requires this input to initiate the charging cycle. Some regulators are mounted on the generator/alternator and others are beneath the sedan's rear seat.

Wiring the Oil-Pressure Warning Light—If your car already has an oil-pressure warning light, transfer the two wires from it to the new warning light. If you're wiring the light from scratch, wire one lead to the 15/54

Dual oil-pressure sending unit. Little one goes to trailer-clearance warning light. Big one goes to oil-pressure gage. Photo by Ron Sessions.

Some short-course racers are weight-conscious. They use light snowmobile or garden-tractor batteries for juice. Photo by Ron Sessions.

terminal on the ignition switch. Wire the other lead to the oil-pressure sending unit on the engine case. Use 16-gage or larger wire.

Oil-Pressure Sending Unit—You can't indiscriminately wire a warning light to an oil-pressure sending unit. Oil-pressure gages and warning lights require different senders.

Ask yourself some questions: Do you want an oil-pressure gage? If not, is the sending unit presently installed on the engine for a warning light or a gage? If it's for a light and you want the light to come on when oil pressure drops to about 7 psi, the original VW sender can be retained.

If a gage was originally used, you'll have to replace the sender with a warning-light sending unit. These are available at most auto-parts stores.

If you want to use *both* a gage and a warning light, or if you want to use a light that indicates at higher oil pressure—recommended for racing applications—you must replace the sender. VDO makes a sender that operates both a warning light and a gage. Or you can "T-off" with VDO's 240850 adapter and install separate light and gage senders as shown in the accompanying photo. This is convenient because you can treat the two units separately.

Also available are senders that are calibrated or adjusted to switch a light on at "higher" low oil pressures. VDO offers one that switches on at 21 psi. Regardless of the instrument manufacturer, these usually have to be specially ordered.

Oil-Pressure Warning Horn—Some racers go one step further in the attention-getting department and wire a horn to the oil-pressure sending unit. Believe it or not, sometimes such a warning horn is almost inaudible, but you can feel it vibrating.

The easiest way to get a horn is from a car or truck at a junkyard. Any late-model car or truck will do as long as it has a 12-volt electrical system. If you use one of these single-wire horns, get the horn *relay* that goes with it. Without the relay, the horn will draw enough current to burn out the power wire.

Single-wire horns must be insulated from the VW chassis by rubber-mounting them. To complete the ground circuit, install a ground wire under a bolt head on the horn body and route it to the oil-pressure sending unit.

In essence, a relay is a mechanical switch that carries the load instead of the sending unit. Off-road, relays are prone to damage from dirt and vibration. If the relay points don't make good electrical contact, the horn won't blow. For this reason, use a warning horn *in addition to, not instead of* a warning light!

If you choose not to use a horn relay, make sure you use a two-wire horn—one that doesn't draw more than 0.1 amps. Phil Reusche of Phil's Inc. in Evanston, Illinois recommends that some motorcycle horns and fork-lift backup alarms draw small amounts of current and are suitable. Just make sure the horn you choose is loud enough to hear over a stinger exhaust! Mount it close enough so you can hear it.

BATTERY

As long as a car is working fine, little attention is paid to the battery. This changes quickly when an ignition-system problem occurs. About the second time you think it's fixed, the battery gives up and won't crank the engine. If you're on a hill you may be able to get rolling and pop the clutch to start the engine. But chances are you'll get stuck in a valley or on a flat area—Murphy's Law again. So, don't buy a "cheapie" battery. Get a good-quality battery: 50—75 amp/hours if it's a 12-volt, or 100—125 amp/hours for a 6-volt. You will be bucks ahead in the long run.

Some racers, especially those running the short-course circuit, are extremely weight conscious. A big, heavy battery is something to be avoided. Lightweight alternatives include garden-tractor, motorcycle and aircraft types. But the energy stored in a lead/acid storage battery is proportional to its weight. Street-driven Baja Bugs and weekend play buggies will

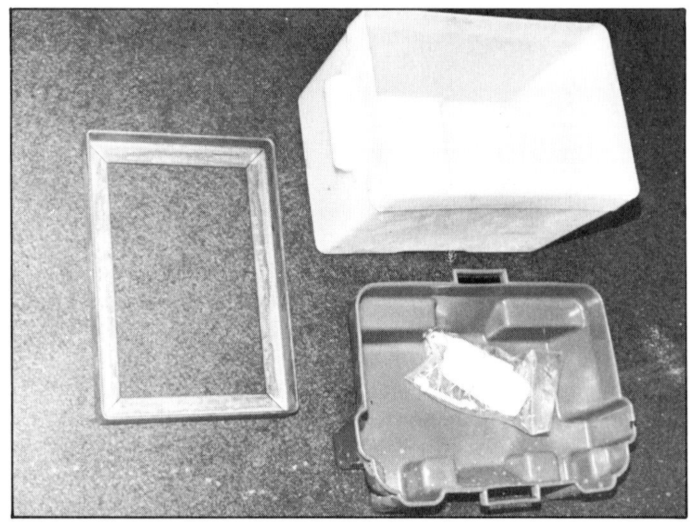

Mounting a battery is a simple job with a battery box and base. Just weld or bolt the metal base in and strap down the battery.

A friend of mine, with whom I won the Mint 400, builds off-road cars for a living. He won't shortcut on workmanship for himself or his customers. This is the kind of clean battery mount he insists on.

be much better off with the reserve capacity offered by a larger—and heavier—battery.

Battery Box/Hold-down—Always mount the battery securely. One way is to build a frame for the battery similar to that I describe for mounting a fuel tank, page 123. To hold the battery in its frame, fabricate a 1-in.-wide, 1/16-in.-thick metal strap that can be bolted or clamped.

Something else: I've just recommended a large-capacity battery for the non-racing off-roader. There's one catch. Make sure the battery is securely mounted. If you go to a larger-than-stock battery, it may not fit in the stock sedan battery tray. Remedy this by doing as I suggested for the racer—build a frame for the battery and secure it with a strap. You don't want to be ducking a battery in a rollover.

One of the first areas to get bent with off-road use is the VW floorpan. For this reason, serious racers move the battery up off the floorpan to a safer location.

If you leave the battery in its stock location, a piece of tire carcass, plywood or something similar should be put under the battery to protect it from sharp rocks, Also, if you retain the sedan rear seat and stock battery mounting, make sure the seat frame or springs do not contact the battery terminals. To minimize the fire hazard, make sure the plastic cover is over the positive terminal.

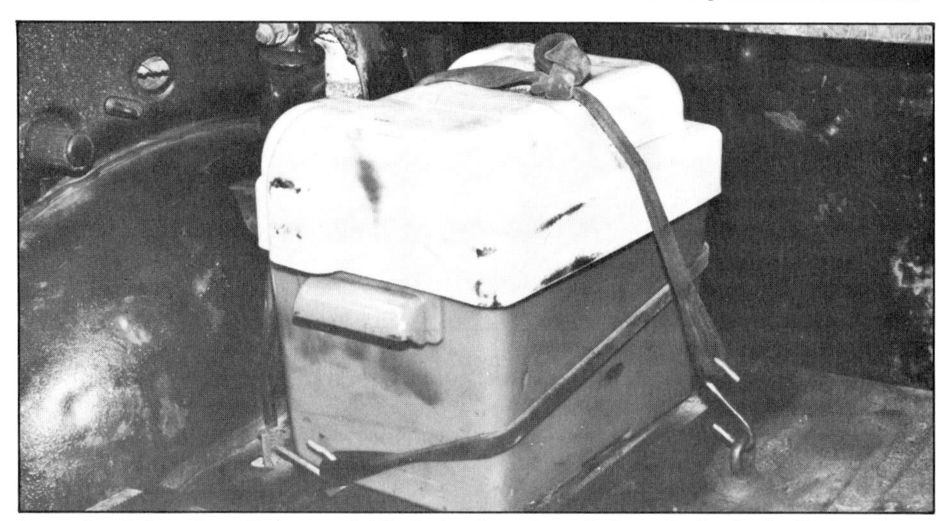

In contrast, here is a "hurry-up job"—trying to get ready for a last-minute fishing trip to Baja. This car has traveled more than 3000 hard off-road miles and that battery stayed put. Nevertheless, do it right. You don't want to worry about the battery coming loose in a rollover.

This brings up the final point about batteries. You don't want to be showered by battery acid if you get upside down. One way to avoid this is by using *anti-spill caps*, originally designed for aircraft use.

When the battery is upright, anti-spill caps vent like conventional battery caps. When the battery is turned upside down, valves in the caps shut the vent holes. This prevents acid from draining out of the battery and onto your head.

Another way to avoid spilled battery acid is to buy a no-maintenance battery. These batteries are vented internally, but are sealed to the atmosphere.

As added insurance against "acid rain," install the battery in a plastic battery box. You can get these at most boat-supply stores. Put the battery in the box, put the plastic lid on, and strap the battery and its box to the frame. Most boxes come with a nylon strap. See the photo.

Transaxle Ground—Even though a battery is grounded to the body, it must also be grounded to the engine. This is done with a ground strap. If you're modifying a sedan, it should already have one. If you're building a tube-frame buggy, you may have to install one.

If you retain the stock rubber engine/transaxle mounts, make sure the strap is attached. Otherwise, the engine will either not run or will ground through the throttle and clutch cables, and eventually weld one of them solid. If you use solid-metal mounts, a transaxle ground strap is not needed. The engine grounds through the solid mounts.

chapter 9 WHEELS & TIRES

Too much rear-tire pressure will cause the rear to bounce in the air like this. Too little air pressure and rocks can cut tires and bend wheels. Photo by Ron Sessions.

When building an off-road car, one of the first things you might think about buying is big, beefy off-road tires. Granted, a Baja Bug or buggy looks a little silly running around on those small, skinny stock Volkswagen tires at the rear. But before you run out and buy gigantic tires for the dirt, remember that bigger is not necessarily better. There is a good reason why VWs come stock with skinny tires. The section on rear tires tells you why.

The advantages of using wider, larger-diameter flotation tires in the dirt are significant. Besides giving more ground clearance, larger-diameter tires roll over smaller holes rather than dropping into them. They also provide a bigger cushion of air between the ground and tire rim for these bumps. This assists the suspension in absorbing shock loads, effectively increasing suspension performance and durability.

The trick is to put the right air pressure in the tires for the terrain. You'll risk bending the rims and cutting the tires with insufficient pressure. This may cause the wheel to spin inside the tire, yanking the stem from the tube—instant flat!

Overinflate them, and you won't be taking advantage of the additional cushioning the tires could provide. When a bump is hit hard enough to compress overinflated rear tires, the tires will kick the rear of the car high in the air. This is especially dangerous in a swing-axle car.

Bending a rim now and then is not all that bad. That's what hammers are for. I've seen cars run races with wheels that were crudely banged back into shapes that wouldn't win any beauty contests. One of the most-often-used tools in an off-road race pit is a small sledge hammer. It can be used to bend a dented rim back in shape, even with the wheel still mounted on the car.

FRONT TIRES

If you're not planning on racing, don't get talked into buying some trick off-road *front* tires. Any used mud and snow tire in the 7.75x15 size range with some tread left works fairly well. Because front tires are not loaded near as much as rear tires, you need be concerned with only two items: ride height and directional control. The taller the front tire, the greater the ground clearance and the easier the car will roll over smaller holes. The more "aggressive" the longitudinal ribbing on the tire, the greater its directional control.

A good-quality *4-ply tire* can resist puncture damage from rocks and cactus reasonably well. Always use 4-ply tires on both the front and rear, *not 4-ply-rated 2-ply tires.* Don't reject the use of blemished tires.

Although farm-implement tires offer excellent directional stability for off-road, they are very heavy and extremely hard to balance. *Never use implement tires on the street;* they are not rated for high-speed use.

An off-road front tire popular with racers for years is the 7.75x15 4-ply Sandblaster Jr., sold by Western Auto. However, it's rapidly being replaced by the 7.00x15 LT Tractionite. Built for mini-trucks, this is another Western Auto tire. Sandblaster Jr. is 27-in. tall with a 5-in.-wide tread. Tractionite is 28-1/2-in. tall with a 7-in.-wide tread.

REAR TIRES

Rear Tire Recommendations—What

Wheel and tire survived 400 miles of rough Nevada desert racing. Tire held air despite the tire and wheel damage. Stamped-aluminum wheel was put back into shape with a sledge hammer.

rear tires to use depends basically on the type of surface you'll be driving on. Available traction and types of hazards vary widely. In the arid Southwest, sand and jagged rocks are major considerations. In the North and East, it's mud, ruts and hard-packed clay. In the inland sand dunes and along the sea coast and Great Lakes, it's deep, loosely packed sand.

Many desert racers use the Western Auto 10.20 x 15 (33-in. tall) Sandblaster 915. This tire has good, but not excessive, traction because of its shallow tread pattern and rounded shoulders. This lets the tire spin and slide somewhat in the rough and in corners. This protects the axle shafts and transaxle against severe impact loads. A tire with lots of traction does not. Unfortunately, because they do a lot of spinning and sliding, Sandblasters wear faster than normal. Serious racers can rarely get more than two races from these tires. Those who can afford it start with new rear tires every race.

For running in the mud, such as in the Midwest and East, the best mud tires have deep, open-tread patterns with big, squared-off cleats and square shoulders. This gives the tire the bite that's needed in the wet stuff. The 10.00 x 15 Formula Desert Dog is a favorite due to its excellent traction. The only problem with this tire is the added stress it places on the driveline components.

Rear Tires As Another Gear—Rear tires act as another gear. The larger

The 7.75x15 Sandblaster Jr. is a popular off-road tire for the front. Aggressive longitudinal tread ribbing offers good directional control. Photo courtesy of Phil's Inc.

Another popular off-road tire is the 10/32-15 Dick Cepek Desert Racer. Note open tread pattern and rounded corners. Photo courtesy of Dick Cepek Inc.

the tire diameter, the "taller" the *effective* final gear. Tires have just as much effect on a car's high- and low-speed performance as the ring and pinion. With a stock engine and transaxle, super-large tires render 4th gear essentially useless on anything but long, level stretches of highway and downhill. The engine will not be able to pull even the slightest grade in 4th gear. You turn your off-road car into a vehicle with a three-speed transmission and a useless overdrive.

The Western Auto 913 Sandblaster. A similar 915 is a popular desert-racing tire for the rear. Both tires feature a shallow tread pattern and rounded corners to let tire spin in rough stuff.

Tectira front tire has similar tread pattern to Western Auto Tractionite. It is often carried as a spare because it is tall, yet sturdy enough to be used at rear wheel. Photo courtesy of Tectira Tire Co.

Worse yet, if you get caught in deep sand, it's possible that your engine won't be able to "pull" the "tall" first gear.

To solve this problem, you can do one of two things: Change the transaxle gearing or increase the engine's torque. If you don't want to do either, but want a usable 4th gear, you'll have to settle for a tire diameter not much larger than stock. A good way to determine whether your car will pull bigger tires is to borrow a pair

Dotted line and white paint highlight the tire-clearance problem areas in the rear fenderwell. There's a box-shaped area on the inner fenderwell above the torsion bar. Indent this area with a large hammer to gain an additional inch of tire clearance.

Distance from torsion-bar housing or fenderwell to center of axle shafts limits how tall a tire can be used. Double this measurement to determine maximum tire OD. Reduction-box setup here permits a tire about 38-in. tall without hitting torsion-bar housing (19 in. X 2 = 38 in.).

from a friend and try them out.

Sedan/Baja Bug Tire Clearance—On the sedan and Baja Bug, large rear tires present a clearance problem. A wide tire will hit a stock rear fender long before it contacts the inner fenderwell. On '58—'66 VWs with Baja-Bug fiberglass fenders, the inner fenderwell becomes the tire clearance problem. On the '57-and-earlier swing-axle and '68-and-later IRS setups, the problem is the longer torsion bar that sticks out through the spring-plate-hub cover.

On '58—'66 Bugs, the tallest tire you can use without hitting the inner fenderwell is about 29 in. OD. On '67—'68 swing-axle cars, or if you put a Type-3 swing-axle transaxle in your car, you can use a tire with an OD slightly over 30 in. This is because their axles and stub-shaft flanges position the tires farther outboard. To determine what size tire will work with early swing-axle cars and IRS cars, read on.

If you don't mind using a sledge hammer on your sedan, you can add about one inch of rear tire clearance. At the front of the rear inner fenderwell is a box-shaped area just above the spring-plate-hub cover. You can reshape this area for added clearance as shown.

To repeat, pre-'58 swing-axle cars and '68-and-later IRS VWs have long torsion bars that stick out and limit clearance for bigger tires. Measure from the center of the axle to the rearward edge of the torsion-bar tube to get the maximum allowable tire *radius.* Knock off 1/2 in. for clearance. Double this figure and you've got the maximum tire OD that can be used without rubbing the torsion-bar tube. For example, if you get a figure of 15-1/2 in., you can use a tire about 30-in. tall.

When figuring out what size tire will fit your car, don't go by advertised OD alone. Why? Tire OD measured at the center of the tread patch—the *crown,* is greater than that measured at the tire's outer edges, the *shoulders.* The difference between these measurements may vary 2—5 in., depending on the tire.

Measure the crown of the inflated tire with a large T-square. If the tire OD measured at the crown is several

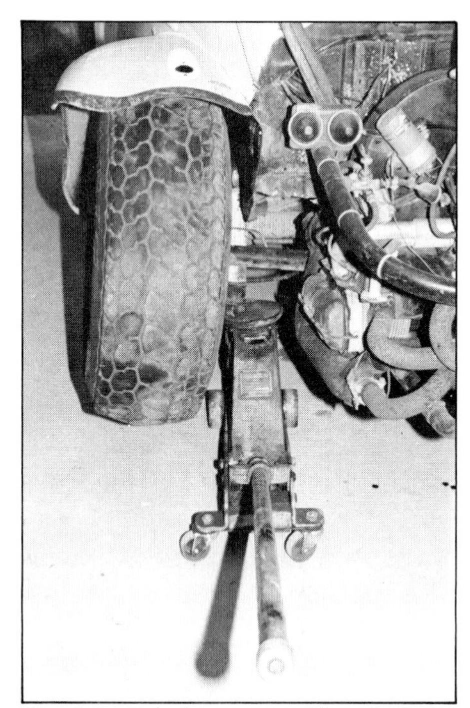

Check for tire-clearance problems by jacking the suspension up into full jounce—against its jounce stop. Jack under the lower shock mount.

Compress suspension fully by chaining car to floor jack. Use strong piece of chain looped over upper shock mount. This will keep car from lifting off ground before suspension reaches full jounce.

A sampler of off-road tires. Left to right they are stock-VW 5.60x15; 7.75x15 Sandblaster Jr.; 7.00x15 Tractionite; 11.00x15 Fun & Mud; 10.20x15 Sandblaster 915; 11.00x15 Positraction. Three tires at left are fronts and three on right are rears. As you can see, these numbers don't tell you much about height and width. Use your tape to check a tire before buying it.

inches greater than the tire OD at the shoulders, you can get away with a slightly taller tire without interference problems. On cars with obvious tire-clearance problems at the forward part of the inner fenderwell, minimize interference by using a tire with more crown—a more rounded tread.

Here again, you can save a lot of time and figuring by borrowing a set of tires similar to those you plan to buy and checking them for fit on your car. If your friend is squeamish about loaning his tires, you can make a quick check without driving the car.

Coat the tread and inner sidewall of the tire with chalk to help you spot interference. Install the borrowed wheel/tire. With a floor jack and strong piece of chain, as recommended for setting rear-suspension preload on page 71, jack under the spring plate until the suspension hits its upper stop. Check clearance with the suspension at full jounce. Most off-roaders will trade *a little bit* of tire rubbing for the advantage of bigger tires.

The reason I've gone into different tire-clearance problems instead of just listing tires and sizes for different models is that some popular off-road tires may not be available where you shop. In addition, tire manufacturers' numbering systems and size codes are confusing. One look at the accompanying photo illustrates what I mean. There isn't always a direct relation between a size code and the height and width, or *aspect ratio,* of a tire. Investigation on your part will tell you what's available and what will fit your car.

Swapping Torsion Bars—There are a few other ways to get added rear tire clearance, but they involve more work. With early swing-axle and IRS setups, you can replace the long torsion bars and hub covers with shorter bars and covers from '58—'68 sedans. While this is OK for a recreational car, I advise against this in a race car because of its effect on spring rate, outlined on page 76.

Lift Kits—Probably the most common method of adding fenderwell clearance is to install a *lift kit.* Lift kits space, or lift, a VW-sedan body away from its floorpan to give additional body-to-tire clearance. Lift kits consist of 2- or 3-in. square tubes that install between the floorpan and the body. For more details, see "Lift Kits," page 128.

Bus Reduction Boxes—Another way to get added tire clearance for swing-axle cars is to install a bus (Type 2) transaxle with the reduction boxes laid back. This setup will move the rear tires back and increase the car's wheelbase about 3 in. For details, see page 36.

Long-Travel Diagonal Arms and Spring Plates—On IRS cars only, you can add tire clearance at the torsion bar by using long-travel diagonal arms and spring plates. These special aftermarket pieces move the wheel center down and to the rear, effectively lengthening the car's wheelbase. This is generally an expensive modification, limited to serious racers only. For details, see page 81.

RECOMMENDED INFLATION PRESSURES

Keep in mind that a lighter car, like a buggy or stripped Baja Bug, requires lower pressures. Heavier street Baja Bugs or sedans need higher tire pressures. Here are some figures:

When running on established roads or trails, most racers use 14—16 pounds per square inch (psi) at the front and 16—18 psi at the rear. For rocky terrain, most run about 15—17 psi at the front and 17—20 psi at the rear. With the latest long-travel race-car suspensions and resulting higher speeds, some racers now run 22—25 psi at both front and rear to protect the rims. On a pre-run vehicle or play car that isn't driven as hard, you can use as low as 10—12 psi at the front and about 14—15 psi at the rear—and seldom bend the rims.

Experiment with pressures lower than this and you risk spinning the rear wheels inside the tires. On a tube-type tire, this rips the valve stem out of the tube, resulting in an instant flat. One way to prevent this is to screw the tire to the wheel. Drill the rim and tire bead. Fix the tire bead to the rim using 3/8—1/2-in.-long, #5 sheet-metal screws. Evenly space the screws around the rim so as not to upset tire/wheel balance. Do not use screws that go completely through the tire bead or the tube may be punctured.

RECAPS

Don't overlook recapped tires. Off-road recaps are about one-half to two-thirds the cost of new tires. Because high speeds and high temperatures aren't encountered off-road, recaps can be used safely. I don't recommend recaps on a dual-purpose recreational Baja Bug or buggy that's driven at highway speeds.

TIRE SEALANT

In the old days, racers put condensed milk in their tires to help prevent leaks from small punctures. The theory was that the milk would stay "rubbery" inside the tire or tube. If a small puncture occurred, the outrush of air would draw this gooey mess into the hole, harden it somewhat, and

Stock-VW front wheel is as good as any steel wheel. Off-road racers have used stock wheels for years with success. And they can be hammered back into shape as well as any other.

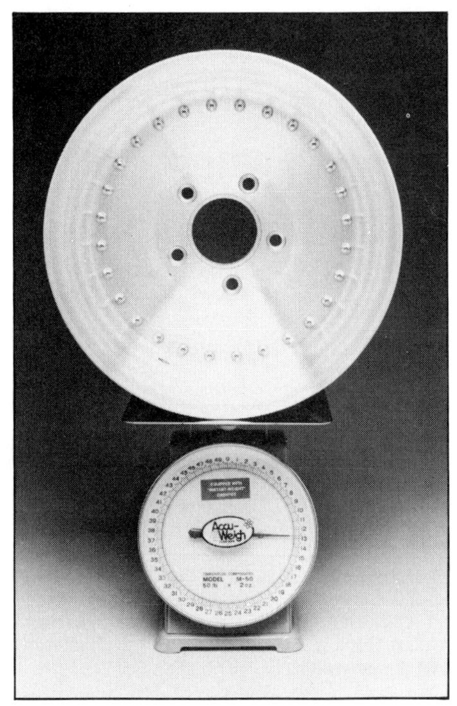

Center Line spun-aluminum racing wheel is very popular in off-road-racing circles. Light weight and strength are a hard combination to beat. Less unsprung weight means better handling. Photo courtesy of Center Line.

plug the leak. Of course, milk is an organic substance with limited shelf life. Needless to say, everyone stepped outside when a condensed-milk-treated tire was dismounted.

Synthetic-based tire sealant was developed a few years back for tubeless tires. It didn't take long for off-road racers to try it. Today, sealant remains one of those things that people just aren't sure of. Most racers put it in their tires hoping that it will work. The trouble is when you get a small hole in a tire and the sealant works, you don't know it. And when you get a bigger hole that it won't fix, it's obvious that it didn't work.

People who swear by tire sealant say it also balances the tire. People who don't like sealant say it gets hard with age and rots the tube. I'm like most racers. I put it in my race-car tires, but I wouldn't put it in a prerunner or play car. It's better to carry a spare tire.

Some tire sealants cannot be used with aluminum wheels because they are corrosive and can pit aluminum. If your car has aluminum wheels, make sure any sealant you use is compatible.

To install sealant, unscrew the tire valve and pour or pump it in. A typical off-road VW front tire takes about 12 ounces (oz) of sealant, and a larger rear tire takes about 24 oz. Reinstall the tire valves, inflate the tires and spin the tires by hand or drive the car to coat the inside of the tube or tire.

WHEELS

Deciding what front wheels to use is easy. Use the stock ones unless you're really interested in appearance or racing. The stock, 15 x 4-in., wheels are strong and were designed to work on the Volkswagen. Period.

You'll need rear wheels compatible with the big tires. Depending on rear-tire size, a 7—10-in.-wide, 15-in. wheel should do. That leaves out the stock rear VW wheels for sure.

Unsprung Weight—I touched on unsprung weight in the suspension and brake chapters. To repeat, it's the weight that goes up and down with the suspension. Unsprung weight is made up of the axle or spindle, suspension components, brakes (unless they're mounted inboard), wheels, tires, tubes and, yes, even tire sealant.

This weight, which moves due to road irregularities, must be controlled by the shock absorbers. Additional torsion bars don't control movement. Unsprung weight increases suspension *momentum,* moving weight. This makes the shock absorbers work harder. This is significant because shock absorbers are taxed to the limit when a car is driven hard off-road.

Almost all race cars use stamped or spun-aluminum wheels to reduce unsprung weight. A typical wrought-aluminum wheel, as opposed to cast, weighs about one-third to one-half of its steel counterpart. Never use cast-aluminum or magnesium wheels off-road. A wrought-aluminum or steel wheel will bend when it encounters a rock, but a cast wheel will break.

Many off-road parts outlets offer flame-cut, steel-spoke wheels suitable for recreational use. Photo courtesy of Johnny's Speed & Chrome.

Keep weight in mind when buying steel wheels, too. Not all steel wheels weigh the same. Aside from diameter and width, wheel weight depends on rim thickness. Thickness varies from about 1/10 in. to 1/8 in. You won't find many steel wheels as light and as strong as the stock Volkswagen wheels.

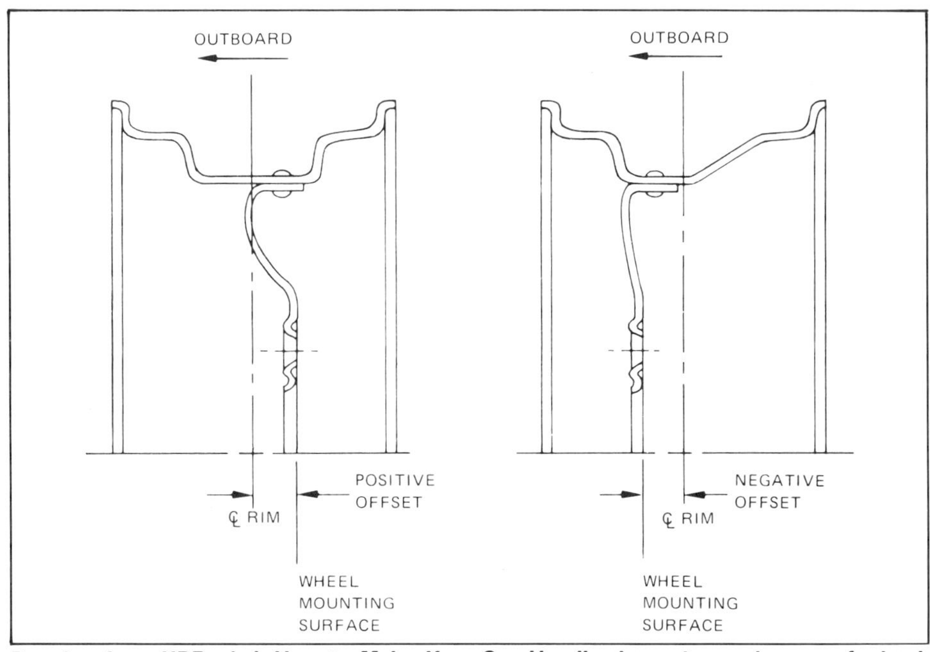

Drawing from HPBooks' *How to Make Your Car Handle* shows two extremes of wheel offset. Positive offset (left) increases track. Negative offset (right) reduces it.

Rim width is measured between the rim flanges. This 4-1/2-in.-wide rim is for the front.

A number of years ago getting a quality off-road wheel was a problem because radial and lateral runout could be excessive. This has changed. If you stick with a known-brand wheel, you should have no trouble. Even so, check the wheels for lateral and radial runout to make sure quality control is up to snuff. No amount of balancing can compensate for a wheel with a serious runout problem.

Wheel Balancing—If your VW is used exclusively off-road, don't bother balancing the wheel/tire assembly. The wheel and the balance the weight(s) will often part company on the first bump. Epoxying or taping on the weight(s) helps, but is no cure-all.

If your car is used for both street and dirt, balancing the wheel/tire assembly can prolong tire life and minimize front-end vibration at highway speeds. If you balance the wheels of your VW, use a crayon to mark the position and value (1/2 oz, 2 oz, etc.) of the weight(s) on the tire. Then any weights that fall off can be replaced in their original locations.

Lug Bolts vs. Studs and Nuts—On some non-stock steel and aluminum wheels, you may have a problem with the stock-VW lug bolts being too short. This occurs if the wheel center is thicker than on the Volkswagen wheel. With fewer threads engaging the brake drum, you run the risk of losing a wheel and/or stripping the threads in the drum. The solution is to install studs in the brake drums and secure the wheels with nuts. For details, see page 97.

Spacers/Wheel Adapters—To match the front and rear tracks, some off-roaders install spacers between the wheel and drum or disc hub. Longer wheel bolts or studs are then used. Although spacers won't hurt the front wheel bearings, the spacers do add unnecessary unsprung weight.

The same criticism applies to wheel adapters. Wheel adapters are available to bolt Ford/Chrysler or GM 15-in. wheels to Volkswagen hubs. I recommend buying wheels with the stock bolt pattern for your off-road car.

Offset—Offset, or *dish,* is the location of the wheel center relative to the rim flanges. If a wheel center is located exactly between the rim flanges, the wheel has zero offset. If the center is moved inboard, effectively moving the rim flanges outboard as in a "deep-dish" wheel, the wheel has positive offset. Moving the wheel center outboard gives negative offset. Got it?

Fancy wide tires on wheels with large positive offsets don't belong on an off-road car. If you must use positive-offset wheels for fender-well clearance for large rear tires, take a second look at those tires. Maybe they're too big.

Positive-offset wheels put a lot more strain on steering components, spindles, link pins and ball joints at the front. Similar strain results at the rear.

Racers using widened front ends and gear-reduction transaxles go to the expense of using wheels with negative offsets. The reason? To get the tire centered over the spindle or axle, where it belongs.

While perusing the aftermarket-wheel catalogs for suggestions, you may be confused by the term *back spacing.* Back spacing is the distance from the wheel-mounting surface to the *inboard rim flange,* where the tire bead seats. This distance will be exactly 1/2 the rim width on a zero-offset wheel.

Width—As a general rule, use wheels that are a couple of inches narrower than the *tire-section width.* Section width is the widest part of the tire, measured from sidewall to sidewall.

Because I'm telling you to use relatively stock-size front tires, use 4-in.-wide front wheels. The rear is another story. Use 7—10 in.-wide rims at the rear. Using a wider rim with a given tire reduces tread crown. Pulling the tire beads together with a narrower rim increases tread crown. Use your tape measure.

Michelin heavy-duty truck-tube valve stem. Tube is thick natural rubber. Stem is bent so it lies flat against the rim. Stem is taped to the wheel to protect it from being snagged by a limb, rock or other object.

Inner liner creates a tube-type tire within a tubeless tire. Using an inner liner requires a special two-way valve.

To increase flotation and traction, use 9-in. or even 10-in.-wide rims. In the rocky desert areas, where rim protection is more important than traction or flotation, use 7- or 8-in.-wide rims. The tire sidewall protrudes past the rim to protect it from damage from rocks.

TUBES

Never run tires without tubes when going off-road, even if the tires are the tubeless variety. A tubeless tire with low tire pressure doesn't take much impact to break loose the tire-bead-to-rim seal. Once this seal is broken, even for a split second, you have a flat tire—unless you use a tube.

When buying tubes, don't cut corners. Heat and friction between the tire and tube is very hard on the tube. Get heavy-duty tubes, preferably natural rubber. My favorite is the Michelin truck tube with a metal valve stem. Using liberal amounts of talc or talcum powder on the tube can cut down chafing against the tire.

If you get stuck and have to let some air out of the rear tires for added traction, don't go below about 5 psi. If you get the pressure too low, you are apt to spin the tire on the rim and tear the valve stem out of the tube.

TIRE INNER LINERS

A liner is nothing more than a small tire that installs inside a tubeless tire. The liner protects against flats. The weekend off-roader needn't bother with them. Considering the cost of an inner liner, it's better to carry a spare. When used, liners are run in the more-critical rear tires only.

Special tubes with two-way valve stems are required with liners. The liner is mounted on the rim with the liner tube inside. By turning the two-way valve one way or the other, you can inflate the tube, or the tubeless tire that fits over the liner. Thus, you have a pressurized outer tubeless tire and an inner tire with a tube at another pressure. If you hit something that flattens the outer tire, it simply deflates and you drive on the pressurized liner. You can't drive as fast, but you can keep going. Pressurize the outer tires as recommended on page 117. Inflate the liner tube to 25—30 psi, regardless of whether it's on the front or rear.

The advantage of the inner liner is that a punctured tire won't go completely flat and stop you. The disadvantages are the liner adds unsprung weight, and flats are more frequent because the outer tire is tubeless.

Even if you don't have a flat during a race, the outer tire may go flat later from pin-hole leaks in the tire or at the bead. Dented or rusted rims aggravate this condition. Because mounting a tire and liner is tricky, and sealing the valve stem on the rim is critical, the use of inner liners is not universally accepted.

Weld Wheel's bead-locking ring helps keep flat tire on rim between pit stops.

Before . . . .

. . . . and after: V-Enterprises can press expensive Centerline aluminum wheels back into shape at a fraction of their original cost.

Chapter 10 FUEL TANKS

If I were this guy, I wouldn't light up a cigarette for a while. Considering the fuel cascading from his tank, I would look into revising the filler pipe and vent system before venturing out again. Photo by George Jirka.

The stock Volkswagen Beetle fuel tank holds a little over 10 gallons (gal). At 15 miles per gallon (mpg) in the dirt, you should have at least a 150-mile driving range. But it doesn't always work out that way. If you're really "cooking," mileage can dip to 10 mpg, or even less if the engine is highly modified. So, unless you keep a close check on how much gas your engine typically uses, you'll find yourself stuck out in the boonies with an empty tank.

Another thing to keep in mind is today's off-again-on-again gasoline supplies. That middle-of-nowhere gas-station attendant might not be on duty on weekends or after hours when you need him. One other spot where gasoline supplies are uncertain is Mexico. The Pemex regular you may find there is sometimes of very poor quality. All these are good reasons for an additional fuel tank.

In a buggy intended for off-road trips, a 15-gal tank should be the minimum. Ideally, a 20-gal tank or larger is the ticket if you have room for one.

On a Baja Bug, you can get 20-gal capacity by simply adding an auxiliary 10-gal tank. But as long as you're going to the trouble of adding an additional tank, you may as well put in a big one. You'll thank me later.

What To Do With the Sedan Front Tank—If you have a sedan, your're in luck. It has several locations for mounting an auxiliary fuel tank. But first, you must decide what to do with the stock front tank.

To shift fuel weight to the rear, some Baja-Bug owners replace the front tank with a larger tank at the rear. If you're building a race car, the front tank and its supporting sheet metal must be removed anyway so you can brace the front torsion tubes and fabricate additional shock mounts. But if you don't have to remove the front tank for suspension modifications, don't.

The disadvantage of removing the front tank is not having the flexibility to carry an extra 10 gal. When you don't need that extra gas, just leave the front tank empty and use the rear tank only.

The difference between a full and empty 10-gal tank is about 70 pounds (lb). On outings where you'll need extra range, simply run off the front tank at the start. Don't try any serious acrobatics until you've burned up most of the front fuel load.

Where To Mount a Sedan Rear Tank—There are two good places to mount a fuel tank in the rear. One is in place of the rear-seat cushion. The other is on the package shelf beneath the rear window. Both have advantages and disadvantages.

A couple of ready-made aluminum tanks are made to fit in place of the rear-seat cushion. They are lightweight, hold about 20 gal and make for a nice, flat area in the rear for the 101 things you end up

Long-range, 24-gal tank fits in rear of VW sedan. Those extra gallons may come in handy some day. Tank is shaped to fit flat on rear of floorpan. Photo courtesy of Johnny's Speed & Chrome.

carrying. Also, because these tanks mount low in the body, they don't raise the car's center of gravity.

A disadvantage with these tanks is the lack of internal baffling. To keep prices down, many manufacturers leave the baffles and supports out. Also, an ice chest, tool box or whatever sitting on top doesn't take long to flex the tank, causing the welds to leak. If you decide to use one of these tanks, cover it with 3/4-in. marine plywood to support your extra baggage without damaging the tank.

Another good spot for that extra rear tank in a sedan is on the rear package tray, immediately below the rear window. I've only seen one tank made especially to fit this area. However, there are tanks built for boats that will fit in this space.

Besides boat-supply houses, larger department stores such as Sears have these tanks. With a tank *securely mounted* in this location, you can retain the rear seat and leave the battery in the same approximate location. A tank here raises the center of gravity of the car, which can affect handling.

Mounting a Buggy Fuel Tank—Unless you've got a special chassis that accepts side-pod fuel cells, the universally accepted place to mount a buggy fuel tank is behind the seat(s), directly over the rear torsion tube. Support the tank in a rubber-isolated cradle attached to the frame.

Make It Removable—Always make sure a fuel tank is removable, especially in a sedan. Otherwise, you may realize after adding a roll bar, upper shock mounts and supports that the tank is now only removable with a cutting torch.

If you're wondering why you would ever need to remove a fuel tank, here are a few instances. There's always the chance of the mounts breaking or the tank springing a leak. Also, a tank installed over the seat-cushion area blocks access to the cover plate for the coupler between the shift rod in the tunnel and the transaxle. To remove the transaxle, you must unbolt the coupler through this opening.

Another thing some tank designs may cover up is the chassis-ID number stamped on the tunnel. A few years ago after a Parker 400 race, my co-driver and his girlfriend were taking their Baja Bug pre-run car home and were stopped by an Arizona Highway Patrolman.

The officer said he wanted to check the registration papers against the engine and chassis-ID numbers. Being a suspicious pair of characters anyway, it didn't suprise me much when my friend told me, "The cop made me take the fuel tank out, right there on the side of the highway, so he could check the numbers." The moral of this story is to make the fuel tank removable—the easier, the better.

GAS TANK ALTERNATIVES

Oval Tanks—These tanks are very pretty. But because of their shape, they won't fit flat on the floor of a sedan. Because they have fewer welds than rectangular tanks, they are more durable and less prone to leakage.

Oval tanks are available in spun aluminum, stainless steel and plastic. Aluminum is susceptible to etching from high-sulphur fuel, such as that found in Mexico. Some plastic tanks are translucent, so fuel level is visible. They are made from seamless molds. Berrien Buggy's fuel tanks are made from polyethylene, which is the same material as used in Army tank fuel pods.

Fuel Cells—The ultimate fuel tank from a safety standpoint is the *fuel cell.* Fuel cells were developed for aircraft use and adapted for automotive applications, first in Indy cars. Not everyone can afford nor justify a fuel cell, certainly not the weekend off-roader. However, they are required by most off-road race organizations.

A fuel cell consists of three main components. From the outside in, these items are a container, a bladder and foam. The bladder is the key to fuel-cell safety. A bladder can be deformed and withstand extreme impact loads without leaking or bursting. A rigid metal tank cannot withstand such loads.

The bladder is contained in a steel "can." It provides the mounting structure for the fuel-cell assembly and protects the bladder from cuts or rips.

Filling the bladder is *open-cell* foam that is 98% void. In simple terms, if the foam occupies 100 cubic inches (cu in.), when filled with fuel, the volume is 2 cu in. of cell material and 98 cu in. of fuel. The function of the foam is to prevent fuel slosh. It does this very well. In fact, fuel-cell foam is the best baffling ever devised for a fuel tank.

A fourth component is the fill plate or "bung," which contains the filler neck, vent and occasionally, the pickup. The pickup is at the bottom of the cell if it's not in the bung.

Fuel cells are a bargain when you consider the possible alternatives—a car burned to the ground and a driver seriously burned or worse. If you're going to race and have a sedan with the fuel tank in the rear-seat area, think of yourself as the wick of a kerosene lantern. Then decide whether or not to install a fuel cell. Common off-road sizes are 8—22 gal.

Fuel cells are stamped with date of manufacture. Race organizations require bladder replacement every 5 years.

Saddle Tanks—FUNCO makes a single-seat buggy chassis to accept 10-gal side-pod fuel cells or saddle tanks. The only type of saddle tanks allowed by most sanctioning organizations are fuel cells. They're expensive. Expect to pay over $250 for each cell.

Fire Walls and Bulkheads—Though not yet required by all racing associations, some racers are installing a metal fire wall or bulkhead between the driver's seatback and the fuel tank. These are custom-made to fit various roll-cage configurations, fuel tank/cell types and body widths. The principle is simple. In a rollover or collision, you don't want to be drenched with gasoline from a ruptured tank or leaky filler.

Flex radiator hose works OK as a filler spout if you don't fill the tank so full that gas sits in the hose. But there's a real danger with this kind of setup. Imagine the aftermath of a rollover; a pulled-off hose and a fire. No thanks!

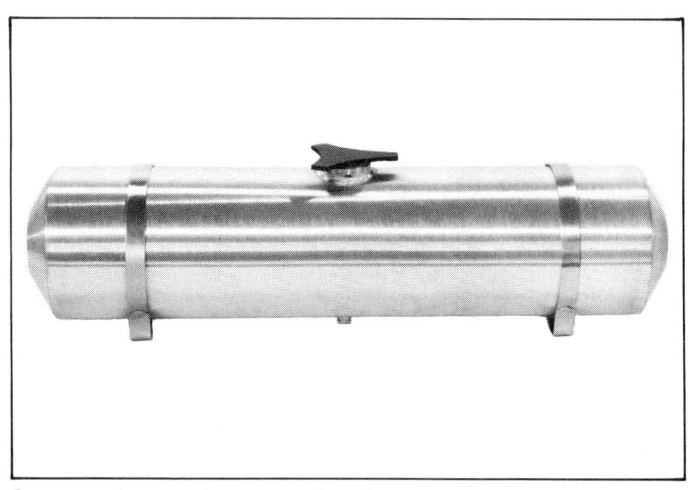

Spun-aluminum tanks are popular with buggy owners. These tanks are attractive and have few seams and welds to crack and leak. Tanks are 8 or 10-in. in diameter and about 2- to 3-ft long. Capacities range from 5 to 12 gallons. Photo courtesy of Bugpack.

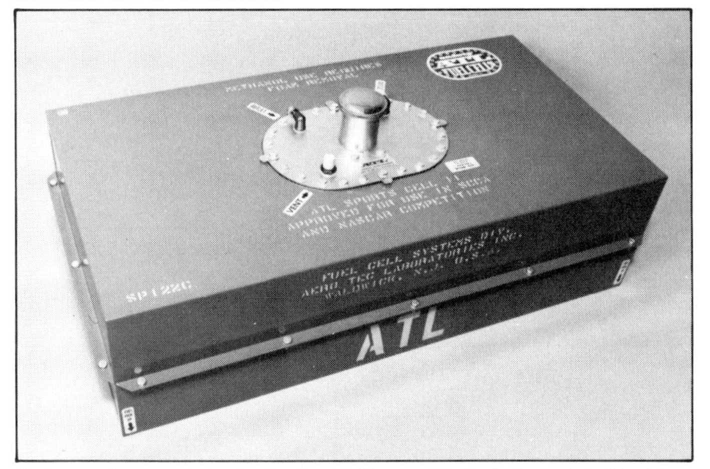

Ultimate fuel tank is the fuel cell. Foam-filled, deformable bladder inside the metal canister is extremely burst-resistant. Foam controls fuel slosh. Photo courtesy of Aero Tec Laboratories, Inc.

FUNCO flush-fit filler side-pod fuel cell. Photo by Ron Sessions.

Venting—As temperatures rise and fall, gasoline expands and contracts. The tank must be vented to prevent it from bulging and splitting at its seams as temperatures rise, or from caving in as temperatures fall. The two most common methods of venting a fuel tank are through the filler cap or via a separate vent and line.

My advice is to use a separate vent and line in your off-road VW. Route the line to a point outside the car and to the rear of the driver.

Direct Fill Or Remote Fill—When you install a fuel tank inside the passenger compartment, it's desirable to have a remote filler. Otherwise, gas spilled during fill up will puddle inside the car. This is both dangerous and offensive to your nose.

Bug Auxiliary-Tank Mounting—A good way to support a floor-mounted fuel tank in the rear of a sedan is with a 1x1-in., 1/8-in.-thick angle-iron frame as shown in the drawing. After making a rectangular frame for the tank to sit in, weld or bolt the frame into place on the floorpan.

You'll also need straps to secure the tank to its new frame. Two 1-in.-wide, 1/16-in.-thick steel straps will do the job. Trial-mount the tank in the frame, and bend and cut the straps to fit. Leave a couple of inches at both ends. Overlap the ends of the straps by about 1-1/2 in. by bending the strap back on itself. Hammer the ends flat. Drill 3/8-in. holes in the ends of the straps so the tank can be secured by bolts.

Before you put away your metal-working equipment and welder, make four tabs to which the straps can be bolted. Again, use the accompanying illustration as a guide. Four 1-in.-long sections of angle welded to the side of the frame work great. Drill 3/8-in. holes in the angles for the bolts.

In the case of an aluminum tank, isolate the tank from its frame and

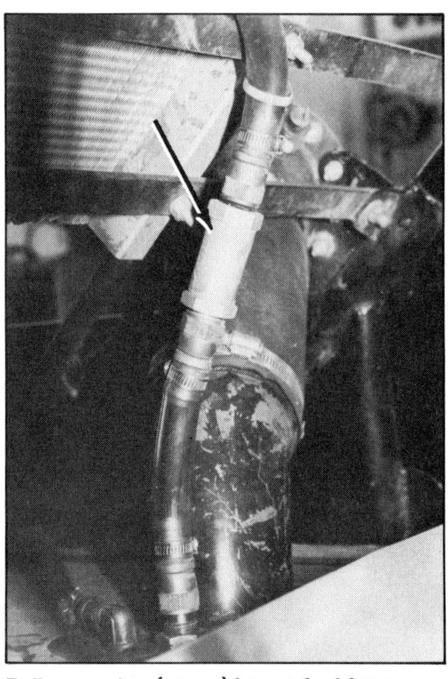

Rollover valve (arrow) keeps fuel from running out of the vent line if you happen to get upside down. Photo by Ron Sessions.

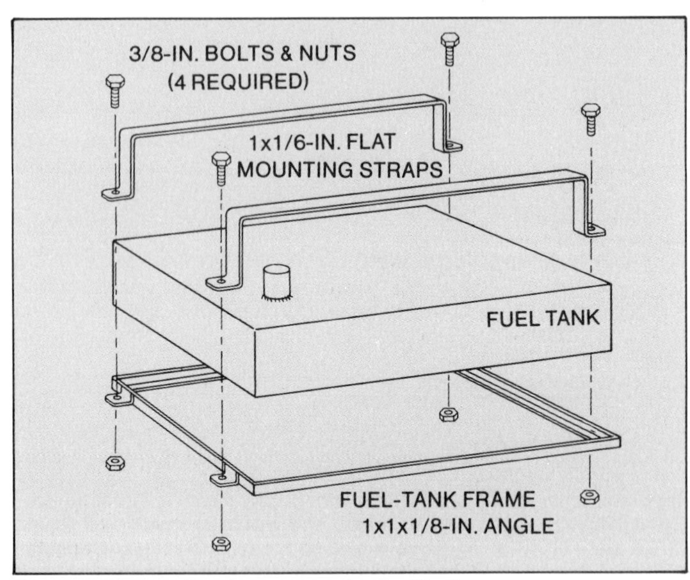

Suggested mounting for rectangular fuel tank.

Most RV-supply houses and all fuel-tank manufacturers sell filler necks and related parts for auxiliary tanks.

Where you mount the remote fuel filler is generally determined by the tank location. Keep the filler hose as short and straight as possible. Otherwise, filling the tank will be time-consuming.

When using two fuel tanks, you need a three-way valve so fuel is drawn from one tank at a time. Three-way valves are available at RV-accessory shops and many auto-parts stores.

straps with strips of carpeting or inner tube. Felt strips also work. Glue the strips of material to the frame and straps using weatherstrip adhesive. This keeps the aluminum tank from chafing or fretting and possibly wearing a hole that might cause a leak as the tank moves ever so slightly in its steel mountings.

A trip to a junkyard will get you a filler pipe and cap. Get a pipe long enough so it can be used for outside fill. Mount the filler pipe through the body side panel, roof pillar or rear quarter window.

If mounting through the body, cut the hole with a hole saw or jigsaw. If mounting through the window, cut one end off the window and fabricate a metal filler panel with a hole for the pipe. Fancy, flush-fit fillers are available from fuel-cell and RV manufacturers. Make the filler pipe fit as flush to the body as possible so it's less vulnerable to damage in a sideswipe or rollover.

Try to get as straight a shot as possible between the filler pipe and the tank. Sharp kinks or bends in the connecting hose make the tank hard to fill. This is especially true with the vapor-recovery nozzles used in California.

Special, gas-resistant hose is available from Gates to connect the filler to the tank. Even a sturdy flex radiator hose will last for years if you don't fill the tank so full that gas sits in the hose. Securely clamp the ends of the hose with worm-drive hose clamps.

Final Plumbing—If this tank is an addition to the sedan's original tank, run fuel lines from both tanks to a three-way valve. Use a small in-line fuel filter between the three-way valve and fuel pump.

Three-way valves are available through RV-supply houses and big auto-parts stores. Use good quality neoprene fuel hose. Secure all fittings with worm-drive hose clamps.

Suspension Considerations—Remember, the addition of a fuel tank at the rear will give more weight over the rear wheels. Figure about 7 lb for every gallon of fuel, plus the weight of the tank itself. Consider this when you're setting rear-suspension preload. If you've already gone through the preload exercise, you'd better recheck it.

Fuel Gage—If your sedan doesn't have a fuel gage for the front tank, check fuel level by simply looking in the filler spout. Next best is a dipstick made from a ruler marked off in quarters of a tank. A dipstick-type gage is incorporated into the filler cap on some direct-fill tanks.

You can fabricate a simple sight glass similar to those used on commercial coffee urns. The gage consists of two fittings welded in one end of the tank with a piece of flexible clear gas hose between them. With one fitting at the top and the other at the bottom, the gas will ride up the clear line to the same level as the fuel in the tank.

Chapter 11 BUILDING A BAJA BUG

A Baja-Bug kit can transform a tired, beat-up sedan into a clean and snappy looker.

The fiberglass kit is what makes a Volkswagen sedan a Baja Bug. Why build a Baja Bug in the first place? Besides the fact that it makes an old sedan look neat, there are some functional reasons. The shortened fenders provide added wheel clearance for larger wheels and tires, and for greater suspension travel. The shortened engine-compartment lid allows for an upswept exhaust system and a tubular engine cage/rear bumper. The smaller fiberglass parts reduce vehicle weight, especially at the front.

If you want a street-legal car, check state laws before going any further. Regulations requiring that the engine compartment and wheels be covered make Baja Bugs illegal in some states.

FIBERGLASS BAJA-BUG KIT

When it comes to Baja-Bug kits, the old saying holds true: "You get what you pay for." Two things are generally wrong with cheap kits: The pieces are thin and flimsy, and they usually don't fit well. Remember, the pieces you buy will have to support headlights, turn signal/parking lights (optional), taillights, driving lights and the occasional drunk who leans on it.

If you intend to do much off-roading, you'll be money ahead starting with a better kit with thicker fiberglass and a good fit. Start with a cheap Baja-Bug kit that doesn't fit and about halfway through the installation job you'll wish you had spent the extra money. Trying to install poor-fitting fiberglass parts is an experience you don't need. And if you manage to install the parts, you'll probably end up with an ugly Baja Bug no matter what you do to it later. Find out which brand works best by asking several Baja-Bug owners who have installed kits themselves. And look at the finished car to see whether you would be happy with a particular brand.

What To Look For In a Kit—It's best to inspect a kit before you pay for it. Do more than look at it. Run your hands over the outside surface to check for smoothness, air bubbles and flaws. If you intend to repaint your car, these defects will require a lot of finishing and maybe some fiberglass repair work.

Make sure there is a large lip or mounting flange on all four fenders. A 2-in.-wide or larger flange is good. Buy the kit if it checks out OK in other respects. A 1-in.-wide or narrower flange will cause problems unless the fender fits perfectly—an

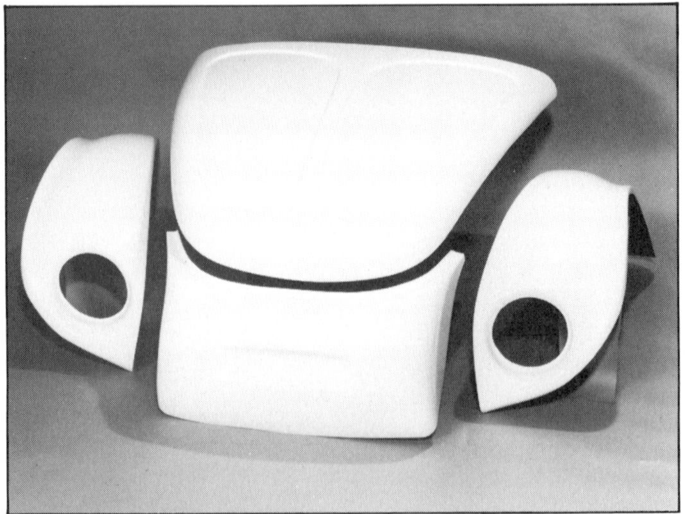

When buying a Baja-Bug kit, don't skimp on quality. Hood, fenders and nosepiece of a 7-piece, wide-eye kit. Photo courtesy of Race-Trim.

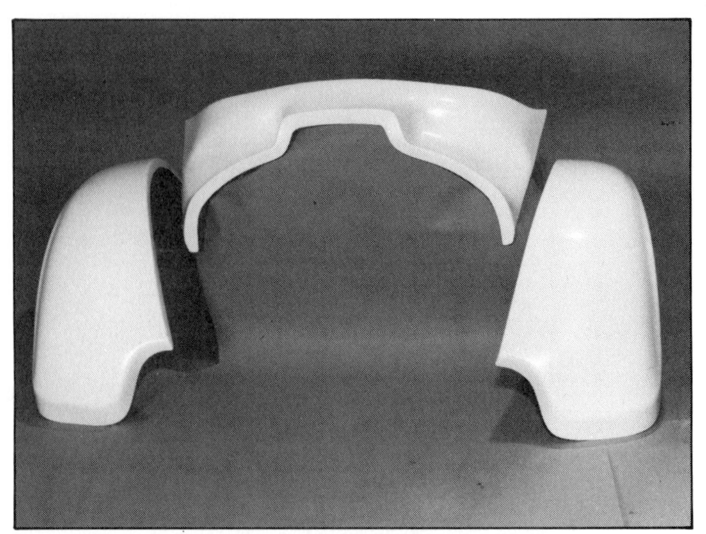

Rear half of 7-piece Race-Trim Baja-Bug kit. Raised section of engine lid is for clearance to large air cleaner. Photo courtesy of Race-Trim.

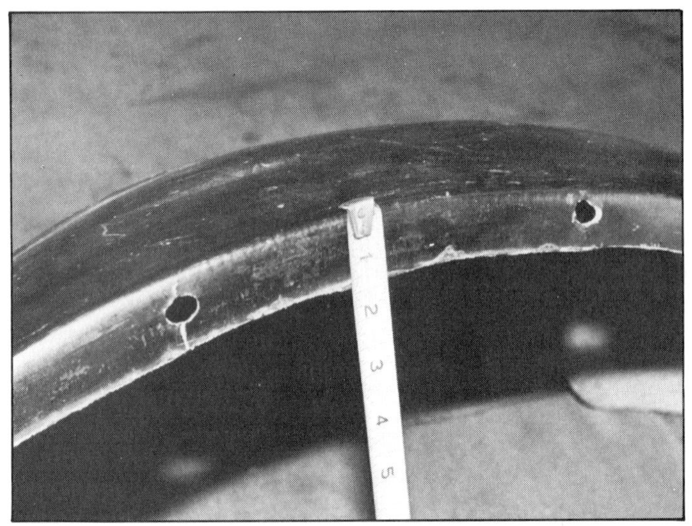

When buying a Baja kit, look for a wide bolt flange on the fenders. This one could be wider. There's already a crack through one of the bolt holes.

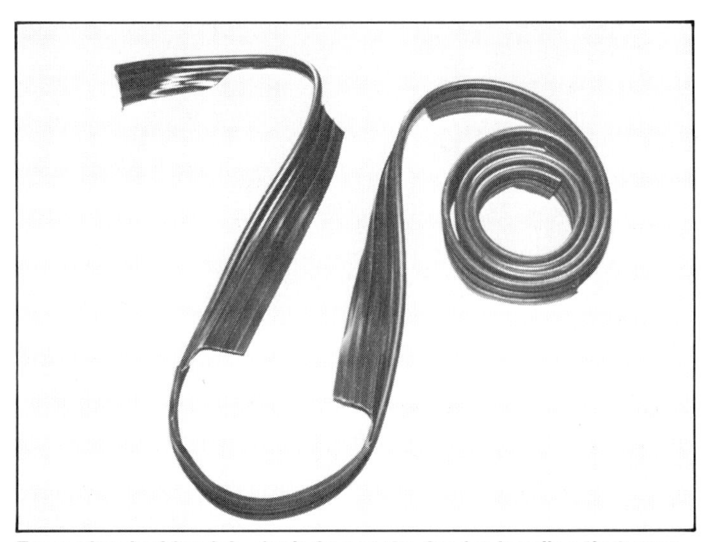

For a nice-looking job, don't forget the fender beading that goes between the body and fenders.

Bugeye kit gets the heavy headlights off the fenders and into the nosepiece. Some people think this presents a weird appearance, but others like it. Check your state laws—the bugeye kit may be illegal because the headlights are too close together.

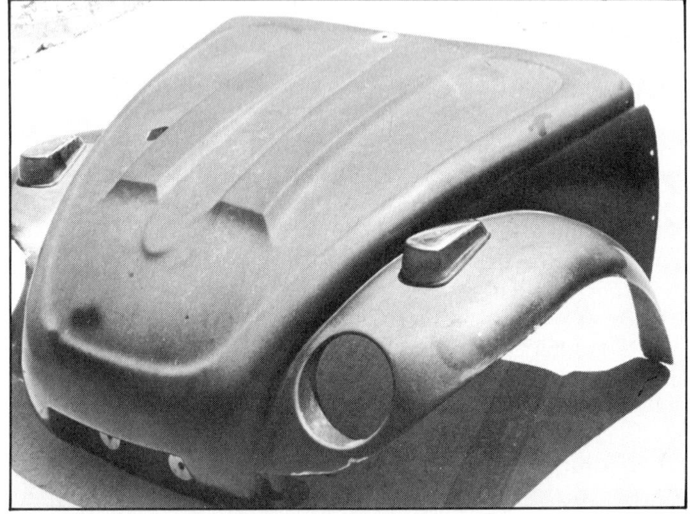

One-piece front end shows pin-mount (front and top) and Dzus-fastener holes on the sides. Quick release fasteners aid serviceability. Note the hole on the top left for access to the the stock fuel-tank filler.

In the desert, the Class-5 Unlimited Baja Bug is the Funny Car of off-road racing. Under that gutted VW shell is an unlimited race car. The one-piece front end on this car is more durable than a multi-piece setup. Photo by Steve Lange.

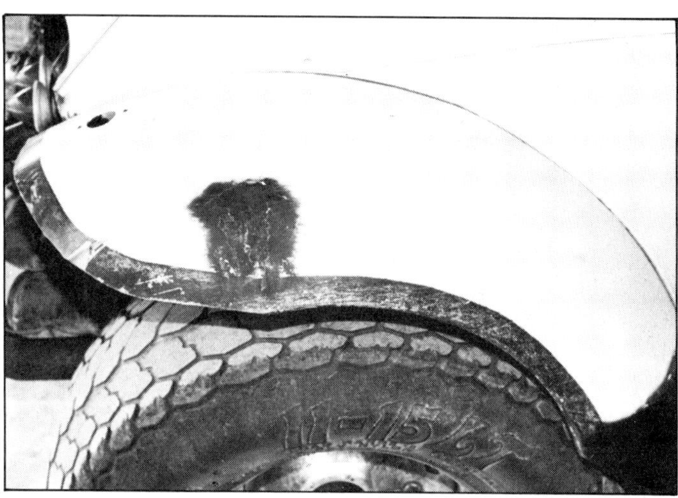

This once was a metal front fender. It was trimmed, turned around and bolted on to become a rear fender. Flexing and cracking fiberglass fender is eliminated. This fender does not cover enough of wheel to satisy some state laws.

unlikely possibility. When mounting fenders, use large washers to keep the bolt or screw heads from pulling through the fiberglass. This will also reduce the chance of the fiberglass cracking or breaking where it bolts to the body.

While you're checking the kit for flaws, remember to check the thickness of the parts. If any look thin and flex easily in your hands, imagine what will happen if you crash through some brush out in the dirt. It's a shame to go to the work of installing a kit, painting the car and then having the fenders fall off. Get a good-quality kit!

Good Baja-Bug kits come with installation instructions that tell you where to cut the body. If in doubt leave a little extra sheet metal for final parts fitting. It's much easier to remove material than it is to to replace it if you cut off too much.

Use fender beading, or *welting*, between the fenders and the body to fill small gaps. Fender beading may be available at the same place you got the kit. If not, try a Volkswagen dealer. Fender beading is also listed in J.C. Whitney's catalog.

Kit Types—There are a number of different types of Baja-Bug kits. Some have headlight mounting holes in the nosepiece—the "bugeye look." Others make provisions for mounting the headlights in the fenders—the "wide-eye look." The style you choose is your business.

The "bugeye" kit has the advantage of getting the weight of the headlights off the front fenders. Headlights mounted in the fenders can cause the fiberglass to flex and crack in use, or cause headlight "flicker" if the fenders aren't rigid.

Most kits consist of seven pieces—four *abbreviated* fenders, a nosepiece, a hood, and a short engine lid. Some kits combine the hood, fenders and nosepiece into a single unit—the one-piece front end.

One-Piece Front End—One-piece front ends are gaining favor in racing circles. They are easy to install, durable and lighter than multi-piece kits. On a Baja built from a sedan with extensive front-end damage, you can avoid much reconstructive surgery.

The drawback to the one-piece front end is poor access to components. Unless you mount the front end so it hinges in front or uses quick-release fasteners for easy removal, you've lost access to the trunk, the dash wiring, and possibly the brake-fluid reservoir.

Most people just cut a hole for gas-tank access. The "tidier" folks add a hinged door or flap. If you've eliminated the front tank, don't worry about it.

Modifying Stock Fenders—One good modification is to move the stock front fenders to the rear and trim off the excess sheet metal as pictured. What you get are a couple of very strong rear fenders—and at little or no cost. The right-front fender goes to the left rear, and the left front to the right rear. Got that?

It may seem too good to be true, but the bolt-hole spacing of the front-fender flanges matches those at the rear. The fenders match the curve of the body, too! Neat, huh?

File the trimmed fender edges smooth for safety's sake to complete the job.

Installing a Baja Kit—Installing a Baja-Bug kit is considerably more involved than just bolting on new fenders, a hood and engine-compartment lid. Your Bug must go "under the knife." You'll be removing several inches of sheet metal from the front and rear of the car.

This can be done in any number of ways. One way is to use an air chisel. You can pick up one of these at a tool rental. Don't forget that you'll also need an air compressor to operate the chisel.

If a compressor isn't available, the job can be done with a jigsaw or torch. An industrial-duty jigsaw is best. I've had good results with the Sawzall jigsaw. With a fine-tooth metal-cutting blade, the job is no problem.

If you use a cutting torch, be careful. Don't burn any wires or blister the paint on metal where it will show. Most people repaint the car after the kit has been installed, but why make extra work for yourself if you don't have to?

Follow the instructions in the kit. A typical kit installation goes something like this: Remove all four fenders, front and rear bumpers, the hood and engine-compartment lid. Measure and mark the front-end cut. With most kits, you cut off part of the front side panels (inner fenderwells), reinforcement plate (forward part of trunk floor) and all of the front valance.

Install the fiberglass hood, leaving the bolts finger-tight. If you're installing a "bugeye-look" kit, cut holes in the fiberglass nosepiece for the headlights. Fit the nosepiece so it matches the contour of the hood. Secure the nosepiece to the front side panels with sheet-metal screws or pop-rivets.

"Transfer" the bolt holes in the stock front fenders to the fiberglass fenders with a marking pen. Drill the holes and bolt the front fenders on. If you're installing a "wide-eye" kit, cut holes in the fenders for the headlights. Mount the headlights. Drill holes in the hood and nosepiece for the hold-downs. While doing this, preload the hood about 1/4 in. by pushing down on it to mark the hold-down holes. Otherwise, the hood will rattle later.

At the rear, remove any remaining engine-compartment insulation. Temporarily mount the rear fenders and the new engine-compartment cover to act as a guide for the rear cut line. After marking the cut line, remove the fenders and cover. On this cut, you'll be removing part of the rear quarter panels and all of the rear apron. Secure the engine cover with sheet-metal screws or Pop rivets. Bolt on the rear fenders.

Now, all that's left is rewiring the lights, smoothing any rough edges and installing bumpers.

LIGHTS

Headlights—Most Baja-Bug kits have fenders made to accept '67-or-later sedan headlight assemblies. You can also use aftermarket driving lights—street-legal sealed beams in a bullet housing—mounted to the bumper. To use the aftermarket lights, you'll have to install mounts on the front-bumper cage.

Locate the lights under the fiberglass fenders so they'll shine out through the openings. Mounting the lights to the bumper keeps the additional weight off the fiberglass fenders, reducing fender flex and the possibility of cracked fiberglass and flickering headlights.

Taillights—Stock taillights can be remounted on the fiberglass fenders. Or you can mount trailer-type lights on the engine cage/bumper. As with the headlights, keeping the taillights off the rear fenders reduces flex and the tendency of the fenders to crack.

After removing the fenders, hood and nosepiece, measure for front-end cut. On this Fiber-Tech kit, top of the cut starts 37-1/4 in. from drip rail at hood seal. Photo courtesy of Dune Buggies and Hot VWs Magazine.

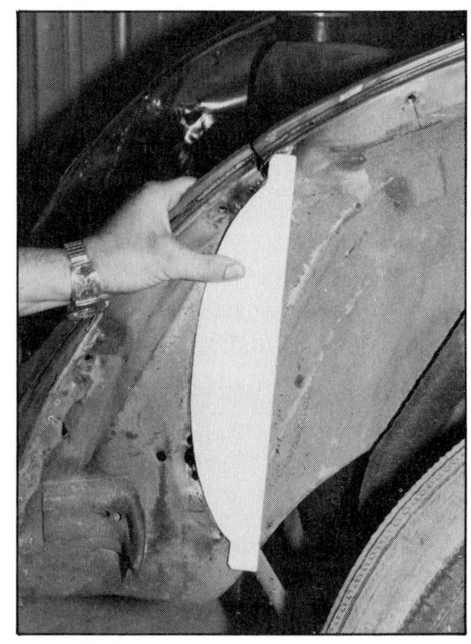

Fiber-Tech provides a template for proper contour to match new nosepiece. Cut on this line and the nosepiece will fit. Photo courtesy Dune Buggies and Hot VWs Magazine.

LIFT KITS

Second to fiberglass Baja-Bug kits, the most popular modification for VW sedans is the body *lift kit.* A lift kit spaces a sedan body 2—3-in. off the floorpan. This provides greater tire clearance for large rear tires. Many people are making this modification to street-driven Baja Bugs just because it's the "in look."

Installing a body lift kit requires unbolting the body from the floorpan and inserting a 2-in. or 3-in. spacer in between. Square tubing is shaped so it follows the contour of the VW sedan floorpan at the sides and rear. Other parts of a typical kit include a cast-aluminum spacer at the front of the floorpan and spacer sleeves and longer body-mount bolts. The cast-aluminum section fits over the front of the tunnel and ties in with the steel tubing at each side of the floorpan.

Be prepared for a suprise if you've never installed a lift kit. The procedure is straightforward, but it involves a lot of time-consuming "grunt" work. Busting loose several dozen rusty or undercoated bolts is no picnic without air tools.

Aside from the normal complement of metric hand tools and the instructions that come with the kit, you'll need two floor jacks, a pair of jack stands, lots of silicone sealer, a hole-saw or jigsaw and a friend or two. I recommend setting aside the better part of a day—a week maybe?—for installing a lift kit.

Installing a Lift Kit—Just so you know what you're up against, I'll briefly outline the procedure. To unbolt the body from the floorpan, you first must find the bolts. There are about 12 bolts along the body-to-floorpan flange under each sill. Get to these bolts from underneath the car.

Inside the car, under the rear seat cushion are two rows of bolts. They are roughly parallel to the rear torsion tube. There are four small bolts on the lower row and two or four larger bolts on the upper row. Access is from above.

At the rear, the body is bolted to the spring-plate supports, with two bolts at each support. Access is from underneath, through each wheel well. Finally, at the front, the body is bolted to the front torsion-bar tubes. Access to these two bolts is from above—after removing the stock fuel tank.

Disconnect the following before lifting the body off the pan: steering shaft at rubber coupler; speedometer cable at left spindle; wiring harness at starter and engine. Mark any wires before you remove them. On cars with a remote-reservoir master cylinder, disconnect the reservoir hose.

Trunk floor is cut along this seam (arrows). Photo courtesy Dune Buggies and Hot VWs Magazine.

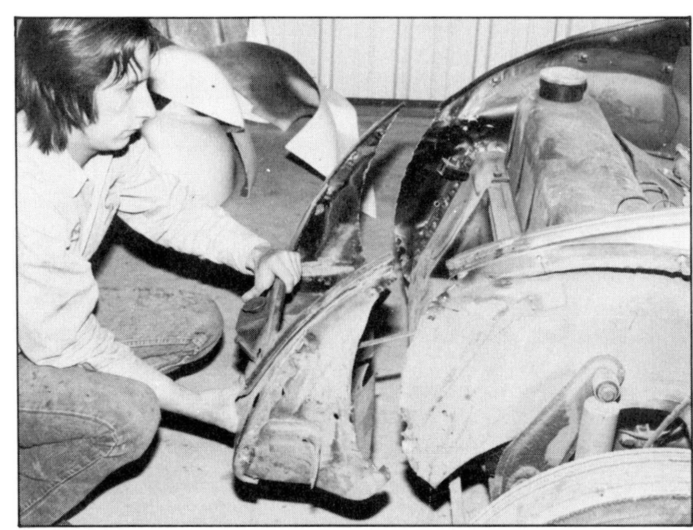

Voila! Hood-latch tube is cut and off comes stock-VW nose. Photo courtesy of Dune Buggies and Hot VWs Magazine.

Leave hood bolts slightly loose so you can align the hood. Alignment is basically a trial-and-error procedure. Take your time and do it right. Fiber-Tech kit supplies new hinge mounts and bolts. Photo courtesy of Dune Buggies and Hot VWs Magazine.

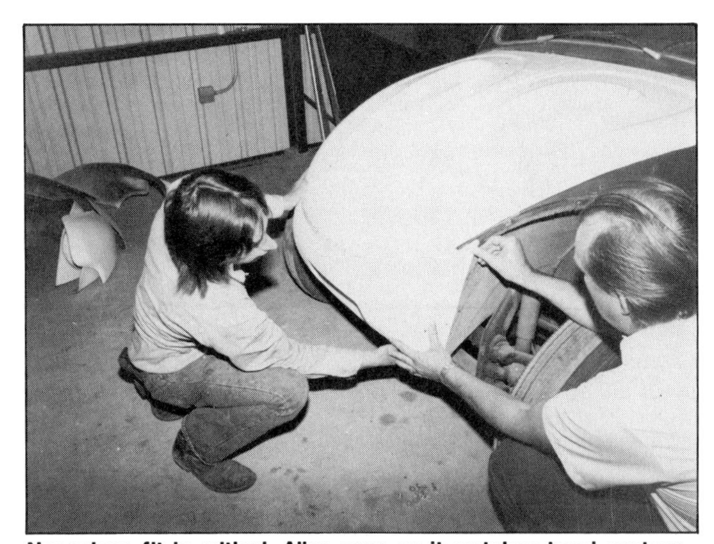

Nosepiece fit is critical. Align nose so it matches hood contour. Secure nosepiece to side panels with sheet-metal screws or Pop rivets. Photo courtesy Dune Buggies and Hot VWs Magazine.

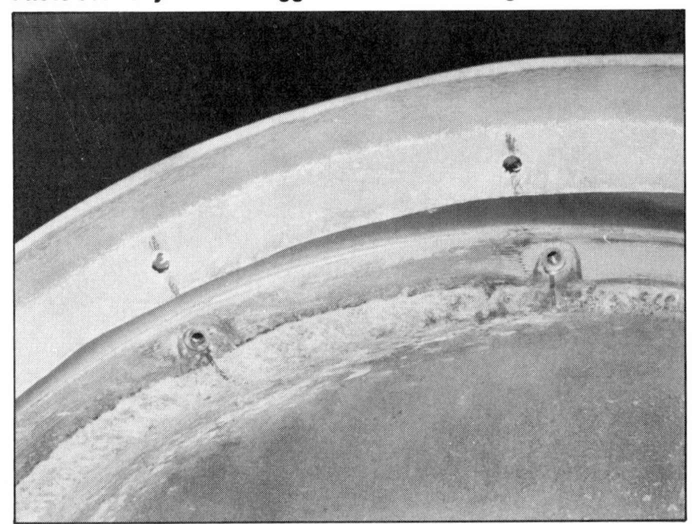

Transfer bolt holes in stock fenders to fiberglass fenders. When mounting fenders, use large washers to keep fenders from cracking at bolt holes. Photo courtesy of Dune Buggies and Hot VWs Magazine.

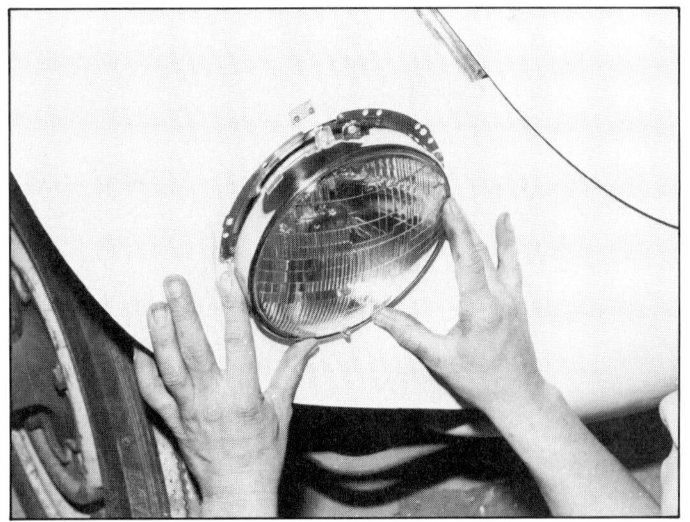

To mount '67-style VW headlights in a "wide-eye" front end, cut a 7-1/8-in. hole in each front fender. Drill holes for the bucket retaining screws. Lightweight nylon headlight-bucket replacements are available. Photo courtesy of Dune Buggies and Hot VWs Magazine.

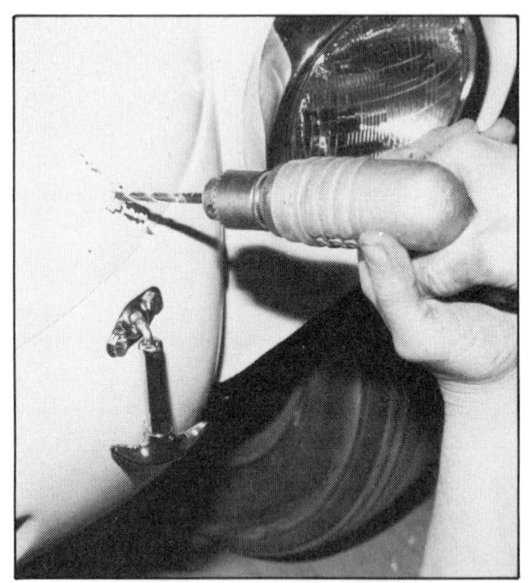

Have an assistant hold hood down against nosepiece when marking hole locations. Otherwise hood will be loose and rattle. Photo courtesy of Dune Buggies and Hot VWs Magazine.

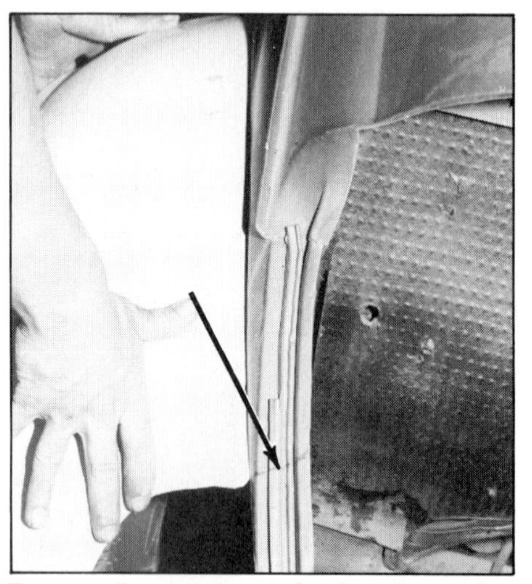

Temporarily mount rear fenders to establish start of rear cutline (arrow). Photo courtesy of Dune Buggies and Hot VWs Magazine.

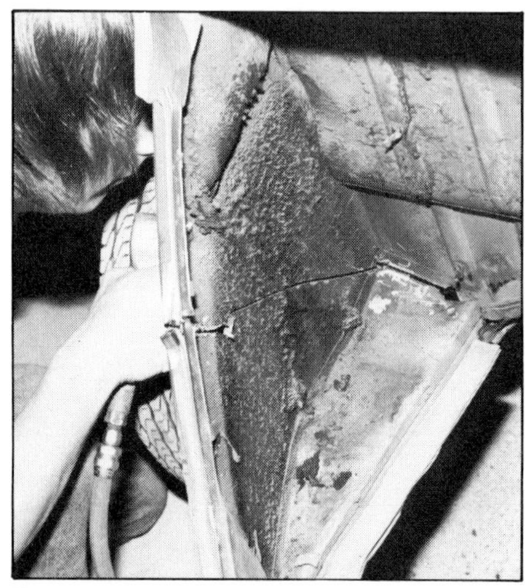

Rear cutline continues diagonally forward to intersection of inner fenderwell and fire wall. Job is easier with engine removed. Photo courtesy of Dune Buggies and Hot VWs Magazine.

Rear apron removed. Now you're ready to bolt on some fiberglass. Make sure insulation doesn't fall down behind engine and clog cooling fan. Photo courtesy of Dune Buggies and Hot VWs Magazine.

All this Baja Bug needs are taillights to complete the job.

After painting, owner of this Baja went one step further and installed plastic welting material over the outside edge of each fender. Photo by Ron Sessions.

RV light, bracket and a little welding solves the taillight/stoplight problem. Mount the lights to the rear bumper/engine cage.

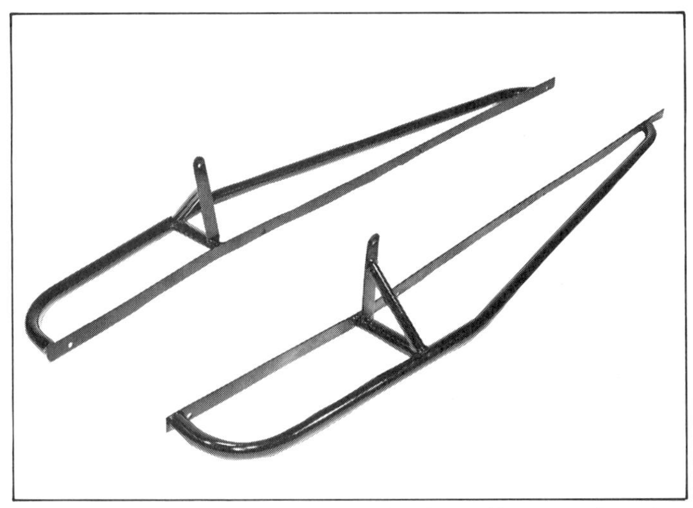

Nerf bars, or *tree bars* as they are called in the East, attach to the sides of the car. The bars will keep you from knocking off a fender or hooking a rear wheel. Photo courtesy of Bugpack.

Sedan has a 3-in. lift kit installed. Note spacer (arrow) between floorpan and body.

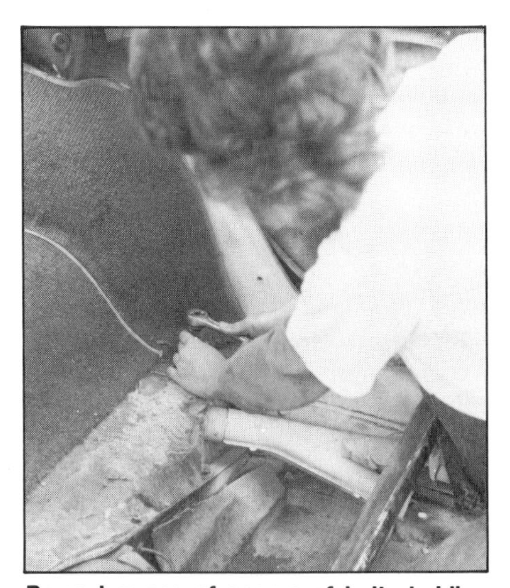

Removing one of scores of bolts holding VW body to floorpan to install lift kit. Originally under rear seat, these are accessible from inside car. Photo courtesy of T-Mag Products.

With all body bolts removed, jack body about 6 in. off floorpan. Support body and install jack stands. Use a wood block between jack and car body. Photo courtesy of T-Mag Products.

Leveling the body to the floorpan. Photo courtesy of T-Mag Products.

After the body is free, raise it 6 in. off the pan. If the stock bumpers remain on the car, put the jack under them. Otherwise, the best procedure is to raise each side separately by jacking under the body-lift point forward of each rear wheel well. Install jack stands under each body-lift point. Then raise the front of the body by jacking under the center of the trunk floor with a floor jack. Use a 2x4 between the jack and trunk floor to spread the load.

With the body safely supported, peel or scrape off the old body-to-floorpan gasket. Check the cast center section and square tubing for fit. If all is well, run a bead of silicone sealer along the bottom of these items and place them in position on the floorpan. Do the same to their top sides.

Carefully lower the body square onto the pan. Align the forward body-to-floorpan bolt holes with a punch or Phillips screwdriver.

Secure the body to the floorpan *using the bolts provided.* There may be fewer bolts in the kit than were originally used to retain the body. This is not good for a body that's going to be banged around off-road. I know; I had a body break loose from the floorpan while running second in the Mint 400. So, if you're going to race your Baja Bug, install additional body-to-floorpan bolts.

Drill through the body, spacers and floorpan. Install the bolts, sandwiching the spacers in between. Install the spacer sleeves and longer bolts supplied with the kit where the the body bolts to the front torsion tubes and rear spring-plate supports.

To complete the job, install a longer hose from the master cylinder to reservoir (remote reservoir only), and bleed the brakes. Elongate the hole in the fire wall for the steering column, and connect the column at

After scraping away old body-to-pan seal, install rectangular tubing and cast center section. Run a bead of silicone sealer along both mating surfaces of tubing and center section. Photo courtesy of T-Mag Products.

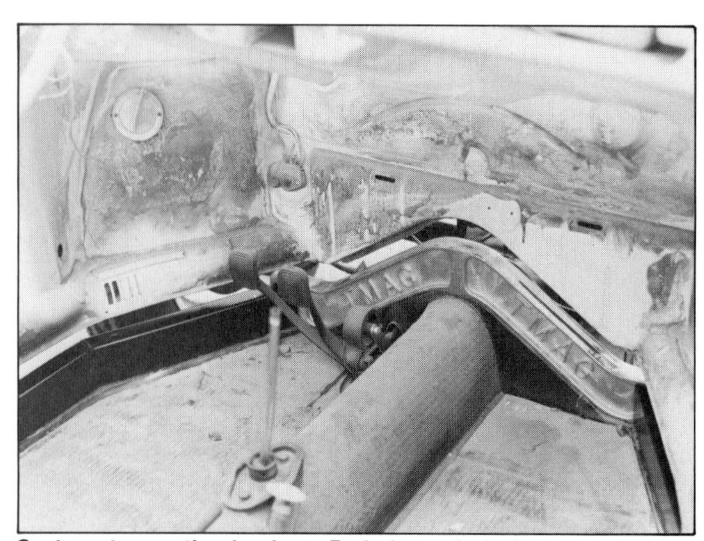

Cast center section in place. Body is ready to be lowered onto spacers. Don't forget the sealer! Photo courtesy of T-Mag Products.

Spacer sleeves go between rear body mounts and perch on spring-plate supports. Photo courtesy of T-Mag Products.

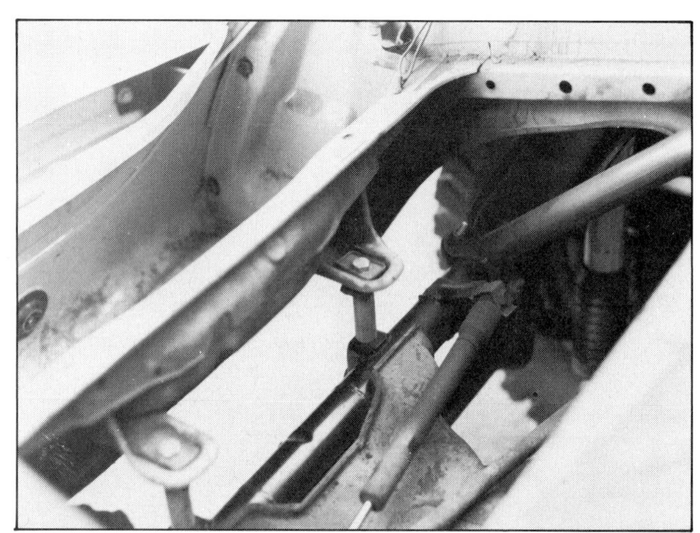

Other spacer sleeves go between body and upper torsion tube. Photo courtesy of T-Mag Products.

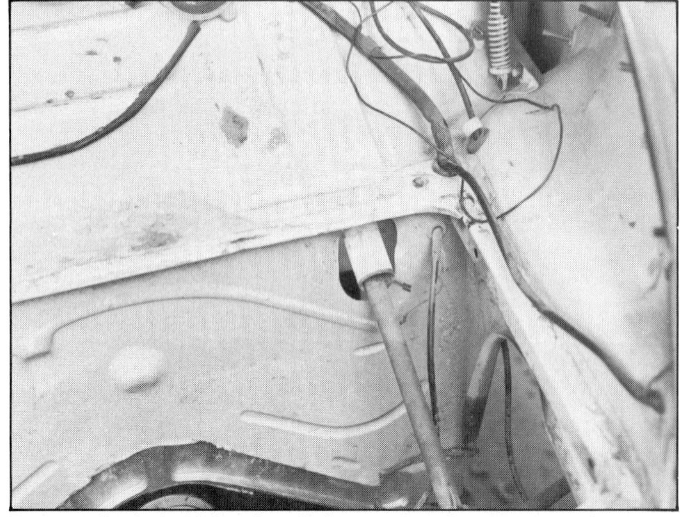

Because body is higher in relation to chassis, you must elongate hole in the fire wall for steering-column clearance. Photo courtesy of T-Mag Products.

With raised body on floorpan, sit inside to determine if seats need to be raised. Note unique split windshield on this Class-5 Baja Bug. Windshield protects driver and open section allows air to flow through car. Photo by Ron Sessions.

If you plan to race your Baja Bug, the body must be reinforced. Here, a support is run off each front hoop and bolted to the door jam. This is a strong part of the body.

Another body reinforcement. This one's run off the rear support and bolted to the body between the rear vents and drip rail. Three body panels come together here, making a sturdy point to tie into.

Yet another reinforcement. This one "captures" the rear hoop and is welded to the strong inner roof panel.

the steering-box flange. You also need to install the fuel tank (if applicable), reconnect the engine wiring, and reconnect the speedometer cable.

Check the seating position now that the body is 2- or 3- inches higher. You might have to raise the seats to retain proper visibility. Usually, all shifter positions can be reached using the stock shift lever. You may find it more comfortable to use a longer or more angled shifter.

SEATS

Stock VW seats are not very comfortable when the going gets rough. Your back is being rubbed up and down against the seat with every bump. This is because the seat cushion is separate from the seat back. If your shirt is not tucked in your pants, it will be rolled into a ball in the small of your back after a few miles of off-road driving. If you tuck your shirt in, the seat eventually pulls it out and rolls it up anyway. Using a simple folding seat cushion from your local dime store will reduce this somewhat.

There are a lot of one-piece, form-fitting, fiberglass or structural-urethane bucket seats on the market. Any will solve this problem and give you a more comfortable ride.

Don't underestimate the comfort needed for going off-road; many beginners do. Padding is very important. After a few hours behind the wheel, you'll be thankful you installed the proper seat. Foam rubber can be added between the fiberglass bucket and the cover for more

Fiberglass, polyethylene and structural-urethane bucket seats are low-cost alternatives. Make sure the one you get is well padded. Photo courtesy of Race-Trim.

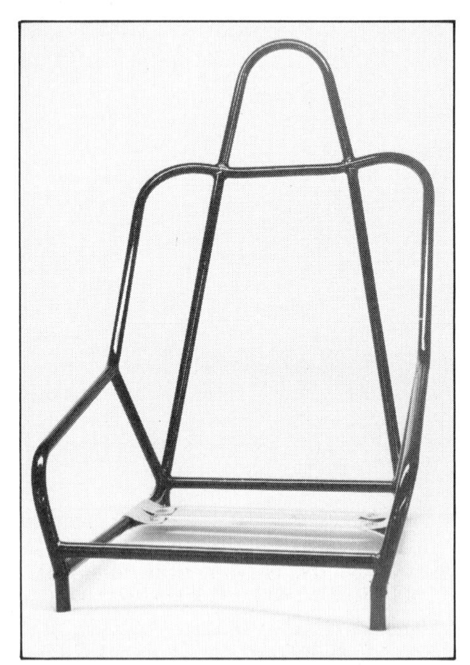
Anatomy of a racing seat. Mastercraft starts out with a 3/4-in., 0.069-in.-wall mild-steel frame. Photo courtesy of Miller Industries.

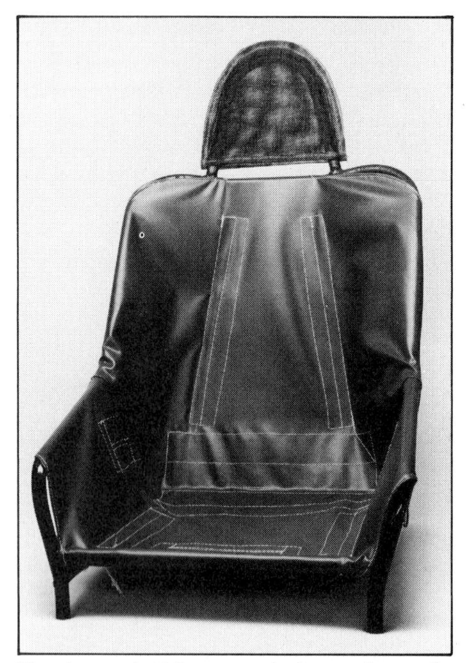
Tough, vinyl-impregnated nylon is stretched over the seat frame. Seat bottom is suspended for comfort. Photo courtesy of Miller Industries.

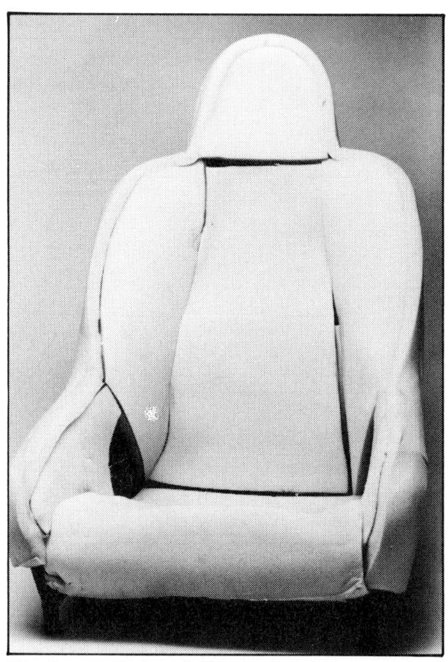
Mastercraft hand-fits polyurethane foam in place to achieve the desired contours. Photo courtesy of Miller Industries.

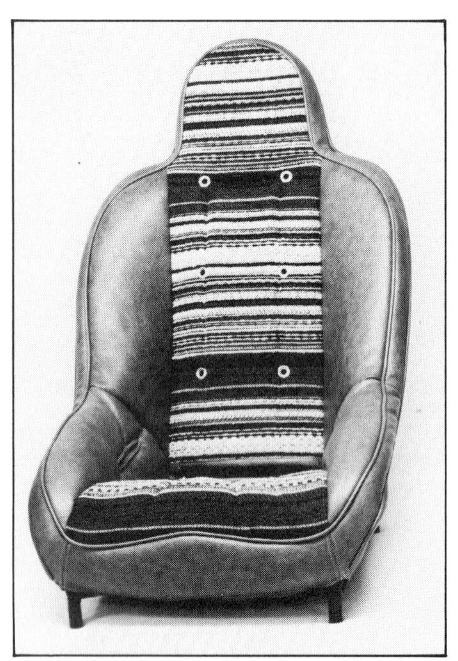
Finished product is a fine piece of furniture. If you plan on racing long distances, don't underestimate the comfort factor. Photo courtesy of Miller Industries.

cushion, if necessary. Just make sure the upholstery and padding is removable for cleaning. If you'll be driving in mud, stay away from cloth-insert seats because they are difficult to clean.

If rear-seat access is not important, consider one of the fixed, tube-frame racing seats. These are by far the most comfortable off-road seats you can buy—more like a piece of furniture than anything you'd find in an automobile.

Those manufactured by Mastercraft are fabricated out of 3/4-in., 0.069-in.-wall mild-steel tubing. Elastic bands are stretched between the sides of the cushion area. The frame is covered with tough, vinyl-impregnated nylon. Polyurethane foam is hand fitted and glued to achieve the desired shape. Over this is fitted the seat cover of your choice.

Adjustable Mounting—If you don't plan on racing, a convenient way to mount one-piece molded seats is to use the seat-adjuster assembly from the stock seats. Using the sliding seat tracks permits easier access to the rear of the car. And it provides fore/aft seating adjustments for different drivers.

Slide the stock seat all the way forward until it comes completely out of its track. Next, tear the seat down to the two loops of tubing that fasten to the slide assembly. Now weld or bolt two pieces of 2-in.-wide, 1/4-in. steel between the two loops to mount your new seats to as pictured. The new seats will now slide back onto the tracks.

There's only one problem with this type of installation. It isn't completely safe. Stock-VW seat tracks are spot-welded to the floorpan. In an accident, these welds could break, causing the seat and its occupant to be tossed around inside the car.

But here's a way to retain the convenience of adjustable seat tracks and make them more secure. Capture the seat tracks to the floor with U-bolts. Get some U-bolts—two for each seat—large enough to straddle the bottom rungs of the seat frame and tracks with the seat installed. Locate the U-bolts so they won't interfere with fore-and-aft seat adjustment. Drill holes in the floorpan for the U-bolts.

With a nut threaded on each U-bolt leg, fit each U-bolt over the seat tracks and in their holes. Adjust the nuts so the U-bolt won't pull down on the seat track. Install flat washers, and lockwashers and two more nuts and washers on the U-bolts from underneath. See the accompanying drawing for details.

Tighten the nuts against one another, sandwiching the floor in between. Check the seat tracks for fore-and-aft movement. Now, if the seat-track welds ever break the U-bolts should hold the seat in place.

Fixed Mounting—If you plan to race, install the seat in a fixed position. This requires building a subframe to mount the seat. The location of this subframe and its mounting tabs determines seat height, rake and distance

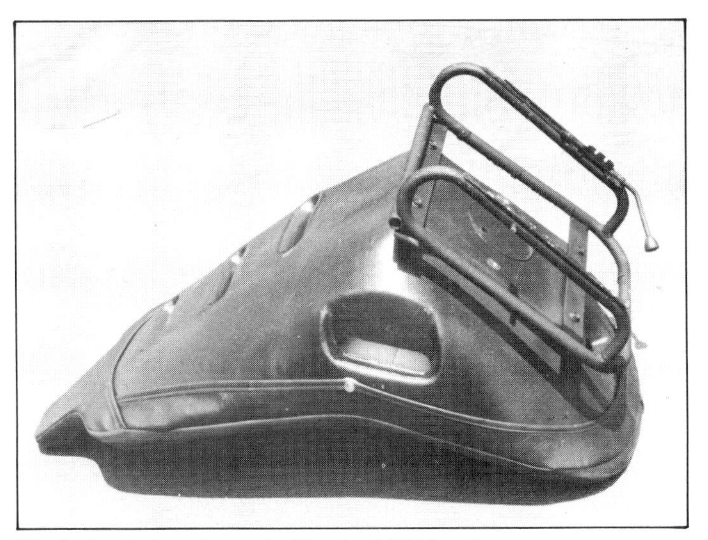

Plastic bucket seat mounted to stock VW tracks.

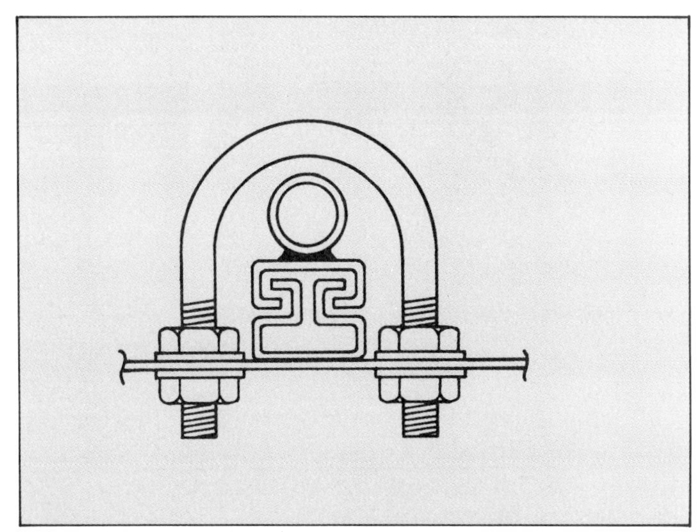

SCORE rules dictate that when sliding seat tracks are retained, seat tracks must be securely clamped to floor with U-bolts. If stock seat-track welds break, U-bolts keep seat in position.

Non-adjustable seat-to-floor mount. Strongest places to tie into are the tunnel and sills—not the center of the floorpan.

You can't work the pedals properly when you're bouncing around. Install a brace for your left foot so you can steady yourself.

from the controls, so make sure you get it right the first time. Do a trial-fit and check position before welding the subframe in place.

The seat subframe can be fabricated from 1-in.-square, 1/8-in.-thick steel tubing. The strongest places to tie into are the tunnel and sill area—much stronger than the floorpan. See the photo for details.

Other Items—Here are some other things that will make a play car more comfortable. Carpets not only look and feel good, they absorb a lot of dust, keeping it from flying around inside the car when you're driving down a fast, dusty road. Carpeting also reduces interior noise. Glue the corners of the carpet in place with weatherstrip adhesive. Don't glue down all of the carpet because you may want to remove it for cleaning some day. You can also use Velcro or snaps to hold the carpet in place.

A foot brace for the driver and one for the rider is a big help in rough terrain. You can use it to hold yourself firmly against the seat so you are not bouncing up against the seat belt. See the illustration for an example.

A grab bar gives the rider something to grab onto and brace against when the going gets rough. It's required in racing. The driver always has the steering wheel to hold but the rider without a grab bar has nothing. The last thing you want is the rider holding onto the roof pillar or door for support. An arm or hand outside the car is just begging for injury, especially in a rollover, or while blasting past a tree or through brush. The grab bar is a good place for the auxiliary horn button used in off-road race cars. This frees the driver to do the driving while the rider "blows for the road" in racing situations.

Give your passenger something to hold onto in the rough. Stock VW "panic handle" does the trick.

chapter 12 SAFETY & FIRST AID

Most off-roaders don't drive like this—only when they're going for the win. You don't have to be a race-car driver to be concerned with safety and first-aid. Knowing what to do in an emergency can save your or someone else's life. Photo by Tom Monroe.

For most of us, off-roading is just for fun. It's a good time to relax and explore mother nature's finest. Or it's great for excitement, climbing steep hills and bouncing over rough terrain. For many, it's a little of both.

You're on your own out in the dirt. There is no one telling you what to do. It's not too tough to figure out who's responsible if you wreck your car or get hurt.

As with many recreational activities, off-roaders consume their share of alcohol. Trying to tell some people not to drink and drive is received about as well as telling them not to get the car dirty.

Fools and drunks either wise up or find another activity, because of the high cost of fixing broken cars and bones. But considering the number of people mixing it up in intimidating terrain, and that natural urge to outperform the other guy, off-roading is a relatively safe sport. Safe until something happens to you.

GLASS

Like in any sport, most injuries can be avoided or minimized with some forethought. Most serious injuries occur in rollovers, but people can get hurt just bouncing around.

Never take a glass container off-road—soda-pop bottle, beer bottle, wine jug, water jar, or whatever. Besides the obvious hazard broken glass presents, dental work is expensive and getting more so. Even the best dentist can't replace your natural teeth. The only glass you should find inside an off-road car are the windows, rear-view mirror and instrument faces.

SHARP EDGES

After an off-road trip, make a note of any part of the car that "bites" you: a cut on the elbow from a sharp edge on the door, a bump on the head from the roll cage, an ankle cut on a sharp corner on the tunnel, or a knee rubbed raw on part of the interior. Most of the time, you're tired when you get home and forget about these things. Consequently, they don't get fixed. Just as sure as there's dust in the desert and mud in the woods, those things that caused minor injuries or irritations will get you again, maybe worse next time.

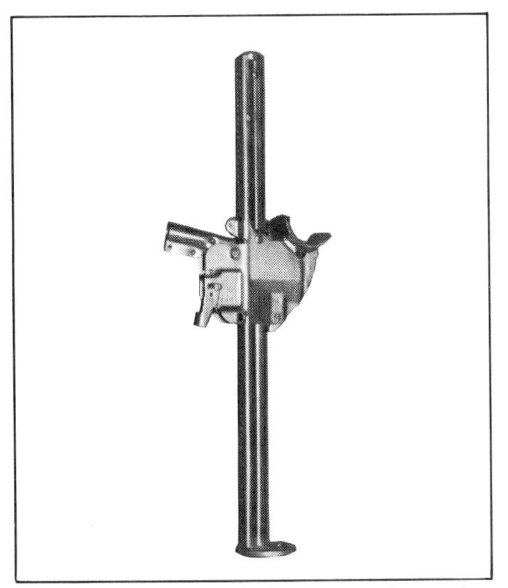

Leave your stock VW jack at home when going off-road. Bumper jack from U.S.-built car can be modified to lift a tubular steel bumper or buggy cage. Pad is welded on end for a solid footing in dirt or sand.

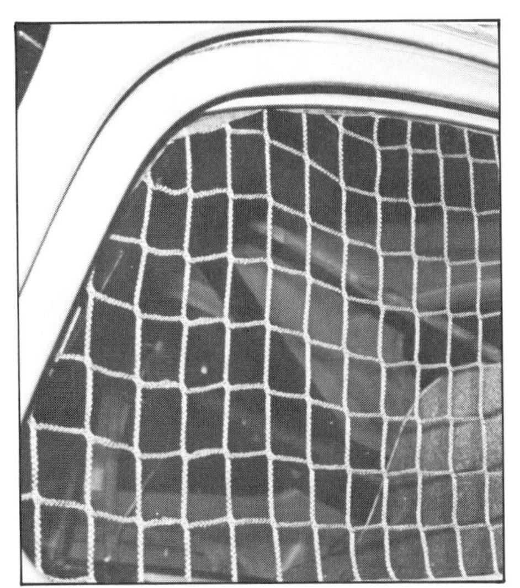

Because of numerous hand and arm injuries, race-sanctioning organizations require window nets. I don't suggest nets are needed for your recreational off-road car, but think about where to put your hands in the event of a crash.

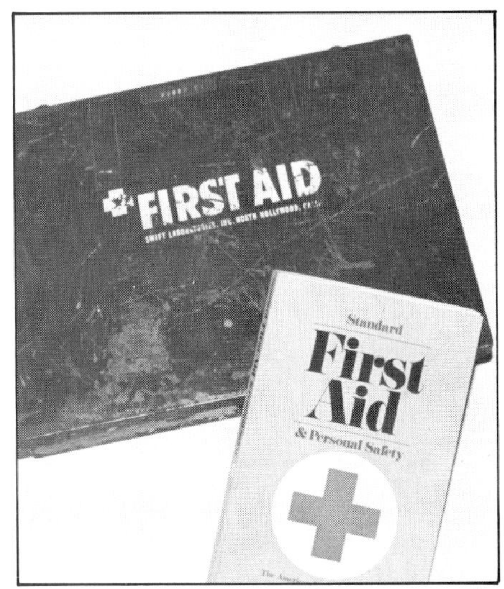

A must in every off-road car. Generally they're nowhere to be found! If you don't carry a first-aid kit or book, at least familiarize yourself with the basics.

JACKING CAUTIONS

The Volkswagen jack was not designed for use on a car with a long-travel suspension and big tires. Neither was it designed for use in the sand or mud. Don't count on using it in the dirt. You won't feel like your VW is a lightweight vehicle if it slips off the jack and lands on you.

Every off-road car needs a sturdy jack to change tires, get unstuck, and so on. One of the sturdiest off-road jacks is the Hi-Lift, or sheepherder's, jack. But this jack is heavy and stands 4-ft tall, making it a tight fit in most buggies. I prefer to use a bumper jack from a U.S.-built car. These are inexpensive and readily available. With a couple of simple modifications so it will fit your car (see photo), you'll have a good off-road jack that's easy to carry.

Other alternatives include a small hydraulic jack or a scissors jack. The only drawback of these is the necessity to get underneath an axle or suspension component that may be buried in the dirt.

ROLLOVER

In a rollover, most injuries result from hands or arms outside of the car. And you can get knocked in the head by something loose inside the car. The driver should keep his hands on the steering wheel. The passenger's hands should be on the grab bar or braced against the dash. Rear-seat passengers, if any, should be strapped in with seat belts.

Other than a few scrapes and bruises, most rollovers are slow enough that only your ego and the roof and fenders of the car are hurt. Heck, that's what roll cages and roll bars are for. But the same rollover can be deadly if a toolbox full of tools and parts is banging around inside the car with you. Anything that isn't tied down— ice chest, spare tire, battery, jack or shovel—can seriously injure you or your passenger. If it's in the passenger compartment, tie it down *securely*.

CARRY A FIRST-AID KIT

A first-aid kit is a must, yet most people don't carry one. Even though all major sanctioning organizations require first-aid kits, I've encountered people prerunning a race in the middle of nowhere without even a Band-Aid. I know—I've tried to borrow one after cutting myself because I didn't have one either! I guess it's natural to think it will happen to the other guy—but it's not healthy.

Most off-road stores and motorcycle shops sell complete first-aid kits. Whichever type of kit you buy, make sure it's in a weatherproof case.

SCORE requires that first-aid kits contain the following items: a 4-in. bandage compress, two 2-in. bandage compresses, a triangular bandage, eight 2 x 3-in. adhesive pads, sixteen 1 x 3-1/8-in. adhesive bandages (Band-Aids), ten treated prep-pads, an eye-dressing packet, ten ammonia inhalants (smelling salts) and an Ace bandage. To this list, add some sort of antiseptic such as iodine or Merthiolate, a tube of Benzocaine-type burn ointment or spray and some medical adhesive tape.

Ideally, you should carry a first-aid instruction pamphlet with your kit. Again, most people don't. So here are some suggestions if you hurt yourself and find you don't have so much as a Band-Aid. A cotton or cotton-blend T-shirt makes a good dressing. Cut a piece to cover the wound and use *duct*—racer's—tape to hold it on.

INJURIES

Broken Bones—Broken bones should be splinted. A splint keeps the broken-bone ends from moving around and doing more damage. A splint can be fashioned from a jackhandle, lug wrench, tree or brush limbs, or heavy cardboard—found on the door panels or covering the wires under the front hood. If none of these items can be found, splints can also be fashioned from some parts of the car that can be unbolted. Examples are the exhaust stinger, steering-brake handle, or a shock absorber.

Hold the splints in place with duct tape or bungee cords. If possible, pad between the body and splints. Use something soft like clothing, paper

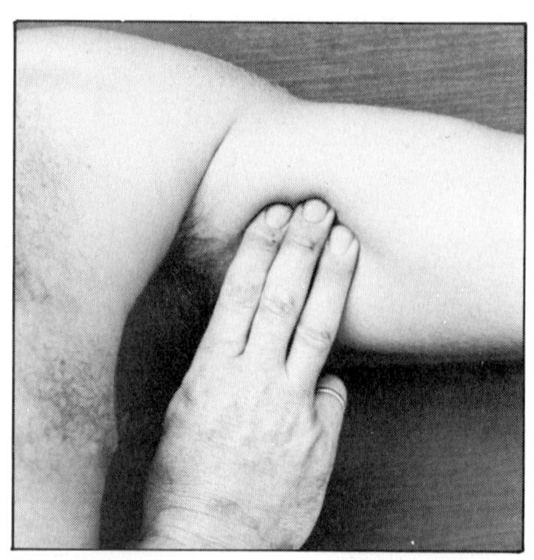

Direct pressure *on the wound* is best for stopping any bleeding. Slow the bleeding by exerting pressure on a major pressure point like this. Compress the artery against the bone.

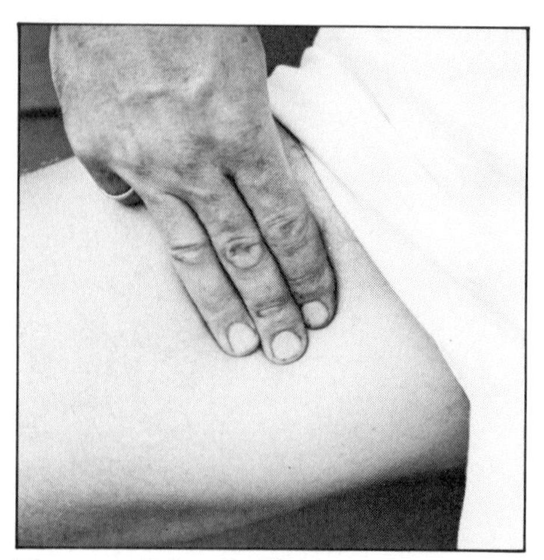

The same is true for the legs. Feel for the femoral artery in the groin area.

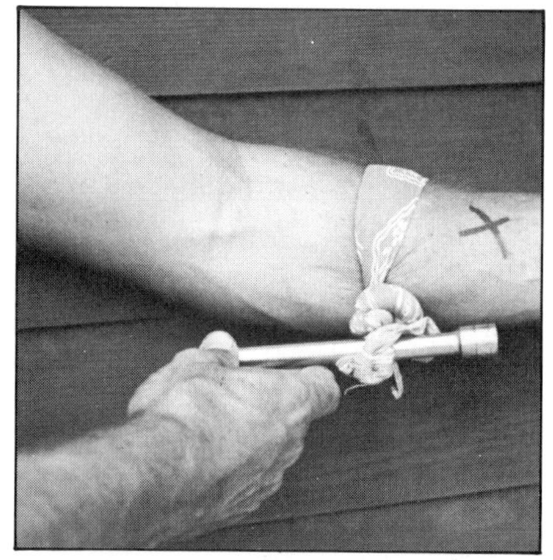

A tourniquet is a last resort! When applying a tourniquet think about sacrificing that limb to save that person's life. The tourniquet goes close to, but above the wound.

towels, or carpeting. Drive out slowly and expect a bunch of squawking from the injured person. There's no way to smooth out the bumps, but make the trip as comfortable as possible.

Heavy Bleeding—If you cut yourself badly enough that direct pressure on the wound with a bandage won't stop the bleeding, find a *pressure point.* On an arm, feel for the artery—it will be pulsing—on the inside of the arm and just below the armpit. On a leg, the best pressure point is in the crease of the groin area. Arm or leg, compress the artery against the bone with your hand.

If heavy bleeding continues, use a tourniquet—but *only if you have to.* A tourniquet is any kind of cloth or plastic band that can be tied around an arm or leg to stop blood flow to the injured area *completely.* **For this reason, the decision to apply a tourniquet is really a decision to risk the loss of an arm or leg to save a life.** If you decide a tourniquet is necessary, place it approximately 2-in. above the wound to allow the maximum amount of blood flow to the uninjured extremity.

If you decide to use a tourniquet, but don't want to risk losing a limb, release it at two-minute intervals for a minute or so. Blood loss will be greater, but there's more chance of saving the limb.

Burns—The best first aid for burns is to cover the area with a water-dampened cloth. This cools the burn and keeps dirt out. Keep it covered until you reach medical assistance. Skin is the body's first resistance to infection. If the skin is not broken, hold ice on the burn. See the following paragraph.

Bruises and Bumps—If you bang into something and get a bruise, bump or lump, try using ice on the hurt. Wrap some ice in a cloth and apply it directly. Not only will the ice numb the irritated nerve endings, but it will reduce swelling as well.

Head, Neck, and Back Injuries—If a person is badly injured, and you suspect a head, neck or back injury, **don't move him.** The best thing you can do, after making sure the victim is as comfortable as possible is to go for professional help.

If there is danger of injury from a vehicle fire, try to move the *car,* not the victim. If the victim is in the car, and a fire starts, try to put it out with a fire extinguisher. If an extinguisher isn't available, carefully drag the victim out of the wreckage by the feet or under the shoulder blades. Keep in mind that without a back-board, you risk doing serious damage to a victim with a head, neck or back injury.

Insect Stings—Although the car provides some protection from most critters, stings from flying insects can be a problem. To most people, a sting from a bee, hornet or wasp is painful but not life-threatening. Yet, over 50 people die every year in the U.S. from flying-insect stings.

If you are allergic to bee venom, get a prescription for a kit to inject adrenalin. Carry the kit with you. Adrenalin can help prevent the body from going into anaphylactic shock after an insect sting.

EQUIPMENT

Seat Belts and Harnesses—An absolute must is a good set of seat belts. For racing, a *5-point harness* is a necessity—and usually required. A 5-point harness consists of a lap belt, shoulder harness and a crotch strap. Wear them. It does little good to tie all your gear down, and leave yourself free to fly around inside the car.

If you've got an old sedan that didn't originally have seat belts, mount anchors through the floor pan. Use large-OD washers on the back side of the sheet metal. For racing, check the rule book for the type of belts and anchors needed.

Fire Extinguisher—Because gasoline is a major fire hazard, especially with an auxiliary tank mounted inside the passenger compartment, always carry a fire extinguisher. Place it within easy reach. Mount it securely so it doesn't roll around and accidentally discharge or hit you. I recommend a dry-powder type, 2-1/2-lb, ABC-rated extinguisher with a capacity gage. If it's a gasoline fire you're putting out, use the entire contents of the extinguisher to make sure the fire is completely out.

The chance of a fuel tank leaking is greater with all the abuse encountered off-road, especially in a rollover. That's the big advantage of fuel cells; they can handle this kind of abuse. Sometimes, leaks can't be stopped until after the fire is put out.

If you ever have the misfortune of not being able to put out a fire, you'll have to walk home. You'll probably start your walk after you watch your

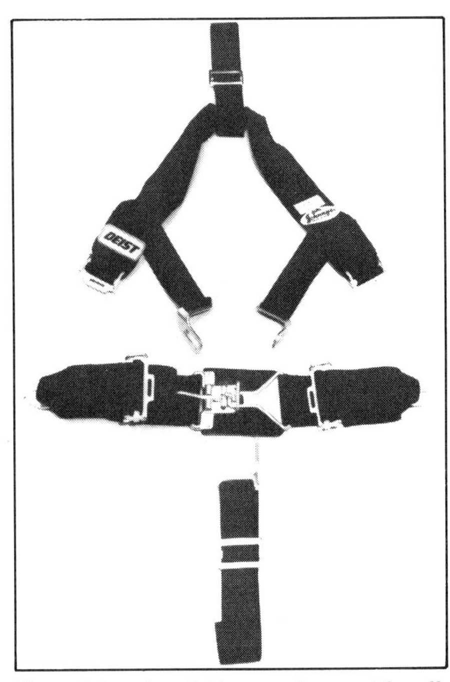
Five-point or six-point harness is a must for off-road racing. Deist harness consists of two shoulder straps, two lap belts and an anti-submarine belt. Photo courtesy of Johnny's Speed & Chrome.

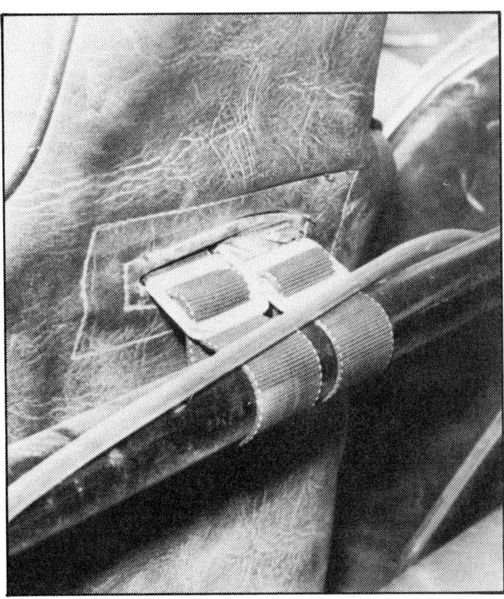
Shoulder straps should be securely anchored to the roll bar or roll cage. Photo by Ron Sessions.

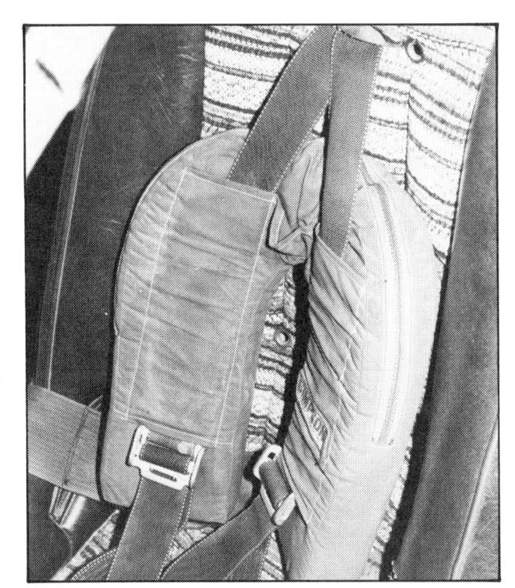
To keep a racer securely in his seat, the straps must be cinched down almost to the point of pain. A padded neck support or "horse collar" helps prevent bruised shoulders and neck. Photo by Ron Sessions.

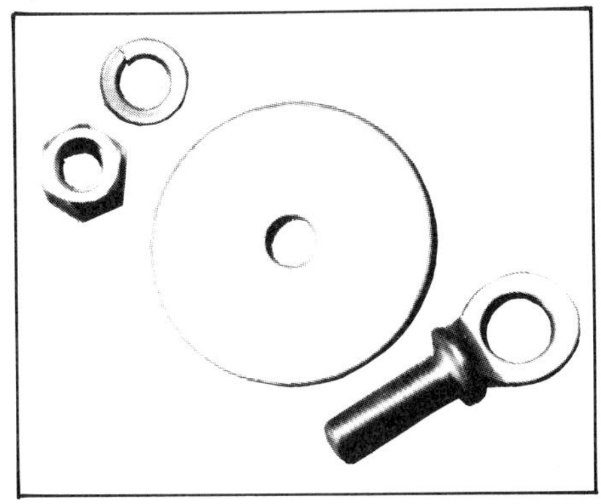
Seat belts are only as good as their anchors. When mounting belts, put a large washer on the underside of the floorpan for each eyebolt. Photo courtesy of Phil's Inc.

A good case for safety harnesses and window nets. Notice how the co-driver has his left hand *outside the car* on the roll cage! It's a good way to lose a finger. Photo courtesy of Bridgestone Tire.

Fire extinguisher is like a first-aid kit: You never think about it until you need it. Carry one!

car and everything in it "burn to the ground."

Flares—One important thing to carry along is at least one emergency roadside flare. If you were stuck or injured so badly you couldn't drive out, a couple of these cheap flares could save your life. Flares can also be used to warn other drivers that your car is disabled and in the way of oncoming traffic.

Emergency flares are inexpensive, light, unaffected by the pounding they'll get off-road, water resistant and can be found at most auto-parts or hardware stores. Stow flares by taping them to the roll bar or cage.

To light one of these flares, pull its cap off, strike the end of the flare like a big match. Position it where it can be seen. It will make a brilliant light for about 10 minutes.

chapter 13 DRIVING TIPS

There is the right way to do it . . . Photo by George Jirka.

The ability to handle a car depends a lot on your natural skill—it's something you're born with. Some off-road racers seem to win no matter what type of vehicle they're in. There are some other drivers who are not as talented but just as successful because they've concentrated on one particular class of car and type of racing. Then there are the rest of us whose success depends mainly on practice and how well we learn from experience. We become better drivers by learning from our mistakes.

Let's take a look at driving off-road. Some off-road driving techniques go completely against what comes naturally on the street.

Advice for Beginners—When I first started racing I asked an old timer if there was any trick to driving off-road cars. He gave me some advice that I've tried to follow ever since. It goes something like this: Only drive "over your head" or at a speed faster than you feel safe when you're in terrain where you can "abandon ship" and leave the road if necessary. Otherwise, drive as fast as you can, but always remain in control. This is especially true in mountainous areas where steep drop-offs loom on either side and in large, rocky and tree-covered areas where leaving the road at speed would spell disaster.

I hope you never find yourself entering a 30-mph turn at 60 mph with nowhere to go except into deep trouble. When you charge over a rise or around a corner and find a sudden change in road direction, the earlier you admit your mistake and look for a way out the more chance for success—survival—you'll have. Ego-wise, it's always easier to explain to your co-driver or passenger why you drove off the road than it is to explain how you caved in the roof to all those waiting to hear.

Use All of the Road—Most dirt roads consist of two tracks, ruts or grooves, a *crown* at the middle and a *berm* along both sides. The crown and berms are built up from loose dirt kicked out of

... and then there's the wrong way! Photo by George Jirka.

Running at night on a well-grooved road. There are more places to drive on this road than down in the grooves. Use all of the road. Photo by Trackside.

the two tracks. The smoothest way to drive is usually—but not always—in the tracks.

Which reminds me of a story. Some friends and I pitted for a team running the Baja. We got a late start and ended up driving my pickup truck for about 36 hours straight to get to the designated pit-stop area ahead of the race car. When my friend drove, I noticed he kept the truck's wheels right in the tracks of the road, regardless of the holes or rocks.

After a number of hours I couldn't take it anymore. Being as polite as possible, I screamed, "You idiot! You don't have to hit all those rocks!"

He took it surprisingly well. After a few snide remarks, he asked, "How do you expect me to miss 'em! They're right in the middle of the road." After talking about it awhile, I realized he didn't see the road the same way I did.

A car's rear tires take up only about 18—20 in. of an average 8—10-ft.-wide road. If there is a rock or hole in one of the tracks, simply steer to the left or right to miss it. This generally requires riding part way up the berm and crown to miss the obstacle. Then the car is steered back in the tracks. With a little practice, this maneuver becomes a smooth and natural reaction.

If the ruts are rough for extended stretches because of water erosion or exposed rocks, you can go one step further. Drive your car to one side so two wheels are atop the crown and the others are on the berm. Driving is a little tricky because all the wheels are in soft material. You'll have to drop back in the groove now and then to avoid rocks and brush on the berm.

Every road and section of road is different. Sometimes these maneuvers aren't worth the trouble and it's better to stay in the tracks and dodge what you can. The point I'm trying to make is this: Don't get locked into driving only in the tracks, or on the berm and crown. Be flexible. You'll drive faster and smoother.

Braking Techniques—One of the most common off-road mistakes is overusing the brakes. You're driving down a dirt road and come around a corner and the road makes a lot sharper turn than expected. A person's normal reaction is to jump on the brakes, scrub off speed and try to steer around the turn.

In the dirt, tire adhesion is poor. The first thing that happens is the brakes lock up and the tires slide. Fact: **A skidding tire won't steer.** When turned, a skidding tire goes straight ahead. The car may slow down somewhat, but it will plow straight ahead.

If you're in the type of terrain that offers a choice, simply drive off the corner through the smoothest spot available. But if there's a big drop-off or a bunch of trees or big rocks on the outside of the turn, you don't have much choice—make the turn or crash! Don't lock the wheels! You won't have any steering control.

Steering With the Rear Wheels—An experienced driver may "dirt-track," or use the throttle to steer with the rear wheels to get through a corner. But never use too much throttle while cornering, or the front wheels will *plow*—push out of the turn.

An experienced driver can also use a steering brake in this situation with success. But unless you're familiar with how they work, you can crash.

Braking and Steering—So, how do you use the brakes? As you approach a turn, apply the brakes to scrub off as much speed as possible *before* turning the steering wheel. Get off the brakes at the last second, turn the wheel and hang on. It's simply not possible to brake hard and steer at the same time. You'll be amazed how quickly a car steers once you let the front wheels roll.

Another time you must force yourself to get off the brakes is right before you hit that big bump. If you're driving along a fairly straight road and are surprised by a washout, hole or a pile of rocks that you'll hit at a speed much faster than you'd like, here's what you do.

As with the unexpected corner, scrub off as much speed as possible while keeping the car straight and under control. Just before the front tires hit the rough, release the brakes. This will let the tires *roll* over the bumps, not plow into them. This also lets you steer and maybe keep the bouncing car on the road.

Another thing that happens when you release the brakes is the front suspension has a chance to extend. With the brakes on, the car dives, or loses

Leaving Mexicali at full song over the canal crossings. Needless to say, getting on the brakes or throttle at this point would have been fruitless.

Knowing when to brake and when not to brake is important. This driver wants that one front tire to be rolling instead of skidding so he can at least aim his car as it lands. Photo by George Jirka.

jounce travel from forward weight transfer. The extra inches of front suspension travel gained by releasing the brakes may be just enough to keep from breaking something.

Another time you can do more harm than good with the brakes is when you're traveling along at a fairly good clip and the car starts to get *crossed-up*—sideways. With most of the weight at the rear end of the car, when one rear tire hits a soft spot, berm or hole, the rear suspension will sometimes hop or swerve from side to side.

This is not the time to jam on the brakes. Let's take a look at why the car sometimes acts like this and how sudden braking affects handling.

Rear-End Hop—A VW off-road car acts similar to a passenger car. When the tires on one side hit a puddle of water, pile of snow or the soft shoulder of the road, the car is pulled or turned in that direction.

The rear of a VW off-road car is most affected because that's where the majority of the car's weight and the large, low-pressure tires are. As the rear of the car is pitched to one side, the rear wheels are no longer pointed in the direction the car is traveling.

The reaction of the tires to this sudden change in attitude causes the rear of the car to swing, or hop back to the opposite side, setting up a fishtail motion, or oscillation. This characteristic is even more violent on a swing-axle car because of its severe camber change. See what I say about camber on page 68.

Another thing that increases this rear-end reaction is rear-wheel toe-out. I can't stress proper rear toe-in enough. There are a lot of good drivers who've gone on their heads when they shouldn't have because the rear tires were toed-out. Make sure you've adjusted it as outlined on page 74.

The novice usually reacts to rear-end hop by getting off the throttle, overcorrecting with the steering wheel and getting on the brakes—all wrong. The combination of these "corrections" causes the hopping and fishtailing to become increasingly violent until it's "adios amigos" and over you go.

Here's the correct method of "saving" a car that is fishtailing: Slow gradually by *easing* off the throttle, staying off the brakes and keeping the front wheels pointed in the direction of travel. *Don't overdo it.* Although this experience will have you shaking in your boots later, you may be able to regain control of the car as it slows.

There's no way to drive in the dirt for any length of time without feeling like the rear end is trying to pass you now and then. Sooner or later the rear will get loose. When this happens, how you react determines whether or not you'll keep the shiny side up.

If you hit something hard enough, all the driving ability in the world won't keep you from performing some of the finer acrobatic maneuvers—items such as the "desert snap-roll" or "slow-motion endo with half twist." If you have the room, don't hesitate to steer off the road to maintain control. Even with your foot on the throttle during the recovery procedure, the car will slow down a good bit with the rear end hopping from one side of the road to the other.

One type of terrain that's great for aggravating rear-end hop is the sand or fine gravel found in dry river beds—called *washes* in the desert. When the rear end starts hopping around and lands in this soft stuff, it takes more than a little power to keep rolling. You must use fairly heavy throttle to maintain speed to keep from bogging.

Don't Lug the Engine—When driving in a sand wash, don't lug the engine. Shift to a lower gear so there is enough power to keep moving in case the tires start sinking in. **Keep engine rpm up!** This also applies any time you're driving fast on an unfamiliar road or one with deep tracks or a soft shoulder. When traveling on a "squirrely" stretch of road that you are unfamiliar with, stay in a lower gear even though engine revs are on the high side. Listen to the engine so you don't overrev it.

Don't Stop in the Soft Stuff—When driving in soft sand, silt, deep gravel, and so on, *don't stop*! A VW will go through most of this stuff without getting stuck *if you keep it moving.* If you stop, the only way out might be on the end of a tow strap.

Here they are folks: Combination "desert snap-roll" and "slow-motion endo with half twist." Photos by George Jirka.

If you have to stop in a sand wash, drive onto a rocky area where the surface is harder. If there's no hard rocky area, head for the brush at the side or up on a slope so you can restart going downhill.

Turning too sharply will also get you stuck in the soft stuff. Never turn your wheels sharply to one side or the other, except maybe when attempting to get out of deep tracks.

Sometimes, especially on heavily used race courses, ruts are so deep and loosely packed that once you get the wheels in them, it's tough to get out without getting stuck. In this case, steer straight ahead and keep moving. If it looks like the road gets worse, maintain speed and look for a good spot to make a move. Don't be hesitant to push on—the ruts were made by a lot of cars that went through ahead of you.

Loose sand is great for getting the rear end of a VW off-road car hopping around—especially if it's a swing-axle car. This can be the end result. Action like this seems to turn everybody into an instant pit crew. But if you're all alone, it may be a long walk home. Photo by George Jirka.

Scouting Ahead—If the car is banged up and you're limping out, or if you come to a particularly bad spot in the road, don't hesitate to stop—but not in the soft stuff! Get out and scout ahead to see if you can make it. If you can, find out the best line to take. If it's on a hill, take some measures to make sure the car stays put before leaving it unattended. Although a handbrake is a handy piece of equipment on any off-road car, don't trust it. Place a couple of rocks under the tires.

Backing Down a Steep Hill—One scary maneuver is backing down a steep hill you couldn't quite climb. If you ever find yourself in this predicament, stop! Take stock of the situation and figure out how to get down before doing anything foolish.

First, look around! Is the dip at the bottom too severe to back through? Is there a place on the hill to turn around? Are there obstacles you definitely want to miss on the way down? Is there an area at the bottom you want to head for if the car gets going too fast? Are there *rain ruts*—ruts caused by erosion—or deep tracks that could help guide you down?

If it's a long, steep hill, you may get only one shot at it once the car starts rolling backwards. Make sure the front wheels are straight. A good way to do this is to use your hand to make a note of the steering-wheel position, once straight, and keep it there all the way down.

If you lose track of the position of the front wheels and the car gets crossed up while sliding down the hill, *you're in big trouble.* The car may come around on you before you realize how to correct. It's like trying to tie a tie in a mirror. Everything is backwards!

Some race-car drivers use a piece of tape on the top or bottom wheel to indicate straight ahead. Sounds stupid,

Alright, fair is fair! Here's your author paying the price for making a "slight driving error." Photo by Trackside.

but you can get confused as to which end is up.

When you're all set, loosen your seatbelt enough so you can turn to get a good view of the rear. Start the engine and shift into reverse. Get the car moving slowly and concentrate on working the brakes.

Ideally, you want to use as much brake as possible without locking the rears. As you're backing down the hill, the front wheels will lock up first. As long as the car is going straight, that's OK. If the car starts to get crossed-up, ease off the brakes so the front wheels roll. This should let you straighten the car without much steering-wheel correction.

If the car gets going faster than the engine can handle, *release the clutch.* Believe it or not, it's better to let the car roll faster than you'd like than to have the rear tires start sliding because of the engine's braking. Get sideways on a steep hill and you'll probably get the car upside down and roll, over and over, all the way to the bottom.

Don't Outdrive the Headlights—When driving at night in unfamiliar terrain, never "outdrive your lights." You may get one heck of a sinking feeling in your stomach. Imagine coming over a rise full blast with the headlights pointed up in the air, then finding that the road turns sharply to the right. You have no choice but to

A hard bump may spin the steering wheel out of your grasp. Don't put your thumbs inside the wheel spokes. The spokes can do nasty things to them.

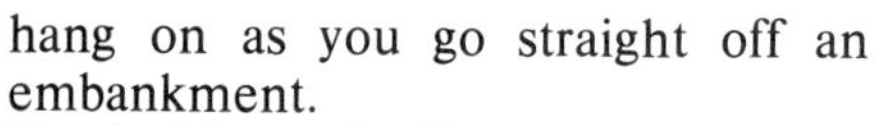

hang on as you go straight off an embankment.

Don't Abuse the Transaxle—As you should already know from the transaxle chapter, the transaxle is one of the more fragile pieces on a bug or buggy. It's also one of the most expensive items to fix.

The best way to keep transaxle expense down is not to break it. And the best way not to break a transaxle is not to abuse it. Simple common sense with the throttle, clutch and shift lever will add many miles to the life of your transaxle. Operate the shifter, clutch and throttle smoothly. For example, there's no reason for full throttle when the rear wheels are turning at only half-throttle speed. Likewise, there's no reason for full power with the rear wheels off the ground.

On rough roads where the rear wheels are bouncing up and down, there is hardly ever a reason for full-throttle operation. While the tires are in the air, they spin faster while the driveline is unloaded and engine revs increase. When the tires come back to earth, they are slowed *instantly* to match the speed of the car. This really shocks the driveline. This will just tear up the transaxle and you won't go any faster.

If you want to go faster, apply only enough throttle to increase speed without spinning the wheels needlessly. Concentrate on steering through the smoothest line. There's no reason to speed or power shift in the dirt except when you may bog down in the soft

Normally, off-road cars run against the clock and intimidating terrain, but not each other. But in the excitement of closed-course or stadium racing, one car can climb up the back of another. Photo courtesy of Bridgestone Tire.

stuff and get stuck. Smooth, positive shifts in the rough will save wear and tear on the whole drive train and keep more money in your pocket.

If it sounds like I'm telling you to baby your car, I'm not. These are simple driving tips that should be made part of your driving style. If you follow them, you'll spend more time driving your car than repairing it.

Some racers are always having transaxle trouble. Generally the problem is not the transaxle, but a short circuit between their brain and their right foot. You can drive just as fast being smooth and positive as the herky-jerky driver who pumps the throttle.

Final Advice—Here's some advice for faster off-road going. Don't practice in the middle of nowhere on a road you're not familiar with. When you are driving all-out to see what you and your car can do, pick a road or an area you're familiar with. It should be within walking distance of help. Remove any heavy, loose objects from the car. Pull your seat belts down tight—like you'd cinch the saddle down on a horse.

Most experienced off-road racers don't grip the steering wheel with their thumbs wrapped around the rim. This is because a hard jar to the front wheels—like a wheel dropping in a hole—can spin the the steering wheel out of their grip. Believe me, those pretty chrome steering-wheel spokes can do nasty things to your thumbs. So keep your thumbs on the wheel rim, not around it.

Chapter 14 SOLUTIONS TO OFF-ROAD BREAKDOWNS

Not the way to start the day. Center torsion-bar adjuster broke loose. The suspension collapsed resulting in gobs of negative camber. No, the car doesn't belong to a low-rider. Photo by Judy Smith.

Although you may not know it by name, I'm sure you know the law—Murphy's Law. It states that, "If anything can go wrong, it will." If your car breaks when driving on the street, you suffer the inconvenience of having to call for help. You may also experience the expense and embarrassment of using a tow truck. When your car breaks down out in the dirt, miles from nowhere, the result can be much more than a simple inconvenience.

Most off-roaders would rather leave the hiking to their "buddies" in the Sierra Club and drive their car out no matter how long it takes. The obvious problem with walking for help is leaving your car unattended. When you're with companions this is not much of a problem. There never seems to be a shortage of volunteers to stay with the car. On the other hand, it's understandable that wives and girlfriends don't like being left alone in the middle of nowhere.

The only time you should walk a long way for help is when you absolutely have to. First try to find out what's wrong. Next, sit down and think of *any way* to fix, replace, bypass or remove the problem so you can limp out. Only when you've eliminated all your options should you consider walking for help.

The best insurance against having to walk home is to travel in pairs. One car can tow the other. All that's needed is a tow strap—preferably nylon. If both cars break down, "rob" parts from one car to get the other one going.

If you must leave one car to get help or parts, push it off the road and out of sight. Some people are better left untempted. Just make sure you will be able to find it. If you leave a car out in the middle of nowhere for any length of time, your chances of finding it untouched when you return are not too good.

Some of the solutions listed in this section are so simple that they seem silly to mention. But you'd be surprised how often simple solutions are overlooked. Also listed are tools and spare parts that are sometimes the only way to solve a problem.

How much equipment you take along depends on the condition of your car and how far you are from civilization. If your car is in top shape and you are not far from help, you won't need many tools or spares.

Let's look at some problems you may encounter. They range from common to most unusual. For each problem I'll give you some options on helpful solutions. I'll also give you tips on any special tools you'll need.

ENGINE QUITS

If your engine was running fine, but it quit all of a sudden, finding the cause should be simple. Most of the time it will be either a fuel or electrical problem.

First look for loose wires at the coil or distributor. If none are found, make some fuel-system checks.

Fuel-System Checks—Check that the carburetor is getting gas. Remove the air-filter hose and/or top of the filter housing. Look inside the throat

Whether the lower shock mount was poorly designed or installed, at least help was close. Photo by Judy Smith.

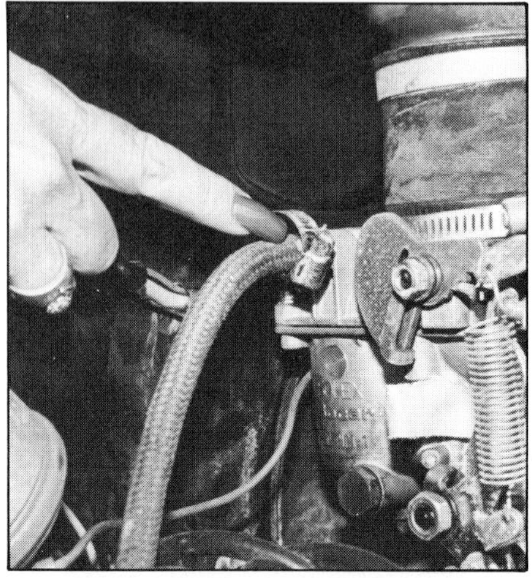

On most carbs, this connection has a built-in filter. A needle valve is also located here. Sometimes these valves stick and can be freed by rapping on top of the float-bowl cover. Manicure job isn't bad either.

Fuel pump used in '66—'70 sedans. Note fuel-screen access plug (arrow) on side. Install this one when replacing a defective pump. Although this pump was not designed to be used on engines equipped with alternators, a little grinding will let it fit.

of the carburetor and work the throttle linkage. There should be a squirt of gas from the accelerator pump every time you open the throttle. If there is, check the ignition. If fuel isn't squirting into the carburetor, make some more fuel-system checks starting with the tank.

Make sure there's gas in the tank. Yes folks, it's been known to happen—frequently! If the tank is empty, you'd be amazed what that long walk can do to prevent your ever running out of gas again.

If there's gas in the tank, work from the carburetor to the tank to isolate the problem. Disconnect the hose at the carburetor and crank the engine. *Be careful not to let gas squirt on the plug wires or hot exhaust.* A safe method is to put the end of the fuel hose into an empty can. If gas squirts out, the problem is the carburetor. The float needle valve is probably stuck or clogged.

A stuck valve can sometimes be freed by striking the top of the carburetor *lightly* where the fuel hose connects. If you don't have a hammer or wrench, a rock will do. If that doesn't work, remove the top of the carburetor and look for the problem.

On Zenith and Solex carburetors, check for a clogged main jet. Access to the main jet is through a large hex bolt. Remove the hex bolt with an open-end wrench. With a screwdriver, unscrew the main jet. Blow through it. *Do not run a wire through it.* If the jet was not clogged, continue working back toward the tank.

CAUTION: When working on fuel lines, fuel pump or carburetor, especially on a hot engine, have a fire extinguisher handy.

If no gas came out the fuel hose when you cranked the engine, check for a clogged filter. I recommend using a see-through fuel filter. But even a see-through filter that appears to be clear can be restricting flow. Disconnect the filter and blow through it. If it's restricted, replace it, bypass it or poke a hole through the filter element and replace the filter as soon as you can. Reconnect the fuel line and be on your way.

If gas is not reaching the filter, go to the fuel pump. Disconnect the fuel line from the tank at the pump. Check for gas. If necessary, suck on the fuel line to check for blockage. Be careful. *Do not swallow any gas!*

If gas is reaching the fuel pump, the pump is the problem. See the section on fuel pumps for details. If there is no gas at the pump, the line is clogged somewhere between the tank and pump, or the screen inside the tank is blocked.

Clogged Fuel Line—If you've retained the stock tank and there's gas in it, but none is getting to the fuel pump, the tank screen may be clogged. Blowing into the fuel line may dislodge the restriction. If this solves the problem, it will only be a temporary fix. Whatever caused the restriction will collect on the screen again. Clean out the tank when you get home.

If blowing through the fuel line doesn't clear the blockage, disconnect the flexible hose from the tank. On sedans, it connects to a metal line at the forward end of the floorpan tunnel. Have a pencil or small bolt ready to plug the hose.

If little or no gas comes out of the tank hose, remove the screen and clean it. This job is easier and you'll spill less gas if you remove the tank.

If gas flows freely out of the tank hose, the blockage is in the line to the pump. On sedans, it's a metal line in the tunnel running the length of the car. Use a throttle cable—preferably a spare—to dislodge any obstruction.

Mechanical-Fuel-Pump Problems—If the mechanical fuel pump quits, first make sure the fuel filter is clean. On '61—'65 and '71—'74 Type-1 VWs, remove the top-cover bolt, cover, gasket and filter screen. On '66—'70 Type-1s, remove the plug on the side, the gasket and thimble-like screen. Check the screen for blockage. If it's clogged, clean it or blow it out.

On '70-and-earlier pumps, if the screen is OK, remove the six screws holding the body together and check the diaphragm. If the diaphragm is not falling apart, the problem is probably in one of the valves. If the diaphragm is damaged, you'll have to replace the diaphragm or pump. This will require a spare, or running without a pump as explained on the next page.

Fuel pump used in '71—'73 sedans equipped with generators. A similar lay-down version with a shorter pump rod is used in '73—'74 models with alternators. Because the pump-body halves are crimped together, the only serviceable item is the filter. If the diaphragm or reed valves fail, it's goodbye pump.

The '71—'74 fuel pumps are crimped together and cannot be disassembled. If the screen is OK on one of these pumps, and it still doesn't pump, the pump is shot.

Behind both the inlet tube (gas-tank side) and the outlet tube (carburetor side) is a small reed valve. If dirt has lodged between either valve and its seat, holding it open, the pump won't work.

Check the valves by blowing gently into the outlet tube. Then suck gently on the inlet tube. Be careful not to swallow any gas. If either valve leaks, you've found the problem. On '66—'70 pumps, you can see both valves from the underside. On '61—'65 pumps, you can see the outlet valve from the underside and the inlet valve from the top.

If dirt is the problem, use a combination of blowing and sucking and working the stuck valve with a small screwdriver or wire to dislodge the dirt. If the valve is broken, forget it. You can't fix it. Replace the pump or run without one.

When replacing the pump cover on '61—'65 and '71—'74 models, put a dab of silicone sealer under the head of the retaining screw. Otherwise, it may leak.

Electric-Fuel-Pump Problems—If the problem is a non-functioning aftermarket electric fuel pump, your options are limited. If you have a continuity tester, check for power to the pump with the ignition on. If you don't have a tester, disconnect the hot wire at the pump. Turn the switch on. Touch the metal end of the wire to the fuel-pump terminal. If you get a spark, it's "hot." If the pump isn't running, the pump is likely jammed or burned out.

Reconnect the fuel line to the carburetor. With the ignition on, try to free the pump by rapping it lightly with a wooden hammer handle or something similar. If nothing happens, read on.

If there is no power at the pump, trace the wiring back to the ignition switch. Don't forget to check the fuse box.

CAUTION: Be very careful when checking for fuel-pump flow. An electric fuel pump will pump gas when the ignition is on whether the engine is running or not.

Running Without a Fuel Pump—If you have a spare fuel pump, no problem. Just put it on and away you go. If not, here's how you get home. Rig up a gravity fuel-feed system.

If you have an auxiliary tank on the shelf below the rear window, you're already set. Run a fuel line from this tank directly to the carburetor. You may have to punch a hole through the sheet metal with a screwdriver to get the line to the carburetor. In this setup, gravity will take the place of the fuel pump. Gravity feed worked for the Ford Model T and still works for motorcycles, so it will work for your VW.

If you don't have a tank in the rear window, find something that will hold gasoline: a water jug—one that's not affected by gas—or a windshield-washer bottle. If you have a sedan, you can move the stock front tank to a location where gravity feed will work. Mount the container or gas tank on the shelf, below the rear window and route the fuel line as I just explained.

CAUTION: When removing the stock gas tank, don't touch the horn connection on the steering shaft with the tank. Pull the horn fuse so the circuit is dead. The resulting sparks can ignite the gas vapors.

You only need about two feet of hose to make this work. Pull the press-fit fuel-outlet tube from the pump with needle-nose pliers. By using this tube as a splice, most cars will have enough neoprene hose and metal fuel line to put together more than the two feet needed.

Any problem you will have depends on how you "jury rig" a temporary fuel system. For example, if you don't

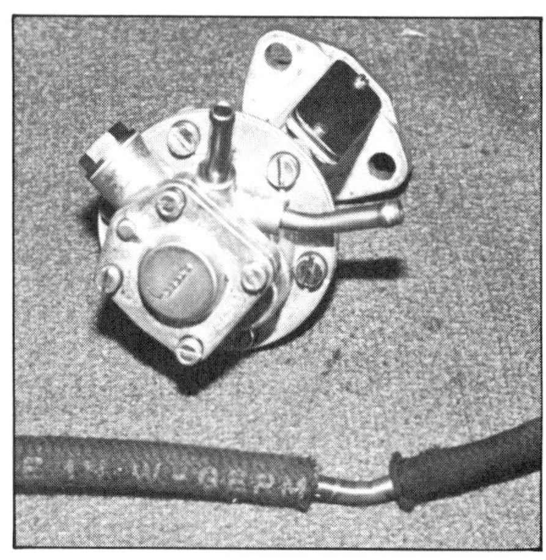

Using one of the fuel-pump tubes, you can splice together enough hose to bypass a defective pump and use gravity feed. 1961—'65 pumps also have a pressed-in outlet tube that can be used to make a splice.

want to punch a hole in the bottom of a plastic water jug and wait for some silicone sealer to dry, you will have to siphon the fuel—draw the gas out the top rather than the bottom—to get gas to the carb. The problem with siphoning is that when the gas gets low, the sloshing around will break the siphon. This just means refilling the jug sooner.

The smaller the container, the more often you'll have to refill it, so figure out an easy way to get gas out of the tank. Unless you can mount the container securely, have someone get in the back and hold it. If you are alone, duct tape the container in place and take it easy.

No matter what you do, it will be slow going home. Remember that with little fuel pressure, flow to the carburetor-float bowl will be slow. Consequently, the carb may run lean at speeds not much higher than idle, so take it easy. Don't use high revs and don't lug the engine.

WARNING: Gasoline is very dangerous! With an open container of gas in the car, keep any windows open for ventilation. If you must smoke, stop and get out. Otherwise, you may end up with bigger problems than a faulty fuel pump.

Ignition-System Checks—If gas squirted into the carburetor throat when you pumped the throttle, the problem is probably the ignition. Before checking for ignition spark, wipe up any spilled gas.

To check for spark at the plugs, pull one of the plug wires off at the plug.

Poor-man's remote starter switch. With a screwdriver at starter solenoid, bridge between positive battery-cable connection and ignition-switch wire to crank engine. Sparks will fly. If ignition is on, engine may start.

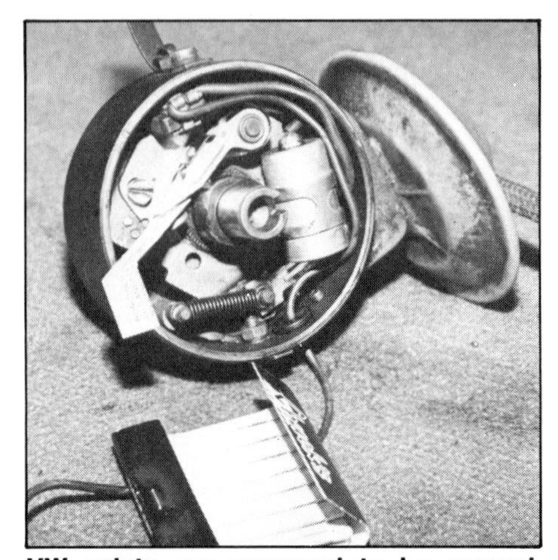

VW points are supposed to be gapped about 0.016 in., but they will work at a few thousandths on either side of that. A common matchbook cover is close and much better than a guess.

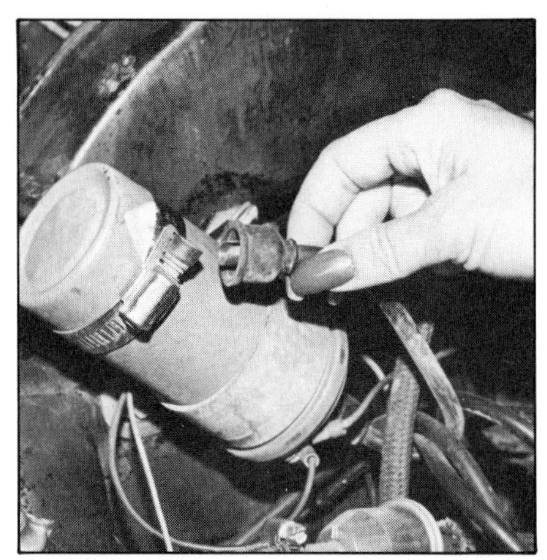

A quick ignition-system check can be made by pulling the coil high-tension wire out of the center of the distributor cap and holding it about 1/4 in. away from a good ground.

Slide a wooden- or plastic-handle Phillips-head screwdriver into the plug socket.

Hold the side of the screwdriver shaft about 1/4-in. from a ground. Any metal part of the engine or body will do. *Don't hold onto the metal screwdriver shaft while cranking the engine. You'll be jolted by 20,000 volts.* Be sure the car is in neutral. Crank the engine.

If you're all by yourself, you can crank over the engine with a homemade remote-starter switch. Make the switch out of a long piece of wire. Wrap one end around the large solenoid terminal and bring the other end up to the engine. With the ignition on, touch this wire to the hot connection on the generator/alternator. The engine should crank over.

If there is no spark, remove the distributor cap. Check the spring-loaded carbon brush in the center of the cap. If it's OK, check the points.

The most frequent off-road ignition problem is dirt between the point contacts. A wipe with a clean piece of cloth, matchbook cover, or business card will usually clean them. If you have an ignition-point file, great. Just get the contact surfaces as clean as possible. A piece of cloth or other foreign material may allow the spark to jump the point gap.

To check if you've corrected the problem, open and close the points with the ignition on. A blue spark should be visible at the points as they open. You may have to shade the distributor to see the spark. If there's spark across the points, check for spark at the end of the coil high-tension wire when you open the points. If OK, install the cap and rotor. Check for spark at the plug leads.

To check if you're getting juice to the coil, pull off the wire from the 15 (or positive) terminal. With the switch turned on, you should get a spark when the wire is touched to the terminal or ground.

If you removed the points to clean them, or just wanted to check the gap and don't have a feeler gage, use a matchbook cover. It will be close to the recommended 0.016-in. point gap. Also, beverage-can pull-off tabs measure 0.014—0.016 in. If you're in a real pinch, "eyeball" it. The idea is to get the engine running well enough to drive out.

When checking the points, if the spark that jumps across the point gap is red or reddish white instead of blue, the coil is probably weak. A weak coil may quit when it gets hot, then work again after it cools.

If you're not getting any spark to the points, check if the coil is getting juice from the ignition switch. With the ignition on, pull the wire off the coil terminal marked 15 (+). Touch the wire to the body of the car or any good ground—just for an instant. If the wire is hot, it should throw a small spark. If you're getting juice to the coil but not to the points, the coil is bad.

A simple test light—continuity tester—will make finding ignition troubles a whole lot easier. Carry one! And the coil problem is easy to fix. Just replace it with the spare you always carry—right?

CARBURETOR PROBLEMS

A disassembled carburetor looks intimidating, but its basic operation is simple. If the carburetor worked fine when you left, the chances of it causing the engine to quit are remote. Usually carburetor problems are just a matter of it not working as well as you'd like and showing up as a sputtering, bogging engine.

$1.98 Tuneup—If you do have trouble with the carburetor, don't tear it apart. Try this first. A good friend of mine, and a fellow racer, runs a local Volkswagen shop in Los Angeles. A number of times I've seen him perform what he calls his "$1.98 tuneup."

Remove the air-cleaner connection at the carburetor. Start the engine. Move the carburetor-throttle lever to its wide-open position. When the engine reaches about 3000 rpm, put the palm of your hand tightly over the carburetor throat. The engine, now without air, will load up and slow down. Just before the engine stops—about 1500 rpm—take your hand off the carburetor but keep the

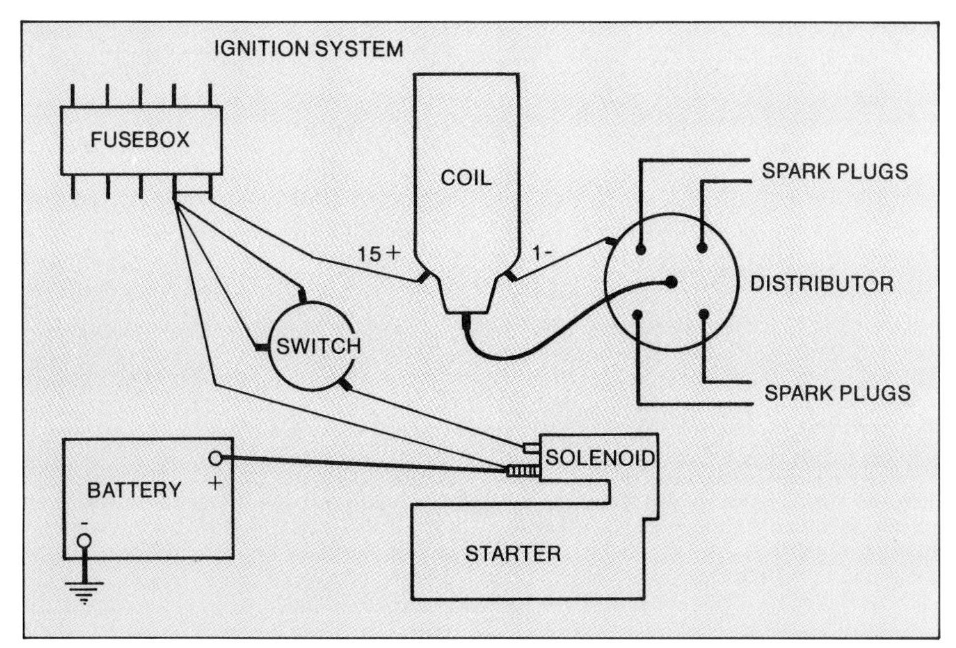

Basic wiring schematic for '67-and-earlier cars. On '68-and-later VWs, the ignition switch is part of the steering column and the electrical connections are not readily accessible. If the switch fails or you lose your key, run the wire from your license plate or parking lights to the 15 side of the coil. Now you can use the first notch on your headlight switch as your ignition switch and jump the terminals on your starter to crank the engine. Neat, huh?

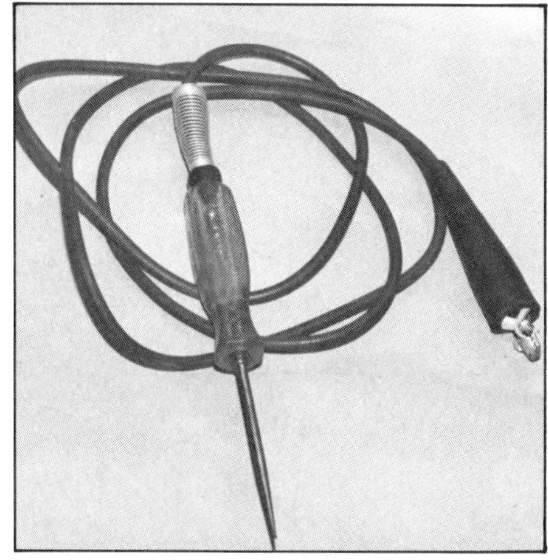

A simple continuity tester is a valuable tool for chasing down electrical problems.

The $1.98 Tuneup. A very small piece of dirt in the right place can cause hesitations and flat spots in engine performance. See the text before you start tearing the carb apart. Don't burn your arm on the exhaust like this guy did.

throttle wide open. The engine will sputter back to life. While keeping the throttle open, let the engine wind back to 3000 rpm and repeat.

This applies all the suction, or vacuum, the engine can produce directly to all the carburetor air bleeds, ports, passages and jets to remove any obstructions.

I'm sure some Volkswagen mechanics will scoff at this procedure. This is because they've never tried it, or would rather rebuild your carburetor for big bucks. All I can say is try it first. I've seen the $1.98 tuneup used successfully on cars running so badly that the owner would gladly pay for a rebuild. I've used it to take flat spots and hesitations out of my own race car on a number of occasions.

DEAD CYLINDER

A Volkswagen engine can go a long way on three of its four cylinders, but it's damaging itself in the process.

Each pair of barrels or cylinders are separate pieces tied together at one end by the case and at the other end by a common cylinder head. With both cylinders firing, heat expands both barrels and sides of the head equally.

But if only one cylinder fires, the "dead" cylinder is being cooled by the unburned air/fuel mixture. The head, case and barrels are all being stressed by uneven *thermal* expansion. The damage being done won't stop you from driving out. But at today's high parts and labor costs, it's worth your time to try to get that fourth cylinder firing.

Check the ignition system from the distributor cap to the plug gap. Check the pushrod tubes for damage. Remove the valve cover to check the rocker arms and adjusters. If the spark plug is oily, clean it off and narrow the gap. You can check things like plugs and plug wires by switching them with another cylinder. Better yet, carry a spare plug and wire.

Don't drive out on three cylinders if you don't have to. Find the problem and fix it, if possible.

FAN BELT

Don't ever go off-road without an extra fan belt. There are a number of different-size VW fan belts, so make sure the spare is for your engine. Match it up with the old belt. Even better, try it on for size and run the engine for at least 10 minutes before venturing out. This will stretch the belt to its working size. You may have to readjust belt tension afterwards to take out excess slack. Remember, if you install a *power pulley*, page 18, you'll need a shorter-than-stock belt.

Without a belt, an air-cooled VW's fan doesn't spin. It will take you a long time to get anywhere without destroying a beltless engine because you have to shut it off about every mile or so to let it cool down. The alternative is damaging your engine if you push it too far without stopping. It won't take long before the engine will seize a piston.

And you'll be running off the battery because the generator/alternator won't be working. At night, most batteries won't last half an hour powering driving lights.

GENERATOR/ALTERNATOR AND REGULATOR

There isn't a whole lot that can be done to fix a generator, alternator or regulator if one quits while you are out in the dirt. However, periodically and before a long trip, you should check generator brushes for excessive wear. Alternator brushes last longer

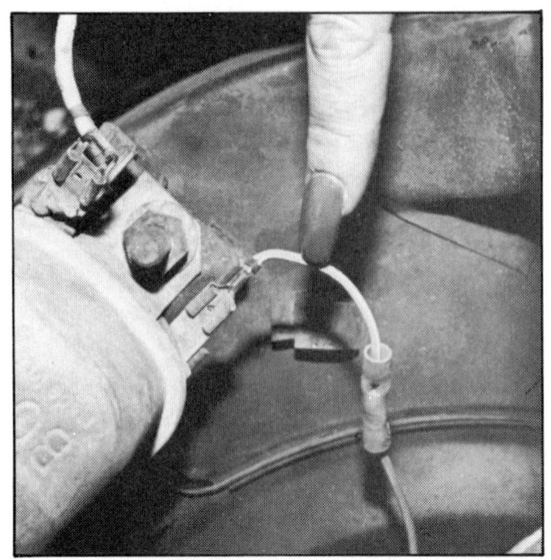

Charging-system idiot light on? Disconnect this little wire at the voltage regulator while the engine is running. If the light goes out, you're probably getting a true reading—probably bad brushes or stuck regulator. If it stays on, you've got a short in the wiring.

Typical race-car arrangement. Two coils mounted side by side so the wires can be rapidly switched from one to the other in the heat of battle.

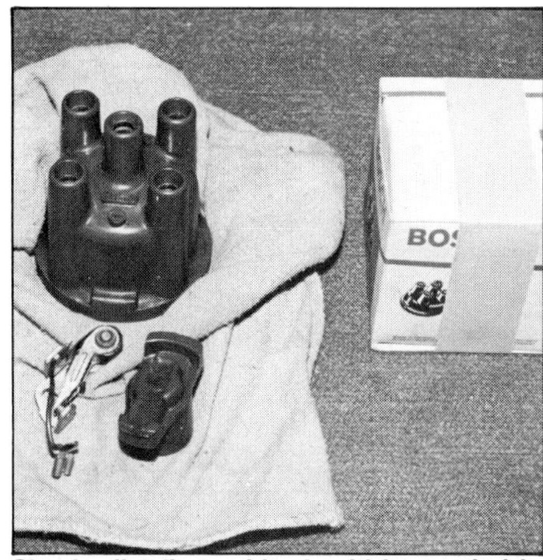

Simple distributor kit should be carried in every off-road car. Wrap the points and rotor in rag, stick the rag inside the cap and put the cap in the box. Tape the box.

and aren't much of a problem, but check them too. Always keep the fan belt adjusted.

If you find yourself off-road with the red generator/alternator warning light on, it's time to ask yourself a few questions. Did you shed the fan belt? Where are you in relation to where you started from? Where are you going? How much daylight is left?

When the charging system quits, you can safely figure on 3—6 hours of daylight running time. This, of course, depends on what shape the battery is in and what juice-robbing accessories, such as driving lights, you can do without.

This includes the starter. A starter requires more current than any other piece of equipment. So if you must shut off the engine, stop your car on an incline so it can be restarted by rolling downhill and popping the clutch.

Before heading back to civilization for repairs, make this quick check to see if your idiot light is giving you a correct reading. With the engine running, disconnect the small warning-light wire from the voltage regulator—terminal 61—and see if the warning light goes out. If the light goes out, you're probably getting a true reading. If it doesn't go out, the light is going on because the wire has shorted to ground somewhere in the car and the charging system is probably OK.

IGNITION COIL

One thing that is almost impossible to patch or substitute is the ignition coil. Most race cars have two coils mounted side by side—one being used, one a spare. If one coil fails, all the driver does is switch three wires to the spare coil and be on his way. Recreational off-roaders need not go to the trouble of mounting an extra coil, but it's a good idea to have one along on a long trip.

DISTRIBUTOR CAP, POINTS AND ROTOR

Most experienced off-roaders, especially ones who have been stranded once, carry a simple distributor kit. Next time you buy a new distributor cap, take the old one and put a rotor and set of points inside it. Stuff the cap with a rag or paper towel to keep these spares from rattling around and put the old cap in the new box. Run a couple of wraps of duct tape around the box both ways and stow your new "distributor kit" in the glove box or toolbox or tape it to a roll bar.

Suppose the distributor cap broke but it was taped—as explained in the engine chapter—and you have all the pieces. You might, with lots of patience, time, glue, tape, etc., patch it up and limp home. But carrying a spare cap is much easier. The same is true of the rotor. Ignition points are next to impossible to fix.

Be careful while working on your distributor cap. If you lose the small, round, spring-loaded carbon brush in the cap that makes contact with the rotor, you're in trouble. This brush delivers the 20,000 or so volts from the coil to the rotor.

Your car won't go anywhere without the brush. If you don't have a spare distributor cap, better start looking for the brush. It's either down in the distributor, somewhere on the engine or skid plate or it has fallen to the ground. The time spent looking for the carbon brush is well spent. If you can't find it, you must improvise. Then you will have the distributor cap on and off several times to fix whatever you substituted for the brush.

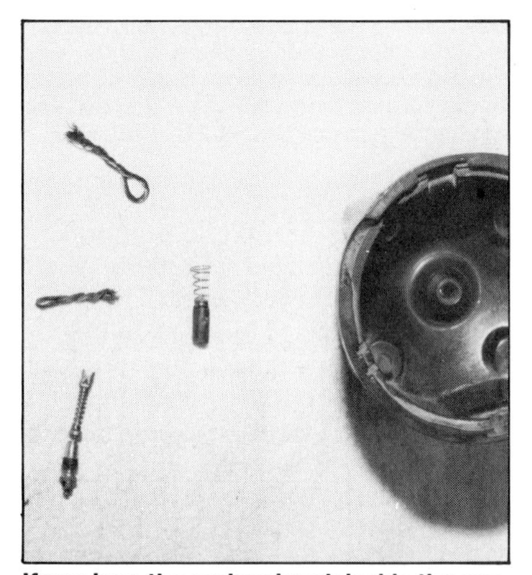

If you lose the carbon brush inside the cap, try using a ball-point-pen spring, a tire valve, a piece of heavy-gage wire or whatever. Anything beats walking.

If you can't find the carbon brush, use its spring if it is still in the center of the cap. Stretch the spring so it will contact the rotor. If the spring is also gone, use a spring out of a ball-point pen or something similar. If you really get desperate, try using a short piece of heavy-gage wire in its place. See the accompanying photo.

Another must is an adjustable pushrod tube. It allows pushrod tube to be replaced in field without removing cylinder head. Photo courtesy of C.B. Enterprises.

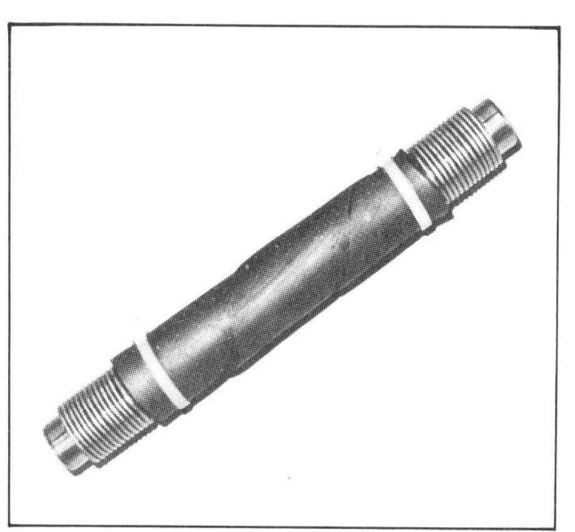

A razor blade, a piece of hose and a couple of cable-ties and you've got good protection for those thin-walled pushrod tubes.

PUSHROD TUBES

Anytime you notice fresh oil around the engine, transaxle or skid plate, check it out immediately. The air-cooled VW engine can't afford to lose much oil. You may be able to catch the problem before too much crankcase oil is lost.

One common problem is a bent or leaking pushrod tube. The stock tubes are made out of very thin material. Even such things as mud or sticks that get packed between the skid plate and the engine can bend these tubes and cause them to leak.

Pushrod tubes are made with a welded lengthwise seam. This weld is the weakest part of the tube. Sharp VW mechanics always install the tubes with the seam at the top so if it opens up, little oil will be lost.

If a pushrod tube is damaged badly enough to cause a leak, the pushrod may be bent. Don't drive the car any farther until it is fixed. Begin by removing the skid plate so you'll have access to do the job right.

Now that you have access to the tube, you can fix the tube and maybe the pushrod one of two ways. If you came prepared, you'll have an adjustable pushrod tube stashed in the car. You should also have a spare pushrod. These adjustable pushrod tubes are available for a few dollars at most foreign-car auto-parts stores.

If you didn't come prepared, you must fix the damaged tube and pushrod as best you can. Let's take a look at both situations.

Adjustable Pushrod Tube—To install an adjustable pushrod tube, remove the valve cover and rocker-arm assembly. Pull the pushrod from the damaged tube. Pry out the tube from underneath and adjust the new tube to fit in its place. Some are spring-loaded and others are threaded to expand into place.

First, visually check to see if it's straight. To double-check, pull the pushrod and roll it against a flat surface such as the car's side-window glass. If it's bent, carefully straighten it as best you can. Reinstall the pushrod.

Replace the rocker-arm assembly and check valve clearance. If the pushrod was bent, adjust valve clearance to about 0.006 in. Make sure the valve is closed when you're making the adjustment. When the valve is closed, you should be able to wiggle the rocker arm and spin the pushrod. Install the valve cover and you're on your way.

Stock Pushrod-Tube Repair—If you don't carry an adjustable tube, follow this procedure. With a gasoline-soaked rag, T-shirt or whatever, clean the oil and dirt off the damaged tube. If it's not bent or split badly, simply wrap the tube several times with duct tape. Silicone sealer or epoxy between the tube and tape will help seal. Then, with a pair of pliers or Channellocks at each end, carefully rotate the tube so the taped-up hole or crack is at the top.

If the tube is badly damaged, or if you suspect the pushrod is bent, do the following. Remove the valve cover and the rocker assembly. Remove the pushrod and check for straightness as described above. Next, reach in through the head with a long screwdriver or something similar. While holding the tube from the outside with your hand, carefully work the bent tube back to something approximating its original shape.

If the tube is so damaged that it almost falls out when the pushrod is removed, you'll have to reinforce it. Tape alone won't do. Use a point file, pencils, small wrench, small sticks or a combination of all if need be. Along with these "splints," use some duct tape and hose clamps, cable ties or wire to get the job done. To give the tube some additional support, you can tape it to adjoining pushrod tubes.

Once you have it so you think it will hold, reinstall the pushrod and check valve clearance. When you're back on the road, stop occasionally to check on how the patch is holding up. If the repair job has greatly reduced the leak, you can keep going. Just keep an eye on the oil level. Probably the worst thing that may happen is that the leak will increase and you will have to put on a fresh patch later.

Pushrod-Tube Covers—One of the best ways to protect the pushrod tubes from damage from brush and gravel is to cover them with resilient material. Get some thin-wall rubber hose and cut the hose slightly shorter than the pushrod tubes. Next, split each hose making a spiral cut, as illustrated.

The hoses can be installed with the tubes in place. Secure the hose to each pushrod tube with a couple of cable ties to keep the hose on the tube.

THROTTLE CABLE

Unless your car sports one of those high-buck hydraulic throttle linkages, you should carry a spare throttle cable. If you don't carry one, at least have some bailing wire in the car. Wire has dozens of uses. When, not if, the throttle cable breaks, it usually breaks at the carburetor or accelerator pedal. Luckily, these areas are easy to get to.

Mend it the best you can with some wire. Expect it to break again after awhile at its splice. When that happens, all you can do is get out and fix it again. If you have no wire, use heavy-gage electrical wire from some part of your car and work the throttle gently. Keep a sharp eye out for a barbed-wire fence to "borrow" from. Do the owner a favor though and don't cut a section out of the fence.

Find an end with some excess wrapped around a pole.

If you have a buggy and the throttle cable breaks at the pedal, here's another alternative. Pull the cable from its housing and route it forward over the front roll-cage hoop. If you have a co-driver, he will operate the throttle by pulling on the cable at your command. If you're alone, tie the cable to the hoop.

Pull down on the cable to operate the throttle. When you let go of the cable so you can shift gears, the throttle will close, but the cable will stay there. Throttle operation won't be very smooth this way but you'll have some semblance of a throttle.

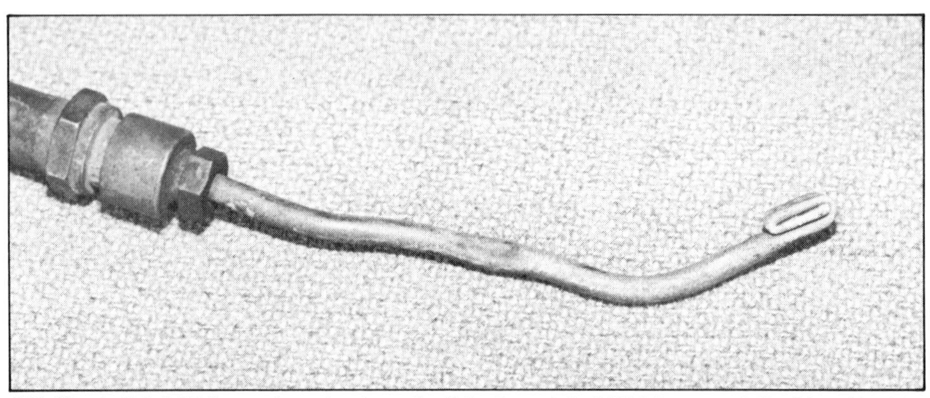

OK, I'll admit I did this one in a vise to make it look neat, but I think you get the idea. I had to do this to a rear brake early in a Mint 400 race. I completed the race using brakes on 3 wheels. I used Vise-Grip pliers to crimp the end shut. It never leaked a drop—Honest!

CLUTCH CABLE

If you break a clutch cable, forget trying to patch it. Because of its high loading, a patch just won't work. For this reason many racers use a hydraulic-clutch setup as detailed on page 44. Actually, the cable doesn't break as often as a crimp breaks and the front eye falls off.

If you don't carry a spare clutch cable, here's what you do. Start the engine with the transmission in gear. Match the gear to the speed you want to drive. You will have trouble starting out in 3rd or 4th gear, so use 1st or 2nd until you find a hill. Drive to the top of the hill, shut off the engine, go to 3rd or 4th gear and try again, this time rolling downhill.

Unless you know what you're doing, and are willing to risk trashing the synchros, don't try matching engine speeds to transmission gear ratios while your moving and "flat shift" or "crash box" the transaxle. The long shift linkage used in the VW makes "feeling" the gears very difficult. Regardless, the VW transaxle just can't take this kind of abuse. If you need more convincing, read about transaxles, page 38.

PROBLEMS WITH WIRING

If a wire burns up or a short develops that you can't find, bypass the problem. Steal a piece of wire, preferably of the same gage, from a component that's not essential to the car's operation. Better, yet, bring along some 10-gage wire for emergency use.

BROKEN BRAKE LINE

If your car experiences brake failure, *don't pump the brakes!* You'll probably just pump valuable brake fluid onto the ground. If you don't carry a spare can of brake fluid, you can't afford to lose any. Slow down and stop, using the handbrake, if your car has one. If you're going to have to run into something, pick something that will give, like bushes.

Nine times out of ten, a leak will be due to a failed steel or flexible brake line. But even if the leak is at a wheel cylinder or caliper, you can close it off.

First pinch the line shut, then cut it with pliers or whatever on the leak side of the pinch. Cut the metal line at a convenient spot between the master cylinder and the leak. As soon as you cut it, bend the end of the line that's connected to the master cylinder upward a few inches so fluid doesn't run out of the line.

Check the master-cylinder reservoir and make sure it has fluid in it. If fluid level is adequate, slowly depress the brake pedal once and hold it there. This will expel any air that may have entered the cut line. If you let up on the pedal, it will suck more air into the system. So find some way to hold the pedal down. Try a big rock.

Next, as illustrated, bend about one inch of the line back against itself and squeeze it flat with Vise-Grip or Channellock pliers. Bend the line back over on itself and flatten it a couple of more times.

Depress the brake pedal to check for any leaks. Remember, once you've crimped the line shut, you can't bleed air out of it. This will eliminate the brake at that wheel, but you'll have brakes at the other three.

If you've lost all the fluid in the reservoir, use water or even urine before resorting to oil. Oil may work, but it will attack all the rubber in the brake system. That includes flexible brake lines and the seals in the master cylinder and wheel cylinders or calipers.

BROKEN FRONT SUSPENSION

With a broken front-suspension component, the solution is about the same no matter what is broken. Replace the broken part if you have a spare and the necessary tools, or run on three wheels if you don't. It's possible to run on only one front wheel because of the VW's high rear weight bias.

The only part that can be straightened in the middle of nowhere is a tie rod. As described earlier, page 46, this job is easier if you can remove the tie rod. If the tie rod can't be removed, it's just a matter of prying it and bending it halfway straight.

If some other part of the suspension is bent so the tires are toed-in or toed-out badly enough that you can't drive, try this. Compensate for the bent part by readjusting the tie-rod ends. If the problem is greater than a bent part and you still have a long way to go, you're going to have to resort to three-wheeling it. Here's what you must do first.

Kingpin Suspension—Jack up the side opposite the broken component. Switch the front tire with the rear tire on the same side. The larger-diameter rear tire on the front and the smaller front tire on the rear will give the front end more ground clearance. Just take it easy over bumps and on the turns.

Jack up the other side and remove the wheel. Take off the complete spindle-and-brake assembly by removing the pinch bolts on the ends of the torsion arms and knocking out the link pins. You'll also have to cut and pinch off the metal brake line using the procedure I just described.

When a tie rod bends, the front wheels toe out. This racer hasn't taken the time to stop and bang the tie rod somewhat straight. Photo by Tom Monroe.

With the right front tire removed to represent a broken front end, the two tires on the left side were switched. Depending on how stiff the front suspension is, or how rough the road is, you might want to remove the left front fender to keep the larger rear tire from breaking it off.

If the lower torsion arm is not broken at the torsion tube, this is all that needs to be removed. If the lower torsion arm is broken at the torsion tube, remove it and put the upper torsion arm in the lower arm's position.

By leaving the lower torsion arm on and hanging down, you'll still have some protection for that corner of the car. The torsion arm will act like a spring-loaded skid when that corner of the car hits the ground—like the tail skid on an old airplane.

Ball-Joint Suspension—The ball joints are pressed into the torsion arms. At the steering knuckle, each joint is attached with a *castellated* nut and cotter pin. Ball-joint studs are always tight in the knuckle. Therefore, removal usually requires that they be pressed out or knocked out with a pickle fork or a ball-peen hammer and a punch.

There's no easy way to take the ball-joint front suspension apart. It all depends on what's broken, and what you can get loose. If all else fails, remove the brake drum and backing plate. Ride on the lower torsion arm.

Once you get the ball-joint suspension apart, do just as described for the kingpin suspension. That is, make sure there is a torsion arm on the bottom torsion bar to act as a skid.

If you put the upper torsion arm at the lower position and you can't press the ball joint out, don't worry about it. At first, the stud-end of the ball-joint will dig into the ground, but when you hit that first big bump the ball joint will pop out anyway. Also switch the tires as previously described for the kingpin suspension.

General Tips—If you break a tie-rod end and you're on a smooth road, the free wheel will generally track parallel to the wheel being steered. If you're out in the rough, treat a broken tie-rod end just like any other broken front-suspension part. Go to three wheels.

If you were careful taking the suspension apart and saved all the nuts and bolts, you might be able to find the part you need in the next town or pit somewhere down the road. Then you could reassemble the suspension and go on your way like nothing happened.

FLAT TIRES

Flat tires are common off-road. Depending on how well you're prepared, a flat can be either a simple inconvenience or a real pain. There are two schools of thought on how to deal with flat tires.

Most people simply carry a small front spare. The advantage is convenience. If a tire goes flat, you just jack up the car and change it. The disadvantages are the additional weight, the lack of space and the possibility of getting a flat rear tire. A front tire will work on the rear, but not very well. Still, it's better than nothing.

Then there is the second group of people who don't carry a spare. Instead, they bring an air pump and tubes or a patch kit. If you are familiar with the procedure and have the proper tire tools, you should be able to fix a flat in less than an hour.

Sparkplug air pumps are handy. Pull out a spark plug, screw in the pump, start the engine and run at idle for a couple of minutes until the tire fills.

If you have another car along, it's easy to break the bead on the flat. Lay the wheel and tire down and run over the tire. Run the driven car's tire right next to the rim of the flat tire. If you're alone, you can use a hammer to try to knock the bead loose. Or slide the tire under the car, under a good jacking point. Set the jack on the side of the tire, next to the bead. Using the car to push against, jack the tire off its bead. Simple!

Once the bead is broken loose, it's easy to push the rest of the tire bead away from the rim. Use a tire tool to pry one side of the tire off the rim and pull the tube out. Using a sparkplug air pump or an electric pump, blow up the tube and find the leak. If the leak is very small, listen very carefully. Spit is great for zeroing in on small leaks.

Breaking the bead loose from a flat tire. Lay a jack against the bead and use the weight of the car to break the bead. This method works better when the tire still has some air pressure.

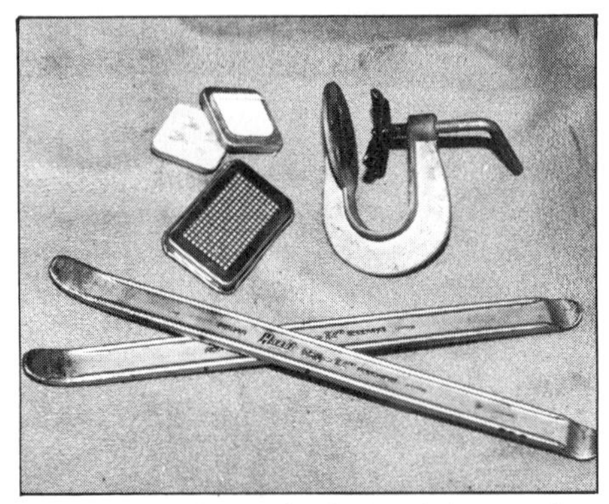

With these tire tools, hot patches and a clamp, you can fix as many flats as you have patches and energy.

A *hot patch*—one that uses a phosphorous charge lit by a match to form a vulcanized bond—will fix most leaks in minutes. Be sure to check for stickers, cactus spines or whatever caused the puncture. Carefully feel all around the inside surface of the tire before putting the tube back in. A rag is great for this because it easily finds stickers and saves your fingers. The advantage of patching is that you can fix any size tire that goes flat, small front or large rear. You can also fix as many flats as you have patches.

If you do find yourself out in the rough with a flat front tire and no spare or patch kit, drive out on three wheels. Switch the tires around as I describe on page 152. During an off-road race, you'd simply drive on a flat front. But if you're not in a hurry, it's foolish to destroy a wheel and tire. Murphy's Law is very clear on this. According to Murphy's Law, you'll drive only a few miles—enough to ruin the tire and tube—only to come across someone with the equipment needed to fix the tire.

The reason you can run with only one wheel in the front is partly the reason why you can't run with only one in the rear—rear weight bias. Locked differentials are not used, so a tire on both rear corners is a must. Otherwise the unloaded drive shaft would spin freely and you wouldn't go anywhere.

If you have a flat in the rear, move the front tires to the rear and the good rear tire to front. The flat goes in the car. This will give you two driving wheels and more ground clearance so the other front brake will hit less.

TRANSAXLE LEAK

If you've installed a transaxle skid plate as I've recommended, the chances of putting a hole through the transaxle case are minimal. Yet transaxle leaks do occur. Common causes are stripped side-cover studs, a broken nosepiece and stress cracks.

If a bad transaxle leak develops, the transaxle will eventually let you know. It will start to howl as bearings and gears run dry. Stop and survey the damage. The location and severity of the leak and your distance from civilization will determine the corrective action.

If the leak is at the side of the transaxle, the fluid level will leak down at least to that point. More fluid will leak as it is splashed or flung off the spinning gears—a good reason to limit your speed once this type of leak is discovered.

If the source of the leak is two or three inches up from the bottom of the transaxle, there should be a pool of oil left in the bottom of the case. Slow the rate of fluid loss by taping or wiring a flattened can and rag over the leak. Or clean the area around the leak and coat it liberally with silicone sealer or weatherstrip adhesive. Anything is better than doing nothing.

If you can keep enough fluid in the transaxle so that some is still being flung about inside by the spinning gears, you can drive a long way if you take it easy. However, if the leak is so bad that all the fluid has leaked out, you have a much bigger problem.

First, slow down the leak as much as you can, using the methods I just described. Then, figure out a way to get oil into the transaxle. As an example, you can bypass the oil filter at the oil cooler and use the extra quart or so of engine oil for the transaxle. Use the oil-filter hose as a filler tube.

Place a container under the leak to catch any precious oil that leaks out. Pour gear oil, engine oil, engine additives—even cooking oil, if that's all you have—into the transaxle.

You're going to recycle this oil as many times as it takes to get back to civilization. When you get as much oil into the transaxle as is possible, drive the car forward about 20 ft. Then, run back, grab the container and catch any additional oil that leaks out.

After coating all the gears and bearings with oil, you can drive more than a few miles before having to repeat the process. Believe me, if it gets you out, it's worth it.

If it's a fast leak and you're far from help, you'll have to do this a number of times so you will need all the oil you can save. If it's a slow leak, just add enough oil to get the gears wet and don't worry about catching any oil. Just drive.

The first gear to pick up oil from the bottom of the case, other than the ring gear, it is the straight-cut gear just to the rear of the fill hole on the left side of the case. Check the illustration.

By jacking up the left side of the car and rotating the left wheel, you can rotate this gear. If it's picking up any oil from the bottom of the case you will see it. On a buggy, the fill hole should be easy to see. On a sedan or Baja Bug, you may need a mirror to see the hole. Looking in from the left wheel well, the fill hole is above the left rear frame horn.

If you take it easy and drive slowly—about 20 mph—the VW transaxle will last a long time with just a little coating of oil on the gears. Heat is what will finally put it out of service. If you think you've run it dry long enough, stop and touch the transaxle case. If it's hot to the touch, let it cool for awhile. Add more oil and go again.

HOW TO GET UNSTUCK

The Volkswagen off-road car gets stuck most often in soft sand. Sand washes and river beds are strange places. You can be breezing down a river bed and for no apparent reason the depth of loose sand changes and

Looking through fill hole on left side of case: Gear shown is first to pick up oil from bottom of case and fling it around inside. If gear is wet, you can keep on driving.

Stuck in a sand wash!

bogs your car down so fast that you're stuck before you know it.

If you ever get stuck, don't sit there and spin your wheels trying to get out. The tires will dig in deeper, making it even harder to get out.

Please read this entire section on getting unstuck before deflating any tires. If you don't have a pump, the air in the tires is pretty valuable, especially if you're crossing any rocky terrain.

The first thing to do is to decide the best way to go once you get unstuck. Does it get worse up ahead? Are there some rocks or hard ground close to either side? Would it be easier to back out, or have you been fighting this soft stuff for a few hundred feet? Remember, unless the shortest distance to hard ground is to go backwards, moving in soft sand is easier going straight ahead.

Once you determine which way you'll be headed, jack up one side of the car. You'll probably have to find something wide and flat, such as a rock, some wood, the spare tire, to support the jack. Otherwise, it will sink into the sand rather than raise the car. Jack each rear wheel up and push sand into the hole it was stuck in.

When the surface is level with the rest of the sand, put some brush or rocks underneath and slightly ahead of the tire. Let the tire down and do the other side. You may have to repeat this several times until you've made it to semi-solid footing.

When you get all the tires up out of their holes, you might try getting out

Put something underneath the jack to keep it from pushing itself into the sand. Now jack the car as far as the jack will go.

With the wheel up out of the hole, push sand into the hole and lower the car. Readjust the jack and repeat until you're out.

once before letting any air out of the tires. If you think you'll have a hard time getting out or if you've already tried once and didn't make it, let some air out of the rear tires and try again.

When deflating the tires, let a little bit of air out at a time. A seriously underinflated tire—less than 5 psi—might slip on its rim and tear out the valve stem. Then you really have problems.

If you don't have a tire-pressure gage, use this rule of thumb. Let out just enough air so the sidewalls start to bulge and are easy to depress with your thumb. Remember: Once unstuck, you'll have to drive on these deflated tires unless you have a pump.

Underinflated tires and tubes are susceptible to damage from rocks and jagged terrain. Otherwise, you can let lots of air out of the tires for maximum flotation, but don't go below 5 psi.

When you're ready to go, try not to spin the tires. This will kick out all the rocks and brush you put underneath to get traction. Instead, get the car rolling, then add the power gradually. You can do this best by revving up the engine and slipping the clutch at first, then letting it out all the way once underway. The trick is to walk the line between letting the tires spin and bogging the engine, without burning the clutch.

If the battery is in good shape and you don't have too far to go to solid ground, try using the starter motor to

Starting a fire without matches. Dampen end of rag in gas and light it with spark plug or coil wire. Use this *only as a last resort* and *only when you're out of matches.* BE CAREFUL!

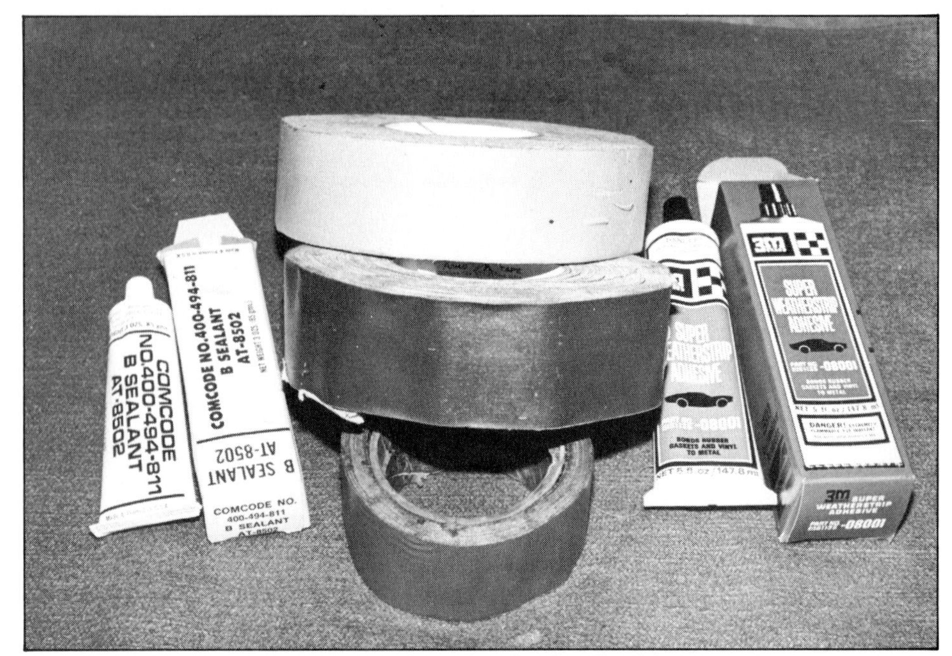

Racer's tape, silicone sealer and weatherstrip adhesive. Don't leave the pavement without them.

move the car. In some terrain, such as mud and silt, the biggest problem is just getting a car moving. When using the engine it's nearly impossible to keep the wheels from spinning. Sometimes a car will "dig in" faster than it moves. In either case, a car can be moved easier with the starter motor.

If the ignition and starter are on the same switch, pull off and ground the coil high-tension wire to keep the engine from starting. Be careful not to drain the battery. Remember, if you're using the old 6-volt starter with a 12-volt battery, using the starter to get the car unstuck may destroy the starter.

NECESSITIES

There are a few things you should have in your off-road car at all times:

Water—You'll never realize how important this is until you really need it. Beer, ice tea and lemonade are popular substitutes. *Never go off-road without fluids.*

Matches—Throw a few packs in the glovebox or tool box. As a precaution, get the waterproof kind. If you get caught without matches, and are desperate for some type of fire, try the following. Put a *very small* amount of gasoline on a rag or paper and light it with a spark from the high-tension wire on the coil. **Be very, very careful.**

Jack—Used to put on the spare, dismount tires, switch tires around and get unstuck. I prefer a bumper jack from a U.S.-built car, modified to fit.

Lug Wrench—For lug bolts, a poor man's hammer, a pry bar and a weapon, should the occasion arise.

Steel Wire or Coat Hanger and Duct Tape—These things will temporarily repair more things on a Volkswagen than anything else you can carry.

Knife—A simple pocket knife can be used to strip electrical wires, cut rubber hoses, as a screwdriver, to clean the points, to pick out splinters or spread mustard and dozens of other things.

Fan Belt—A must as explained earlier.

Hammer—The bigger, the better. A ball-peen is best.

Wrench—A Crescent or adjustable wrench and Channellock or Vise-Grip pliers are musts. They can be used on anything you can get a grip on: nuts, bolts, wire or sheet metal. You will also need a wrench to change a fan belt.

Tow Strap—A nylon line, a piece of chain, a heavy rope, anything strong enough to tow a Volkswagen. You will find that most people will tow you out if you have a tow strap. Almost no one will attempt to push you out if you don't have one. Fiberglass bodywork is fragile.

Flashlight—It beats setting a fire to see what you're doing. One with a magnetic base is best. Make sure the batteries are fresh. A spare set of batteries and a spare bulb are worth considering.

Tire-Pressure Gage—So you can adjust pressure for different terrain. Sure beats guessing.

Silicone Sealer and 3M Adhesive—Although these take a while to set up, they may fix some things you'll have no other way of fixing.

DON'T LOSE YOUR CAR

Last, but not least, if you suffer the supreme off-road embarrassment of having to walk for help, don't compound the problem by losing your car. Where your car is on some dirt road is very obvious when you are standing next to it. But if you walk five miles to a paved road and hitch a ride into town for help, you will be surprised at how all those dirt roads look strangely the same when you return. You will also be surprised how quickly a seemingly flat, open desert can swallow up a Volkswagen.

Stop often when walking and get your bearings from any landmarks available. Mark roads and intersections with cans, rocks or brush. If you leave someone with the car, have them light a fire, wave a flashlight or mirror, or set off a signal flare about the time you expect to be headed back. Smoke signals really do work. Sunlight reflected from a mirror can be seen from as far as 20 miles away.

APPENDIX

AIR FILTERS, BREATHERS

Donaldson Company, Inc. (Cyclopac)
P.O. Box 1299
Minneapolis, MN 55440
(612) 887-3414

K & N Engineering, Inc.
P.O. Box 1329
561 Iowa Ave.
Riverside, CA 92502
(714) 684-9762

La Mar Industries
1500 Daisy Ave., P.O. Box 2589
Long Beach, CA 90801
(Tri-Phase Air Cleaners) (213) 437-6363

Uni Filter (Pro-Comp II)
13522 Newhope St.
Garden Grove, CA 92643
(714) 530-6101

BRAKE COMPONENTS

AMS
(see suspension components)

Edco Disc Brakes
3411 W. MacArthur Blvd.
Santa Ana, CA 92704
(714) 540-6536

Hurst Performance, Inc.
50 W. Street Rd.
Warminster, PA 18974
(215) 672-5000

JFZ Engineered Products
7851 Alabama Ave., Unit 3
Canoga Park, CA 91304
(818) 887-6776

Summers Brothers
(see driveline components)

Wilwood Engineering
9666 Owensmouth Ave. Unit S
Chatsworth, CA 91311
(818) 998-3535

BUS REDUCTION BOX ADAPTERS

John Johnson Racing Products
P.O. Box 81
Lemon Grove, CA 92045
(714) 583-2054

CAMSHAFTS

Engle Racing Cams
1621 12th St.
Santa Monica, CA 90404
(213) 450-0806

Ed Iskenderian Racing Cams
16020 S. Broadway
Gardena, CA 90248
(213) 770-0930

Norris Performance Products
14762 Calvert St.
Van Nuys, CA 91411
(818) 873-4846

Sig Erson Cams
1588E Chemical Lane
Huntington Beach, CA 92649
(714) 898-9604

Web-Cam, Inc.
1663 Superior Ave.
Costa Mesa, CA 92627
(714) 631-1770

CHASSIS, ROLL CAGES

Appletree Automotive, Inc.
Box 310, Silver Lake
Mears, MI 49436
(616) 873-4830

Berrien Buggy Inc.
U.S. 31 South
Berrien Springs, MI 49103
(616) 471-1411

Chenowth
943 Vernon Way
El Cajon, CA 92020
(619) 449-7100

FUNCO Race Cars
8847 East 9th St.
Cucamonga, CA 91730
(714) 985-6813

Hi-Jumper Chassis
San Fernando Bug Center
1533 Truman St.
San Fernando, CA 91340
(818) 361-1215

Pro-Tech Engineering
707 South 10th
Blue Springs, MO 64015
(816) 229-3272

Sandhawk Chassis
(see Station One)

COOLERS

Earls's Performance Products
825 E. Sepulveda Blvd.
Carson, CA 90745
(213) 830-1620

Lockhart Consumer Products
15555 Texaco St.
Paramount, CA 90723
(213) 774-2981

Perma-Cool
671 E. Edna Place
Covina, CA 91723
(818) 967-2777

Rapid Cool Division, Hayden Inc.
1521 Pomona Rd.
Corona, CA 91718
(714) 371-0450

DRIVING LIGHTS

Robert Bosch Sales Corp.
2800 S. 25th Ave.
Broadview, IL 60153
(312) 865-5200

Duramex Inc./Zelmot
414 W. Rowland Ave.
Santa Ana, CA 92707
(714) 540-5872

General Electric Co.
Lighting Business Group
Nela Park
Cleveland, OH 44112
(216) 266-2121

KC Hilites, Inc.
P.O. Box 155
Williams, AZ 86046
(602) 635-2607

Lucas Industries Inc.
5500 New King St.
Troy, MI 48098
(313) 879-1920

Per-Lux Inc.
1242 E. Edna Place
Covina, CA 91724
(818) 331-4801

Racemark International, Inc. (Hella)
P.O. Box 82
Burnt Hills, NY 12027
(518) 399-9106

SEV Division, Valeo Automotive
34360 Glendale Ave.
Livonia, MI 48150
(313) 522-7320

DRIVELINE COMPONENTS

AMS
(see Suspension Components)

Dura-Blue Manufacturing
14911 "B" Moran St.
Westminster, CA 92683
(714) 891-7013

Henry's Machine Works, Inc.
1220 Knollwood Circle
Anaheim, CA 92801
(714) 761-2152

Summers Brothers
530 S. Mountain Ave.
Ontario, CA 91761
(714) 986-2041

Super Boot
1649 W. Collins
Orange, CA 92667
(714) 997-0766

DRIVER/SANCTIONING ORGANIZATIONS

American Closed Course Racing Association, Inc.(ACCRA)
15802 Springdale St.
P.O. Box 36
Huntington Beach, CA 92649

American Motor Sports Association (AMSA)
P.O. Box 5473
Fresno, CA 93755

High Desert Racing Association
961 W. Dale Ave.
Las Vegas, NV 89119
(702) 361-5404

SCORE International
31332 Via Colinas, Suite 103
Westlake Village, CA 91362
(818) 889-9216

EXHAUST SYSTEMS

Ermie Immerso Enterprises, Inc.(Thunderbird Products)
18700 Susana Rd.
Rancho Dominguez, CA 90221
(213) 537-1800

Phoenix Engineering
3963 Alamo St.
Riverside, CA 92501
(714) 788-2361

S & S Enterprises
1401 E. Ball Rd.
Anaheim, CA 92805
(714) 758-0355

Tri-Mil Industries
2912 S. Compton Ave.
Los Angeles, CA 90011
(213) 234-9014

FABRICATED METAL PARTS

Unique Metal Products
8745 Magnolia Ave.
Santee, CA 92071
(619) 449-9690

FIBERGLASS BODY PARTS

Fiber-Tech Engineering, Inc.
10809 Prospect Ave.
Santee, CA 92071
(619) 448-0221

Leslie Enterprises, Inc.
298 E. Bellevue Rd.
Atwater, CA 95301
(209) 358-7141

HYDRAULIC CLUTCH AND BRAKE CONTROLS

Ja-Mar Off-Road Products
42066 C Avenida Avarado
P.O. Box 1538
Temecula, CA 92390
(714) 676-2066

Neal Products, Inc.
7170 Ronson Road
San Diego, CA 92111
(714) 565-9336

INSTRUMENTS

Stewart-Warner Corp.
Consumer Products Div.
1826 Diversey Parkway
Chicago, IL 60614

Sun Electric Corp.
Instrument Products Div.
1560 Trimble Rd.
San Jose, CA 95131
(408) 946-7500

VDO-Argo Instruments, Inc.
980 Brooke Rd.
Winchester, VA 22601
(703) 665-0100

LIFT KITS

Ideal Metal
8524 Ablett Rd.
Santee, CA 92071
(619) 449-6116

T-Mag Products
1320 E. St. Andrews, Unit 1
Santa Ana, CA 92705
(714) 966-0706

Up Your Bug
356 N. Marshall, Suite H
El Cajon, CA 92020
(619) 561-4080

PARTS SUPPLIERS/OUTLETS

Bow-Wow of Seattle, Inc.
2922 6th Ave. So.
Seattle, WA 98134
(206) 382-1707

Bug-E Warehouse
1915 So. Presa St.
San Antonio, TX 78210
(512) 533-8056

Buggy Bunker
2120 W. Magnolia Blvd.
Burbank, CA 91506
(818) 841-0660

Buggy Shop
11433 Concord Village Ave.
St. Louis, MO 63123
(314) 842-0200

Bugformance Inc.
944 W. El Camino Real
Sunnyvale, CA 94087
(408) 245-3591

Bugpack Performance Products (Dee Engineering, Inc.)
3560 Cadillac Ave.
Costa Mesa, CA 92626
(714) 979-4990

Bugstuff
709 Jefferson Ave.
W. Brownsville, PA 15417
(412) 785-7000

C.B. Performance Products
28813 Farmersville Blvd.
Farmersville, CA 93223
(209) 733-8222

Crown Auto Parts and Marine
218 So. King St.
Gloucester, NJ 080030
(609) 456-0157

D & D Buggy
1435 Swift
N. Kansas City, MO 64116
(816) 471-3066

Deal Automotive
4584 Columbus Rd.
Macon, GA 31206
(912) 474-9292

Dune Buggy Supply Co.
717 E. Excelsior Ave.
Hopkins, MN 55343
(612) 938-8877

Dyna-Motive
941 W. 80th St.
Bloomington, MN 55420
(612) 888-2012

Florida VW & Porsche
5409 NW 161st St.
Hialeah, FL 33014
(305) 625-7032

Gene Berg Enterprises
1725 N. Lime St.
Orange, CA 92665
(714) 998-7500

House of Buggies
7302 Broadway
Lemon Grove, CA 92045
(619) 589-6770

Bob Hoy
1122 Airway Blvd.
El Paso, TX 79925
(915) 778-7788

J.M.Performance Racing Products
11549 Charles Rock Rd.
St. Louis, MO 63044
(314) 344-0022

Johnny's Speed and Chrome
6411 Beach Blvd.
Buena Park, CA 90620
(714) 994-4022

Larry's Off Road Center
4156 Wadsworth Rd.
Dayton, OH 45414
(513) 275-9501

McKenzie's Incorporated
12945 Sherman Way
North Hollywood, CA 91605
(818) 764-6438

Mr. Bug Inc.
1202 W. Struck Ave.
Orange, CA 92667
(714) 633-1093

Off-Road Guide
13237 Sierra Hwy.
Canyon Country, CA 91351
(805) 252-5180

Off-Road Outfitters
10061 Valle De Paz
El Cajon, CA 92021
(619) 561-3400

Oreo Off-Road Center
3600 S. Palo Verde Ave.
Tucson, AZ 85713
(602) 791-9386

Phil's Inc.
2204 Ashland Ave.
Evanston. IL 60201
(312) 869-2434

SCAT Enterprises, Inc.
P.O. Box 1220
Redondo Beach, CA 90278
(213) 370-5501

Small Car Specialties
618 E. Ball Road
Anaheim, CA 92805
(714) 635-6620

Speed Unlimited
621 Allen Ave.
Glendale, CA 91201
(818) 841-0502

Station One
3101 W. Thomas Rd.
Phoenix, AZ 85017
(602) 272-9333

VB & W
12133 E. Valley Rd.
El Monte, CA 91732
(213) 444-5519

OFF-ROAD PREPARATION AND FABRICATION

Off-Road Engineering
9720 Cozycroft St.
Chatsworth, CA 91311
(818) 882-2886

V-Enterprises
32817 Crown Valley Rd.
Acton, CA 93510
(805) 269-1279

PUBLICATIONS

Off-Road Advertiser
P.O. Box 340
Lakewood, CA 90714
(213) 860-7007

Off-Road Action News
9371 Kramer Suite J
Westminister, CA 92683
(714) 893-0953

Off-Road America
6637 Superior Ave.
Sarasota, FL 33581
(813) 921-5687

SCORE News
SCORE International
31332 Via Colinas Suite 103
Westlake Village, CA 91362
(818) 889-9216

Dune Buggies and Hot VWs Magazine
P.O. Box 2260
2949 Century Place
Costa Mesa, CA 92626
(714) 979-2560

VW & Porsche Magazine
Argus Publishing
12301 Wilshire Blvd.
Los Angeles, CA 90025
(213) 820-3601

RACE ENGINES

Auto Craft
1050 K Cypress
La Habra, CA 90631
(714) 870-9797

Bear's Bug Service
3540 S. 11th
Beaumont, TX 77705
(713) 842-1599

Lyle Cherry
7808 Maplewood
Fort Worth, TX 76118
(817) 498-0565

Fat Performance
1450 N. Glassell
Orange, CA 92667
(714) 639-2833

Giese Racing Engines
1705 Monrovia
Costa Mesa, CA 92627
(714) 631-5877

Jeff's Bug Shop
5236 Elm St.
Houston, TX 77036
(713) 665-4582

Dave Midgett
c/o Beetle Works
1012 W. Main St.
Lebanon, TN 37087
(615) 444-0036

Drino Miller Enterprises
942 Sunset Dr.
Costa Mesa, CA 92627
(714) 642-3690

Small Car Repair
3589-1/2 "G" St.
Tacoma, WA 98408
(206) 473-2474

RACING TRANSAXLES

A & A Transmissions
12623 Sherman Way, Unit B
North Hollywood, CA 91605
(818) 765-3566

B.C. Engineering
860 Pico Blvd.
Santa Monica, CA 90405
(213) 390-4471

BK Transmissions
7652 Slater
Huntington Beach, CA 92647
(714) 847-2681

C&S Transmissions
17416 Clark Ave.
Bellflower, CA 90706
(213) 804-1282

Fat Performance
(see Race Engines)

Giese Racing Engines
(see Race Engines)

Mike Leighton
11435 Santa Fe Ave., East
P.O. Box 1216
Hesperia, CA 92345
(619) 244-3584

Transaxle Engineering
9833 Deering Ave., Unit H
Chatsworth, CA 91311
(818) 998-2739

SAFETY EQUIPMENT

Aero Tec Laboratories Inc.
Spear Road Industrial Park
Ramsey, NJ 07446
(201) 825-1400

Diest Safety, Ltd.
641 Sonora Ave.
Glendale, CA 91201
(818) 240-7866

Filler Product Inc.
9017 San Fernando Rd.
Sun Valley, CA 91532
(818) 768-7770

Simpson Safety Equipment, Inc.
22630 S. Normandie Ave.
Torrance, CA 90502
(213) 320-7231

SEATS

Mastercraft by Miller Industries
1165 Walnut St.
Chula Vista, CA 92011
(714) 423-0272

Leslie Enterprises, Inc.
(see Fiberglass Body Parts)

Super Seats
208 4th Ave.
Buckeye, AZ 85326
(602) 386-2592

SHOCK ABSORBERS

Bilstein Corp. of America
11760 Sorrento Valley Rd.
San Diego, CA 92121
(619) 453-7723

KYB Corp. of America
901 Oak Creek Drive
Lombard, IL 60148
(312) 620-5555

KYB Shocks
Buck Bradley (West Coast Distributor)
3605 W. McArthur Blvd., Suite 707
Santa Ana, CA 92704
(714) 540-3884

Phoenix Phactory
P.O. Box 725
Campbell, CA 95009
(408) 379-0446

Rough Country, Inc.
19007 S. Reyes Ave.
Compton, CA 90221
(213) 639-6211

SUSPENSION COMPONENTS

AMS
1180 N. Fountain Way
Anaheim, CA 92806
(714) 632-9410

JT Machine Products
Star Route 2, Box 49
Boulevard, CA 92005
(714) 766-4762

Sway-A-Way Corp.
7840 Burnet Ave.
Van Nuys, CA 91405
(818) 988-5510

The Wright Place
9420 Flinn Springs Ln.
El Cajon, CA 92021
(619) 561-4810

Woods Off-Road Products
2733 W. Missouri
Phoenix, AZ 85017
(602) 242-0077

TIRES AND WHEELS

Center Line Wheels
13521 Freeway Dr.
Santa Fe Springs, CA 90670
(213) 921-9637

Dick Cepek, Inc.
1700 Kingsview
Carson, CA 90746
(213) 217-1059

Faas Wheels
6695 Amah Parkway
Claremore, OK 74017
(918) 342-4270

Formula Tire Division (Desert Dog)
Armstrong Rubber Co.
500 Sargent Dr.
New Haven, CT 06516
(203) 562-1161

Great Western Tire Co.
2866 Commercial St.
San Diego, CA 92113

Jackman Wheels
1035A Pioneer Way
El Cajon, CA 92020
(619) 440-1014

Sand Tires Unlimited
2990 Grace Lane
Costa Mesa, CA 92626
(714) 540-8473

Teltira, Inc.
701 Charleston
Lee's Summit, MO 64063
(816) 525-3555

Weld Wheel
933 Mulberry
Kansas City, MO 64101
(816) 421-8040

Western Auto Supply Co.
2107 Grand Ave.
Kansas City, MO 64108
(816) 346-4209

TRANSAXLE COMPONENTS

Crown Manufacturing
P.O. Box 2860
858 Production Place
Newport Beach, CA 92663
(714) 642-7391

Speed Unlimited
(See Mail-Order Houses/Retail Outlets)

VW ENGINE MACHINE WORK

Gene Berg Enterprises
(see Mail Order House/Retail Outlets)

Pauter Machine Co.
367 Zenith St.
Chula Vista, CA 92011
(714) 422-5384

RIMCO
520 East Dyer Rd.
Santa Ana, CA 92707
(714) 549-0357

Small Car Specialties
(see Mail Order House/Retail Outlets)

ACKNOWLEDGEMENTS

Over the years I've picked up a great deal of information from a number of off-road racers. In the course of writing this book, I found their help invaluable. A special thanks goes to Bob Vanegas, Jim Weber, Bill Apple, John Howard and Bill Varnes.

This book wouldn't be what it is without many of the exciting action shots provided by photographers and friends, including Judy Smith, Jean Calvin, George Jirka, Mike Rehler, Steve Lange, Anthony J. DePalmer, Jere Alhadeff, Centerline Photo, Trackside Photo and Action Photo.

When it came to getting product information from businesses involved in VW off-roading, there was no shortage of cooperation. What precedes this is a listing of those businesses that contributed to this book. There are many other businesses not listed here that offer quality parts and competent advice.

This book is dedicated to Linda, whose typing and patience was much appreciated.